NEW FRONTIERS IN OPEN

Henry Chesbrough is Faculty Director of the Garwood Center for Corporate Innovation at the Haas School of Business at UC Berkeley. His research focuses on managing technology and innovation. His first book, *Open Innovation* (2003), articulates a new paradigm for organizing and managing R&D. His second book, *Open Business Models* (2006), extends his analysis of innovation to business models, intellectual property management, and markets for innovation. His third book, *Open Services Innovation* (2011), explores open innovation in services businesses.

Wim Vanhaverbeke is Professor at the University of Hasselt. He is also Visiting Professor at ESADE Business School and the National University of Singapore. He has published in leading international journals including *Organization Science, Research Policy,* and *Strategic Management Journal.* He was co-editor with Henry Chesbrough and Joel West of *Open Innovation: Researching a New Paradigm* Oxford, (2006). His current research is focusing on open innovation in SMEs, innovation ecosystems, and on the implementation of open innovation practices.

Joel West is a Professor and Program Director at the Keck Graduate Institute, one of the seven Claremont Colleges, and is also Professor Emeritus at San José State University. He co-edited a 2014 special issue of *Research Policy* on open innovation and co-founded the World Open Innovation Conference; he also created and edits the Open Innovation Blog (oiblog.net). His research focuses on firm management of external open innovation networks, including platforms, ecosystems and innovation communities.

New Frontiers in Open Innovation

EDITED BY
HENRY CHESBROUGH,
WIM VANHAVERBEKE, AND JOEL WEST

OXFORD
UNIVERSITY PRESS

OXFORD
UNIVERSITY PRESS

Great Clarendon Street, Oxford, OX2 6DP,
United Kingdom

Oxford University Press is a department of the University of Oxford.
It furthers the University's objective of excellence in research, scholarship,
and education by publishing worldwide. Oxford is a registered trade mark of
Oxford University Press in the UK and in certain other countries

© Oxford University Press 2014

The moral rights of the authors have been asserted

First published 2014
First published in paperback 2017

All rights reserved. No part of this publication may be reproduced, stored in
a retrieval system, or transmitted, in any form or by any means, without the
prior permission in writing of Oxford University Press, or as expressly permitted
by law, by licence or under terms agreed with the appropriate reprographics
rights organization. Enquiries concerning reproduction outside the scope of the
above should be sent to the Rights Department, Oxford University Press, at the
address above

You must not circulate this work in any other form
and you must impose this same condition on any acquirer

Published in the United States of America by Oxford University Press
198 Madison Avenue, New York, NY 10016, United States of America

British Library Cataloguing in Publication Data

Data available

Library of Congress Cataloging in Publication Data

Data available

ISBN 978–0–19–968246–1 (Hbk.)
ISBN 978–0–19–880399–7 (Pbk.)

Links to third party websites are provided by Oxford in good faith and
for information only. Oxford disclaims any responsibility for the materials
contained in any third party website referenced in this work.

Foreword: Open Innovation and Knowledge Creation

INTRODUCTION

Open Innovation is a new management paradigm which originally grew out of the practices and research of the high technology industry in the U.S. and Japan. As the world entered into the century of the "Knowledge Society" (Drucker, 1993), utilizing ideas and knowledge from both inside and outside of the firm's boundaries is becoming more important than ever. Accordingly, the topic of Open Innovation can be analyzed and discussed from various perspectives in management: strategic positioning, value chain, business model, core competence, knowledge creation and management, etc. Henry Chesbrough, the leading scholar in this field, has actively been conducting research and promoting the concept to practitioners and academics alike. I have been interacting closely with him over the years, as our interests intersect in the area of innovation and knowledge creation.

The knowledge-based view of the firm originally emerged from the studies of product development in Japanese firms. Takeuchi and I have described the importance of the "scrum" approach to product development process and used this rugby metaphor to show how the process can be made more agile and how knowledge can be shared within the team. These studies led to the development of the knowledge creation theory, which focused on the spiral process of making tacit knowledge explicit (and vice versa) among individuals, groups, and organizations (Nonaka and Takeuchi, 1995). Sharing knowledge in our view does not deteriorate the value of knowledge, but rather, promotes knowledge creation. In this sense, the knowledge creating process needs to be open; that is what "Open Innovation" is about.

In this foreword, I would like to explain briefly how Open Innovation is different from the existing theories of innovation, and why it is important to practitioners and academics.

HOW OPEN INNOVATION IS DIFFERENT FROM THE EXISTING THEORIES OF INNOVATION

Open Innovation is defined in Chapter 1 of this book as "a distributed innovation process based on purposively managed knowledge flows across organizational boundaries, using pecuniary and non-pecuniary mechanisms in line with the organization's business model." (Chesbrough and Bogers, this volume: 17) In other words, one of the main characteristics of Open Innovation is the utilization of knowledge, both inside and outside of the firm, to innovate something new. That something can be a new product, a new service, or even a business model; in this sense, Open Innovation yields the same results as conventional innovation. What makes Open Innovation different is that it is open, not closed at the firm's boundaries—it transcends the firm's boundaries. Because Open Innovation transcends the firm's boundaries, both the knowledge utilized in the innovation and the knowledge resulting from the innovation are able to permeate both inside and outside of firms' organizational boundaries.

Most conventional management theories and business practices have pursued a different path: they have placed knowledge inside a "black box" and kept it within the firm's organizational boundaries, assuming that such knowledge is indeed the source of their superior sustainable performance, that it adds value within the firm's value chains, maintains the core competence of the firm, and differentiates the firm from other firms. This kind of thinking leads to phenomena such as the "Not Invented Here" syndrome, by which firms regard knowledge not generated within the organization as inferior. The truth is that, during the time of technical innovation, i.e., until the 2000s, the "black box" strategy worked well: it gave high return on invested capital by keeping the knowledge—mostly explicit knowledge in the form of know-how and intellectual properties—within the firm's value chains.

However, since the 2000s, there has been a shift in innovation from products to services. As businesses became more globalized and customer needs became more complex, both were accelerated by the penetration of ICT. Products were not enough to meet value propositions; they had to be accompanied by services within a business model. We call this movement a shift from thing-oriented to event-oriented; or from *mono* (thing) to *koto* (event). Therefore, firms needed to expand their value chains from vertical to horizontal integration; from a closed and linear system to an open and complex ecosystem; from extraction to inclusion. However, conventional management theories cannot fully explain such a movement because they are based on neo-classic economic theory, which presumes perfect competition and market equilibrium. This is where the concept of Open Innovation comes in. The concept of Open Innovation explains the processes and consequences of opening up the black box: how the shared knowledge can promote innovation beyond the firm's boundaries and realize event-based value creation. It can also explain the co-creation of customer values between/among firms and customers, academia, and government. There are no longer five pressuring forces in the market, but rather, cooperative forces for creating relationships and achieving common interests and goals. We can call this a business ecosystem.

WHY OPEN INNOVATION IS IMPORTANT TO PRACTITIONERS AND ACADEMICS

As described above, we live in a world of global complexity, and we need a theory and practice that can realize a win-win situation. We need an inclusive business model, in which all the stakeholders may gain value. We believe Open Innovation is the solution. Open Innovation integrates diverse knowledge within the business ecosystem and seeks to create new knowledge which benefits multiple stakeholders. The ecosystem we live in today is very complex; where events and things interact with each other in multiple layers with constantly changing relationships. All things flow without end. In such a context, ideas and knowledge need to be shared and combined among and beyond the stakeholders in order to be effective and efficient in that specific context. Open Innovation can explain such processes and consequences. Therefore, for practitioners, research on Open Innovation can offer the theorization of such processes and consequences; practitioners can utilize the theory to institutionalize such processes in practice. For academics, research on Open Innovation will open up a new field of study in various areas such as strategy, organizational behavior, marketing, finance, and knowledge management.

In fact, the theory of Open Innovation can explain recent developments in the field, such as "Living Labs." Promoted mainly among European countries, Living Labs is one of the movements of Open Innovation. Living Labs embody both the research and the innovation process and are operated as a public-private-people partnership, often in a single city or a town. A concept and/or technology is tested, designed, and created in a real-world situation, in order to create new user-centric products and services. In this context, knowledge is co-created among the public-private-people partnership. The theory of Open Innovation can explain the concept and practices of "Living Labs."

CONCLUSION

Open Innovation is not a buzz word—it is happening in the real world. To understand and pursue Open Innovation, we may face a number of questions such as: How can a firm overcome the internal knowledge silo and transcend its organizational boundary? How can a firm obtain knowledge inclusively from the outside? How can a firm offer its knowledge to the outside, and what are the issues with and consequences of doing so? And, what kind of individual capabilities are needed at multiple levels of the firm? By answering these questions, this book will provide the reader with a clear understanding of what Open Innovation is, and how to pursue it.

August 2013
Ikujiro Nonaka, Professor Emeritus, Hitotsubashi University

Contents

List of Figures — xi
List of Tables — xiii
List of Contributors — xv

Part I: Open Innovation: Ten Years Later

1. Explicating Open Innovation: Clarifying an Emerging Paradigm for Understanding Innovation — 3
   Henry Chesbrough and Marcel Bogers

2. Firms, Users, and Innovation: An Interactive Model of Coupled Open Innovation — 29
   Frank Piller and Joel West

3. A Classification of Open Innovation and Open Business Models — 50
   Wim Vanhaverbeke and Henry Chesbrough

Part II: Analyzing Open Innovation at Different Levels of Analysis

4. Challenges of Funding Open Innovation Platforms: Lessons from Symbian Ltd. — 71
   Joel West

5. Open Innovation and Industrial Dynamics—Towards a Framework of Business Convergence — 94
   Jens Frøslev Christensen

6. Exploring Open Innovation at the Level of R&D Projects — 115
   Wim Vanhaverbeke, Jingshu Du, Bart Leten, and Ferrie Aalders

Part III: New Application Fields for Open Innovation

7. Exploring Open Innovation in Small and Medium-Sized Enterprises — 135
   Sabine Brunswicker and Vareska van de Vrande

8. Open Innovation in Multinational Corporations: New Insights from the Global R&D Research Stream — 157
   Kazuhiro Asakawa, Jaeyong Song, and Sang-Ji Kim

9. Open Social Innovation — 169
   Henry Chesbrough and Alberto Di Minin

Part IV: Managing and Organizing Open Innovation

10. Open Innovation and Intellectual Property: A Two-Sided Market Perspective ... 191
    Henry Chesbrough and Roya Ghafele

11. Managing Inside-Out Open Innovation: The Case of Complex Ventures ... 208
    Henry Chesbrough and Chris Winter

12. Patterns of Implementation of OI in MNCs ... 223
    Letizia Mortara and Tim Minshall

13. Getting Help From Innomediaries: What Can Innovators do to Increase Value in External Knowledge Searches? ... 242
    Nadine Roijakkers, Andy Zynga, and Caroline Bishop

14. Theories of the Firm and Open Innovation ... 256
    Wim Vanhaverbeke and Myriam Cloodt

Part V: Conclusions

15. Surfing the New Wave of Open Innovation Research ... 281
    Wim Vanhaverbeke, Henry Chesbrough, and Joel West

Reference List ... 295

Index ... 337

List of Figures

1.1	Annual citations to open innovation in Google Scholar	5
1.2	Growth of publications on open innovation in Web of Science	7
1.3	Word cloud based on abstracts, top-30 words	14
1.4	Word cloud based on author keywords, top-50 words excluding "open" and "innovation"	15
1.5	The open innovation model	18
2.1	Two forms of coupled open innovation	39
4.1	Symbian global smartphone unit sales and market share, 2002–2012	83
4.2	Overall market share of leading handset makers, 1997–2012	87
5.1	Conceptions of business environment	103
5.2	The convergence life cycle	109
6.1	Graphical representation of Philips open innovation process	121
6.2	Factors affecting the organization of R&D projects and their outcome	128
9.1	Social innovation as seen by NESTA and the Young Foundation	170
9.2	Ashoka's Hybrid Value Chain	177
9.3	CHAMPS2: Transforming public services	182
10.1	The many possible paths for IP Creation	192
12.1	An open innovation implementation framework	237
13.1	Potential value added by innomediaries in different stages of external knowledge searches, value-adding actions by innovating companies, and key success factors related to each phase	245
14.1	The open innovation funnel	258

List of Tables

1.1	Books with open innovation in the title from 2003–2012	6
1.2	Fields represented by journals in sample	8
1.3	Top 20 most cited articles in sample with forward citations, main topics, and levels of analysis	10
1.4	Possible units of analysis and research objects for open innovation research	26
2.1	Contrasting open and user innovation	31
2.2	Multiple dimensions of coupled open innovation processes	38
2.3	A process model for coupled open innovation projects	40
3.1	A classification of combinations of open innovation and open business models	54
4.1	Categories of Symbian ecosystem members ca. 2002	78
4.2	Shareholders of Symbian Ltd., 1998–2008	80
4.3	Changes in Symbian Ltd. capital structure, 1998–2008	81
4.4	Licensees of smartphone operating systems	85
4.5	Successful open innovation platforms	89
4.6	Examples of successful chaining from a cash cow platform to a new platform	92
7.1	Overview of empirical studies on open innovation in SMEs	138
8.1	Common characteristics of open innovation research	161
8.2	Cross fertilization of open innovation research and global R&D research	168
9.1	Open social innovation in the three non-profit organizations	184
12.1	Key areas for future research	239

List of Contributors

Ferrie Aalders is responsible for the Business Excellence approach and implementation at Philips Research worldwide. He started his career executing scientific research in physics at Utrecht University, The Netherlands, concluded in 1985 with a Ph.D. in Mathematics and Natural Sciences. He has been working for Philips since, initially in Research on CAD and computer simulation of electron-optical devices, and as of 1990 in Advanced Development at Philips Display Components, later LG.Philips Displays. In 2001, he returned to Philips Research, where he combines his responsibilities for Business Excellence with the position of secretary of the Research Management Team and Export Control Officer for Philips Group Innovation. Throughout his career he has worked on innovation on the interface between technology and organization. He is member of the Strategy Committee of the European Foundation for Quality Management (EFQM) and of the Scientific Advisory Board of IPERF (Luxemburg).

Kazuhiro Asakawa is Mitsubishi Chaired Professor of Management at the Graduate School of Business Administration, Keio University, Japan. He received his Ph.D. from INSEAD and his MBA from Harvard Business School. His research interests are focused on innovation and R&D management of multinational corporations. He is an Associate Editor of *Global Strategy Journal* and serves on the editorial boards of the *Academy of Management Perspectives, Journal of International Business Studies, Journal of International Management*, and *Asia Pacific Journal of Management*. He chairs the Japan Chapter of the Academy of International Business. He was a visiting scholar at MIT Sloan School of Management and adjunct faculty fellow at the research institutes of the Ministry of Economy, Trade and Industry and the Ministry of Education, Japan.

Caroline Bishop Following an early career with the NHS, Caroline spent 20 years at the University of Birmingham where she took on a variety of progressively influential roles. Highlights of her career include setting up an MBA program in Hong Kong, and developing and implementing a £2 million, five-university partnership across the East and West Midlands to provide training in technology transfer to medical students. Caroline was particularly influential in the development of IXC UK, securing £3.6 million of grant funding to set up the initial pilot project. Guiding IXC UK through its early years, she oversaw its spin-out from the University of Birmingham as a "limited by guarantee company," and continues to drive the company as Managing Director.

Marcel Bogers is Associate Professor of Innovation and Entrepreneurship at the Department of Food and Resource Economics at the University of

Copenhagen. He obtained a combined B.Sc. and M.Sc. in Technology and Society (Innovation Sciences) from Eindhoven University of Technology and a Ph.D. in Management of Technology from Ecole Polytechnique Fédérale de Lausanne (Swiss Federal Institute of Technology). His main interests center around the design, organization, and management of technology and innovation. More specifically, he has studied areas such as open innovation, business models, family businesses, users as innovators, collaborative prototyping, entrepreneurship, improvisation, and university-industry relations.

Sabine Brunswicker is an innovation researcher and advisor with a particular focus on open innovation and innovation ecosystems. She is an Associate Professor of Innovation and Director of the Research Center for Open Digital Innovation (RCODI) at Purdue University in West-Lafayette, United States. She is also affiliated with the Innovation and Knowledge Management Institute (IIK) at Esade Business School, at Ramon Llul University in Spain and the School of Information Systems, Queensland University of Technology (QUT) in Australia. Before joining Purdue in fall 2014, she was Head of Open Innovation at the Fraunhofer Institute for Industrial Engineering of the Fraunhofer Society. She holds a Master in Mechanical Engineering and Management Science (dual degree), a Master of Commerce with a specialization in Marketing and Entrepreneurship, and a Doctorate of Engineering in the area of innovation management.

She maintains an active network with industry partners through joint action research projects, both with SMEs as well as global firms such as SAP, Philips, and Porsche. Further, she closely interacts with the European Commission and is a member of the Open Innovation Strategy and Policy Group (OISPG) of DG Connect.

Henry Chesbrough is Faculty Director of the Garwood Center for Corporate Innovation at the Haas School of Business at University College Berkeley. Previously, he was an assistant professor of business administration, and the Class of 1961 Fellow at the Harvard Business School. He holds a Ph.D. in Business Administration from the University of California Berkeley, an M.B.A. from Stanford University, and a B.A. from Yale University, *summa cum laude*.

His research focuses on managing technology and innovation. His book, *Open Innovation* (2003), articulates a new paradigm for organizing and managing R&D. His second book, *Open Business Models* (2006), extends his analysis of innovation to business models, intellectual property management, and markets for innovation. His third book, *Open Services Innovation* (2011), explores open innovation in services businesses.

His academic work has been published in *Harvard Business Review, California Management Review, Sloan Management Review, Research Policy, Industrial and Corporate Change, Research-Technology Management, Business History Review*, and the *Journal of Evolutionary Economics*.

List of Contributors

Jens Frøslev Christensen is Professor of Management of Innovation at Copenhagen Business School. He has published several books and numerous articles in international journals such as *Industrial and Corporate Change, Research Policy, Industry and Innovation,* and *Managerial and Decision Economics*. His main research interest has focused on the interplay between management of innovation, corporate strategy, industrial dynamics, and, more recently, the challenges of sustainability.

Myriam Cloodt is Assistant Professor of Entrepreneurship and Innovation at Eindhoven University of Technology, the Netherlands. Her research interests mainly include open innovation, (corporate) entrepreneurship, mergers and acquisitions, and strategic technology alliances. Her work has appeared, amongst others, in *Research Policy, R&D Management, International Entrepreneurship and Management Journal, European Journal of Innovation Management, Business History,* and *Business History Review*. She co-authored the chapter "Open innovation in value networks," in H. Chesbrough, W. Vanhaverbeke and J. West (eds), *Open Innovation: Researching a New Paradigm* (Oxford University Press, 2006).

Alberto Di Minin is Associate Professor of Strategy at the Istituto di Management—Scuola Superiore Sant'Anna (Pisa, Italy—http://www.sssup.it), and Research Fellow with the Berkeley Roundtable on the International Economy (BRIE, University of California Berkeley http://brie.berkeley.edu/). Alberto's research and teaching deals with the appropriation of innovation. In particular he focuses on open innovation and new business models. He also works on technology transfer, intellectual property, and R&D management. His latest publications include *California Management Review, Journal of International Business Studies, R&D Management Journal, Research Policy*.

Alberto regularly contributes to the daily *Il Sole 24 Ore* covering case studies of innovative Italian companies. He has received a Ph.D. from University College Berkeley, and a M.S. in Public Policy from the Georgia Institute of Technology.

During 2013 he served as Advisor for Innovation Policy to the Italian Minister of Research and Education. http://www.diminin.it.

Jingshu Du is Assistant Professor of Strategic Management and Innovation at VU University Amsterdam and Research Fellow at Vlerick Business School. She received her doctoral degree in applied economics from Hasselt University in 2013. Jingshu's research interests cover open and collaborative innovation, IP strategies, and new product development. In her research, she investigates the effects and organization of open innovation at the project level by focusing on how R&D collaborations affect innovations in R&D projects, what are the contingencies of R&D collaborations, and how to manage collaborations successfully. Jingshu's research is conducted in close collaboration with European firms in manufacturing and the pharmaceutical industries. Both empirical and case-based research approaches are used to address her research questions.

Roya Ghafele is an Assistant Professor with the School of Law of Edinburgh University. She holds three Fellowships with the University of Oxford, among them at the Said Business School and the Oxford Intellectual Property Research Centre. In addition to that she is a Founding Member of Oxfirst Limited, a boutique consulting firm specializing in the Economics of IP.

Prior to that, she was an Assistant Professor with the University of Oxford and a Research Scholar with the Haas School of Business, University of California at Berkeley. From 2002 to 2007 she worked with the U.N's World Intellectual Property Organization (WIPO) and the Organization for Economic Cooperation and Development (OECD). In 2000 she started her career with McKinsey & Company.

Her Ph.D. was awarded the Theodor Koerner Research Prize by the President of the Republic of Austria. Dr. Ghafele was trained at Johns Hopkins University, School of Advanced International Studies, the Sorbonne, and Vienna University.

Sang Ji Kim is a visiting researcher at Seoul National University (SNU). She received her B.A. in Economics in 2002 from SNU and Ph.D. of strategy and international management (2014).. Before going back to study, she worked for Samsung Corning Precision Glass (current Samsung Corning Precision Materials), a joint venture company of Samsung Electronics and Corning Inc. Sang Ji Kim's current research interests cover competitive strategy based in knowledge management and learning, global strategy of multinational corporations, and innovation strategy; knowledge sourcing and sharing of global R&D networks of a firm, catching-up strategy of venture firms, and open innovation.

Bart Leten is Associate Professor of Innovation Management at the KU Leuven and the University of Hasselt. He is Associate Editor of the journal *Industry and Innovation*. His research focuses on innovation and international business strategies of large firms and has been published in journals such as *Journal of International Business Studies, California Management Review, Journal of Product Innovation Management, Regional Studies and Environment and Planning A*. Topics of current interest include R&D location choices, R&D collaborations and open innovation, technology licensing, intellectual property management, and design of innovation tournaments.

Tim Minshall is a Reader at the University of Cambridge Centre for Technology Management. He researches, teaches, writes, and consults on the topics of open innovation, technology enterprise, the financing of innovation, and university-industry knowledge exchange. He is a Fellow of Churchill College, Cambridge, a non-executive director of St. John's Innovation Centre Ltd, Cambridge, and a Visiting Professor at Doshisha University Institute for Technology, Enterprise and Competitiveness in Japan. He has a BEng from Aston University, and a Ph.D. from Cambridge University Engineering

Department. Prior to joining the University of Cambridge, he worked as an engineer, teacher, consultant, freelance writer, and project manager in the U.K., Japan, and Australia.

Letizia Mortara is a Senior Research Associate at the Centre for Technology Management at the Institute of Manufacturing, University of Cambridge (U.K.). She is also a By-fellow at Churchill College, Cambridge and an Associate Editor for the journal "R&D Management". Her interests and expertise include open innovation, additive manufacturing, and technology intelligence. In these topics she researches, teaches, and consults. Letizia has a first degree in industrial chemistry from the University of Bologna (Italy). Prior to joining the University of Cambridge, she worked as a process/product manager for the IVM group specializing in coatings, stains, and primers for wood, and then she gained her Ph.D. in processing and process scale-up of advanced ceramic materials at Cranfield University when she moved to the U.K.

Frank Piller is a professor of management and the director of the Technology & Innovation Management Group at RWTH Aachen University, Germany. He also is a faculty member of the MIT Smart Customization Group at the MIT Media Lab, U.S.A. His research focuses on value co-creation between businesses and customers/users, strategies to increase the productivity of technical problem solving by open innovation, and the market of open innovation intermediaries. He also has a continuous interest in mass customization and customer co-design. Frank Piller's research is supported by grants from the European Commission, the DFG, BMBF, and other institutions. He has consulted and delivered executive workshops for many Dax30 and Fortune500 companies. As an investor, member of the Board of Directors, or as a scientific adviser of several technology companies, he transfers his research into practice.

Nadine Roijakkers wrote her Ph.D. thesis at the United Nations University/MERIT on inter-firm collaborative innovation in the pharmaceutical biotechnology industry. During the period 2002–2007 she held a number of positions in research and academia. From 2007 to 2009 she was a senior strategy consultant to companies operating in various sectors of industry at KPMG Consulting. As of November 2009 she has held the position of Assistant Professor of Strategy and Innovation Management at Hasselt University in Belgium. Nadine Roijakkers has published numerous articles and book chapters on alliance management and innovation management. Outlets for her work include among others *Long Range Planning, Research Policy, Business History Review, British Journal of Management, European Management Journal, Technological Forecasting and Social Change, Small Business Economics*, and *California Management Review*.

Jaeyong Song was a professor at Columbia and Yonsei before joining Seoul National University (SNU). He serves as vice president of Korea Academy Society of Business Administration (KASBA) and president of Association of Korean Management Scholars. He won best dissertation awards from the Academy of Management and European International Business Association. He won Chazen Teaching Innovation Award at Columbia Business School, SNU Teaching Award, and the best performing professor award from Yonsei University. He received the best paper awards from KASBA, Korean Academy of Management, and Korean Academy of International Business. His papers were published in *Management Science, Harvard Business Review, Strategic Management Journal, Organization Science, Journal of Economics & Management Strategy, Journal of Management, Journal of International Business Studies*, and *Research Policy*. He serves as an editor of *Journal of International Business Studies*.

Vareska van de Vrande is an Associate Professor of Strategic Management at the Rotterdam School of Management, Erasmus University. Vareska joined RSM in 2007 after completing a Ph.D. in Industrial Engineering and Management Science at the Eindhoven University of Technology. Vareska's expertise focuses on the areas of corporate entrepreneurship, external technology sourcing, corporate venture capital investments, strategic alliances, joint ventures, and mergers and acquisitions. Other research interests include open innovation and corporate venturing. She has published articles in leading journals, including the *Strategic Management Journal*, the *Journal of Product Innovation Management, IEEE Transactions on Engineering Management*, and the *Journal of Business Venturing*.

Wim Vanhaverbeke is professor at the University of Hasselt. He is also visiting professor at ESADE Business School and the National University of Singapore. He published in several international journals such as *Organization Science, Research Policy, California Management Review, Journal of Management Studies, Small Business Economics, Journal of Business Venturing, Technovation*. He was co-editor with Henry Chesbrough and Joel West of the book *Open Innovation: Researching a New Paradigm* (Oxford University Press, 2006). He is a dedicated open innovation researcher collaborating with different partners in universities and companies around the globe. His current research is focusing on open innovation in SMEs, innovation ecosystems, and on the implementation of open innovation practices. He established the European Innovation Forum with Henry Chesbrough in 2012. He is a frequently asked speaker at leading international conferences and an adviser for several globally operating companies. He has consulted and organized workshops for multinationals and technology companies.

He was recently recognized by the International Association of Management of Technology (IAMOT) as one of the top 50 authors of technology and innovation management over the last five years (2008–2012).

Joel West is Professor of Innovation & Entrepreneurship at the Keck Graduate Institute School of Applied Life Sciences, after holding a similar position at the San Jose State University College of Business. His research on open innovation includes co-editing both *Open Innovation: Researching a New Paradigm* (Oxford, 2006) and a 2014 special issue of *Research Policy*; he also created and edits the Open Innovation blog (blog.openinnovation.net). Other research has examined firm platform strategies in open source software and mobile telecommunications. He has published articles in *Industry & Innovation, Information Systems Research, Journal of Management Studies, Journal of Product Innovation Management, Journal of Technology Transfer, R&D Management, Research Policy*, and *Telecommunications Policy*, among other journals. He has a Ph.D. from UC Irvine, an S.B. from MIT and more than 20 years' experience in the software industry.

Chris Winter is a Venture Partner at New Venture Partners, where he has worked since 2003. He is also Chairman of Health Enterprise East, a part of the U.K. National Health Service charged with exploiting IP developed in the NHS. Previously, Chris co-founded Brightstar, BT's corporate incubator in 2000, and acted as its CTO in identifying and building companies spun out by BT. Prior to co-founding Brightstar, he was the Vice President of Engineering at Cyberlife Technologies Ltd., a software start-up based in the United Kingdom. Chris spent the previous 12 years at BT in various positions, including the Head of Futures Research at BT, where he ran teams of scientists and engineers in researching and developing optical switches, optical computing, next generation operational support systems, AI software and mobile network management.

Chris holds a B.A. in Biochemistry from Oxford University, a Ph.D. in Physics from Lancaster University and is a Visiting Professor of Technology Entrepreneurship at UCS, Suffolk, U.K.

Andy Zynga got a doctoral degree from the RWTH Aachen (Germany) in 2015. He joined NineSigma in February 2008, establishing and growing the European company presence rapidly. In September 2009 Zynga was appointed the Global CEO of NineSigma Group, expanding business across all regions. During his career, Zynga has built four high tech and service businesses successfully in both Europe and the U.S.A. He also spent six years at KPMG Consulting (now Bearingpoint) in Germany and London where he successfully grew a division of the Information, Communication, and Entertainment (ICE) Consulting Practice into a major international player. Prior to joining NineSigma, Zynga spent four years as a member of the Executive Committee of Telindus, a global network integrator and outsourcing company with more than $1B in revenues and 3,000 employees in 14 countries. While there, he turned around a region from loss-making to more than 10% of EBITDA, and growing revenues by more than 30% to over $150M. Zynga earned a dual Masters degree in Business Administration and Mechanical Engineering from the Technical University of Berlin, Germany.

Part I

Open Innovation: Ten Years Later

1

Explicating Open Innovation

Clarifying an Emerging Paradigm for Understanding Innovation

Henry Chesbrough and Marcel Bogers

1.1 INTRODUCTION

In this chapter, we survey the considerable academic literature that has arisen since the publication of *Open Innovation* (Chesbrough, 2003a) in April 2003. According to Google Scholar, there have been over 6,000 citations to this book since its publication a decade ago. Moreover, a number of academic and practitioner journals have organized special issues that were inspired by the book and the subsequent scholarship it motivated. These journals include *R&D Management, Technovation, Research Policy*, and *Research-Technology Management*. Stimulating more academic scholarship in open innovation was furthermore the explicit intention of an edited volume by Chesbrough, Vanhaverbeke, and West (2006).

This academic scholarship has been more than matched by the response of industry to the book and its subsequent work. Google now reports millions of page links in response to the term "open innovation." Many consulting firms now feature an open innovation practice area in their work. Job titles like Manager or Director or VP of Open Innovation are becoming more common in many industries.

In this chapter, we seek to provide an overview of the work that has led to this broad acceptance. Informed by an analysis of academic citations, we show the growth of academic research in the domain of open innovation (including the growing breadth of open innovation research in other fields), identify key areas of significant progress of open innovation research, and consider understudied areas. By highlighting the more general themes and development of the broad body of research related to open innovation, we aim to complement existing literature reviews in this area (e.g., Bogers and

West, 2012; Dahlander and Gann, 2010; Elmquist, Fredberg, and Ollila, 2009; Huizingh, 2011; van de Vrande, Vanhaverbeke, and Gassmann, 2010; West and Bogers, 2014).

Our other goal for this chapter is to further clarify the concept of open innovation, which Chesbrough (2006b) defined as "the use of purposive inflows and outflows of knowledge to accelerate internal innovation, and expand the markets for external use of innovation, respectively." (p. 1) In discussing the underpinnings of open innovation, we also aim to link it more clearly to the pre-existing literature in economics and management of innovation. We particularly address the notion of Research and Development (R&D) spillovers and propose that purposively structuring mechanisms to manage such spillovers is a uniquely defining characteristic of the open innovation concept. Accordingly, we conclude that open innovation ought to be conceptualized as a distributed innovation process that involves purposively managed knowledge flows across the organizational boundary. We also link this definition to the main types of open innovation, namely Outside-In (inbound), Inside-Out (outbound) and the combined Coupled type, as well as the associated mechanisms, including pecuniary and non-pecuniary flows.

Building on our aim to foster greater consistency around the definition of open innovation within the broader innovation literature, we moreover discuss divergent views on open innovation, we pay particular attention to the confusion that has emerged between "open innovation" and "open collaborative innovation," and we discuss the differences and complementarities among these perspectives. In this way, we hope to encourage our academic colleagues to adopt a consistent definition of the concept, so as to retain coherence of the research in this area (and in other areas as well).

Next, we summarize and address some of the critiques of open innovation that have emerged in the literature. Such scholarly criticism is important to the development of any domain of academic work, and so we respond to the most prominent critiques in hopes of advancing the debate around the open innovation concept. Finally, we consider the progress this research has made, relative to the research agenda identified in Chesbrough et al. (2006).

1.2 A REVIEW OF TEN YEARS OF OPEN INNOVATION RESEARCH

1.2.1 Growth of Open Innovation

Open innovation has been growing rapidly since the term was coined by Chesbrough (2003a). When that book appeared, a Google search on the term *open innovation* yielded roughly 200 page links, most of which merely had the two words near each other in a sentence, such as "company X opened its

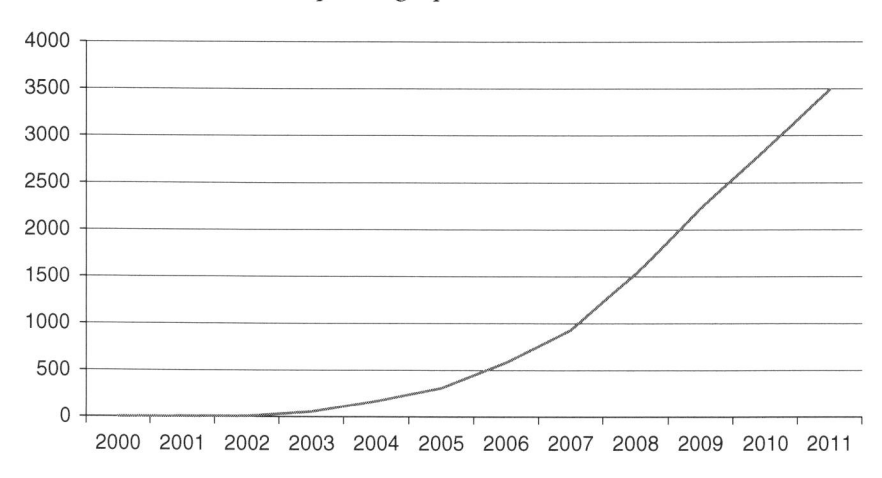

[Figure: line graph showing annual citations rising from near 0 in 2000 to ~3500 in 2011]

Figure 1.1 Annual citations to open innovation in Google Scholar
Note: Search criteria: "open innovation," Chesbrough

innovation office at location Y." There was not yet any specific meaning to the two words together as a phrase. By contrast, a search on the same search engine in April 2012 generated 483 million links, which, by scanning through some of the links, generally seemed to address this new and very different model of innovation. This suggests that open innovation has become widely known.

Within academic research, the citations to "open innovation" in Google Scholar as a term and/or to the Chesbrough (2003a) book have also been growing, as shown in Figure 1.1. However, as can also be seen in the figure, this growth in annual citations (the second derivative) has begun to slow down in the past couple of years. Nonetheless, the figure shows several thousands of new scholarly contributions each year.

Due to the framing of the 2003 book, the audience for *Open Innovation* has included both academics and managers from its inception. In contrast to the slowing rate of new scholarly articles year over year, the number of books on open innovation is accelerating year over year (Table 1.1). This suggests that open innovation is being translated beyond academic research into industry practice at an increasing rate.

1.2.2 The Scope and Breadth of Open Innovation Research

Next, we more specifically explore the academic research on open innovation that has been published in peer-reviewed journals so far. Our aim here is to provide an overview of the importance of open innovation as a research area by investigating the growth of related academic research, the areas in which open innovation has attracted most attention, and the breadth of open innovation in

Table 1.1 Books with open innovation in the title from 2003–2012

Author(s)/Editor(s)	Title	Year Published
Chesbrough	Open Innovation: The New Imperative for Creating and Profiting from Technology	2003
Chesbrough	Open Business Models: How to Thrive in the New Innovation Landscape	2006
Chesbrough, Vanhaverbeke, and West	Open Innovation: Researching a New Paradigm	2006
Gaule	Open Innovation in Action: How to Be Strategic in the Search for New Sources of Value	2006
OECD	Open Innovation in Global Networks	2008
Fasnacht	Open Innovation in Financial Services: Growing Through Openness, Flexibility and Customer Integration	2009
Hafkesbrink, Hoppe, and Schlichter	Competence Management for Open Innovation	2010
Lindegaard	The Open Innovation Revolution: Essentials, Roadblocks, and Leadership Skills	2010
Bingham and Spradlin	The Open Innovation Marketplace: Creating Value in the Challenge Driven Enterprise	2011
Chesbrough	Open Services Innovation	2011
Dahlander, Frederiksen, and Rullani	Online Communities and Open Innovation: Governance and Symbolic Value Creation	2011
Kinoshita	Service Entities in Open-Closed Innovation	2011
Rahman and Ramos	SMEs and Open Innovation: Global Cases and Initiatives	2011
Sloane	A Guide to Open Innovation and Crowdsourcing: Advice from Leading Experts	2011
Brem and Tidd	Perspectives on Supplier Innovation: Theories, Concepts and Empirical Insights on Open Innovation and the Integration of Suppliers	2012
de Pablos Heredero and Lopez	Open Innovation in Firms and Public Administrations	2012
Lyons, Coronado Mondragon, Piller, and Poler	Customer-Driven Supply Chains: From Glass Pipelines to Open Innovation Networks	2012
Spithoven, Teirlinck, and Frantzen	Managing Open Innovation: Connecting the Firm to External Knowledge	2012

Note: Based on search on amazon.com for books up to 2012 with search term "open innovation" in the title. Inclusion of books is determined by relevance to the practice and research of open innovation, while largely excluding pure "academic publishing," such as theses or dissertations.

Explicating Open Innovation 7

influencing other areas than business and management within social science as well as other areas outside of social science.

In order to identify relevant research, we searched for articles within the Thomson Reuters (formerly ISI) Web of Science[1] that have "open innovation" in the title, abstract or keywords and/or cite the original Chesbrough (2003a) book. Compared to some other reviews (e.g., Dahlander and Gann, 2010; West and Bogers, 2014), we opted for a broad and general approach by also including articles (1) that did not explicitly address or adopt the definition of open innovation as suggested by Chesbrough (2003a, 2006b) and (2) that were listed as editorial material, review or meeting abstract. Moreover, we searched for relevant articles not only in the Social Science Citation Index (SSCI) but also in the Science Citation Index[2] (SCI) and the Arts and Humanities Citation Index (A&HCI).

Figure 1.2 shows the growth of the number of publications related to open innovation with a total number of 941 articles[3] within all fields of SCI, SSCI, and A&HCI (shown as All). Open innovation research began in the business and management journals, and soon expanded to other social science journals in 2005, with a growing number of publications in the broader social science journals emerging by 2006.[4] By 2008, a further growth in journals publishing open innovation research emerged in the natural and physical sciences (and to a lesser extent in the arts and humanities).

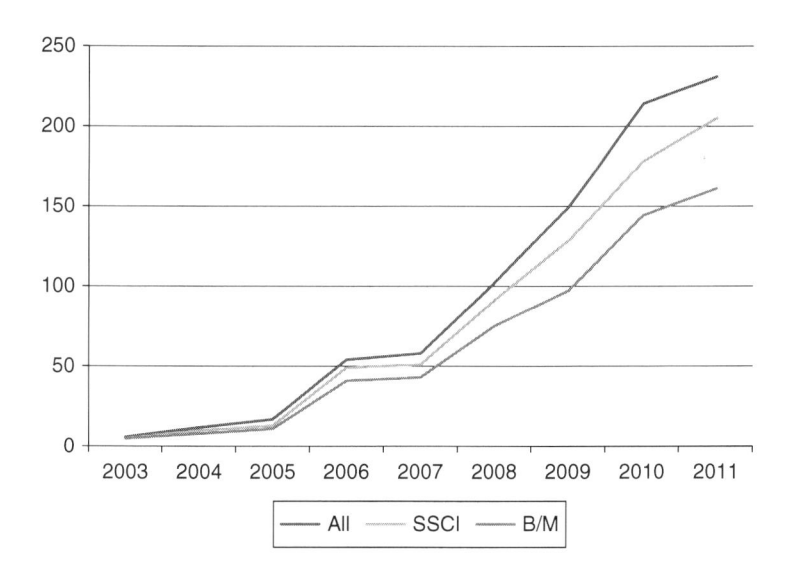

Figure 1.2 Growth of publications on open innovation in Web of Science

Notes: Search criterion: "open innovation" in title, abstract or keyword or citing Chesbrough (2003a); All = SCI, SSCI and A&HCI; SSCI = Social Science Citation Index; B/M = Business or Management category (within SSCI)

Table 1.2 Fields represented by journals in sample

Field (Web of Science Category)	Number of Articles
Management	603
Business	338
Industrial Engineering	173
Operations Research & Management Science	129
Planning & Development	123
Economics	72
Engineering, Multidisciplinary	47
Information Science & Library Science	43
Geography	32
Environmental Studies	29
Computer Science, Information Systems	26
Urban Studies	22
Chemistry, Multidisciplinary	20
Multidisciplinary Sciences	20
Engineering, Electrical & Electronic	15
Computer Science, Software Engineering	14
Public Administration	14
Computer Science, Interdisciplinary Applications	13
Pharmacology & Pharmacy	13
Computer Science, Artificial Intelligence	11
Engineering, Manufacturing	11
Chemistry, Medicinal	10
Computer Science, Theory & Methods	10
Biotechnology & Applied Microbiology	9
Food Science & Technology	9
Telecommunications	9
Education & Educational Research	8
Social Sciences, Interdisciplinary	8
Applied Psychology	7
Sociology	7
Art	5
Biochemistry & Molecular Biology	5
Engineering, Chemical	5
(Topics with four publications or fewer)	105
Total	**1965**

Note: One journal may be listed in multiple categories, which explains why the total number of categories is higher than the total number or articles in the sample.

Table 1.2 shows this breadth of the influence of open innovation by presenting the number of articles in the various Web of Science Categories, which are used to classify journals and thereby the articles in them. The table shows that management and business are the most prominent categories. At the same time, other areas such as engineering, geography, computer science, and chemistry are also well represented in open innovation research. Such publications generally include case studies or more general discussions of open innovation in particular industries, such as the chemical, pharmaceutical, and food industry (although we may note that low-tech industries are still understudied). Also represented are proposals for how open innovation can solve larger social problems, such as neglected diseases.

1.2.3 The Impact of Open Innovation Research

We now consider the most cited open innovation-related research. Table 1.3 shows the 20 most cited articles that are in our sample; such citation counts are biased towards earlier publications and cannot identify more recent work that will be influential in the future.

The table lists articles that we would consider to be within the domain of open innovation as well as those that do not address open innovation per se. For the latter, there are a number of articles that do not advance our understanding of open innovation itself but instead rely on the open innovation concept to further advance the understanding of other areas. The most cited article, Teece (2007), is much more focused on advancing theories of business strategy than it is about theories of industrial innovation. Teece argues that embracing open innovation is an important part of firms' dynamic capabilities—thus bringing open innovation explicitly into the domain of theories of firm strategy. A similar argument, although less pronounced, could be made for Rothaermel and Hess (2007); other articles in this list also apply open innovation to advance the understanding of business strategy. Table 1.3 classifies these 20 articles into seven categories, based on our analysis from reading each of the articles in detail.

While some articles specifically address product development models—such as Cooper (2008) who discusses reinventing his stage-gate model for use with open innovation—most articles address some aspect of the innovation process. This includes Laursen and Salter's (2006) article on the (optimal) use of external knowledge for innovation performance, the most cited journal article that is at its core about open innovation. Other articles that specifically deal with the role of universities—such as university-industry relationships or openness from the university's perspective—and environmental context have a less explicit focus on the core open innovation concept. It is moreover important to note that only a few of these influential articles specifically

Table 1.3 Top 20 most cited articles in sample with forward citations, main topics, and levels of analysis

Author (year)	Title	Journal	Citations	Strategy	Product development	Innovation process	Toolkits/ users	Limits/ risks/ costs	University	Environmental context	Individual/ Group	Firm/ Organization	Network	Industry/ Sector	National/ Institutional
Teece (2007)	Explicating dynamic capabilities: The nature and microfoundations of (sustainable) enterprise performance	Strategic Management Journal	324	•								•			
Laursen and Salter (2006)	Open for innovation: The role of openness in explaining innovation performance among UK manufacturing firms	Strategic Management Journal	284			•		•				•			
Chesbrough (2003b)	The era of open innovation	Sloan Management Review	214		•	•			•			•			
Rothaermel, Agung, and Jiang (2007)	University entrepreneurship: A taxonomy of the literature	Industrial and Corporate Change	101			•			•						
Chesbrough and Crowther (2006)	Beyond high tech: Early adopters of open innovation in other industries	R&D Management	86		•	•						•		•	

Gassmann (2006)	Opening up the innovation process: Towards an agenda	R&D Management	85					•			•	
Cooke (2005)	Regionally asymmetric knowledge capabilities and open innovation exploring "Globalization 2": A new model of industry organization	Research Policy	70									
Perkmann and Walsh (2007)	University–industry relationships and open innovation: Towards a research agenda	International Journal of Management Reviews	70	•				•			•	
Henkel (2006)	Selective revealing in open innovation processes: The case of embedded Linux	Research Policy	68	•		•		•				
Rothaermel and Hess (2007)	Building dynamic capabilities: Innovation driven by individual-, firm-, and network-level effects	Organization Science	65	•	•			•			•	
Piller and Walcher (2006)	Toolkits for idea competitions: A novel method to integrate users in new product development	R&D Management	64		•		•					

(continued)

Table 1.3 Continued

Author (year)	Title	Journal	Citations	Strategy	Product development	Innovation process	Toolkits/ users	Limits/ risks/ costs	University	Environmental context	Individual/ Group	Firm/ Organization	Network	Industry/ Sector	National/ Institutional
Christensen, Olesen, and Kjaer (2005)	The industrial dynamics of Open Innovation: Evidence from the transformation of consumer electronics	Research Policy	63	•		•						•	•	•	
Dodgson, Gann, and Salter (2006)	The role of technology in the shift towards open innovation: The case of Procter & Gamble	R&D Management	63		•	•	•			•					
Thrift (2006)	Re-inventing invention: New tendencies in capitalist commodification	Economy and Society	62			•						•			•
Nieto and Santamaria (2007)	The importance of diverse collaborative networks for the novelty of product innovation	Technovation	61		•	•	•	•					•		
Fleming and Waguespack (2007)	Brokerage, boundary spanning, and leadership in open innovation communities	Organization Science	56			•					•		•		

Article	Journal										
West and Gallagher (2006)	Challenges of open innovation: The paradox of firm investment in open-source software	R&D Management	55	•		•		•		•	•
Jacobides, Knudsen, and Augier (2006)	Benefiting from innovation: Value creation, value appropriation and the role of industry architectures	Research Policy	55	•		•		•	•	•	•
Cooper (2008)	Perspective: The Stage-Gate (R) idea-to-launch process-update, what's new, and NexGen systems	Journal of Product Innovation Management	54		•	•	•			•	
Chesbrough and Appleyard (2007)	Open innovation and strategy	California Management Review	51	•		•			•	•	
Total for all 20 articles:			1951	6	6	17	6	4	4	5	16	9	3	2

address the limits, risks, and costs of open innovation. Finally, a number of articles consider the role of technology and users in innovation. An interesting example here is the Fleming and Waguespack (2007) article on what they call "open innovation communities," although they do not explicitly refer to the Chesbrough (2003a) notion of open innovation.[5]

1.2.4 Topics in Open Innovation Research

We now address some of the main themes that have emerged in the literature on open innovation. To complement our general understanding of this literature and what we know from other recent reviews, we scanned through the articles in our database to iteratively construct patterns of recurrent themes and we developed a word cloud as a tool to find the most frequently recurring words in the articles' titles, abstracts and keywords.[6]

Figure 1.3 provides a word cloud with the most commonly occurring words in the abstracts of the 941 articles in our sample. It is no surprise that "open" and especially "innovation" occur frequently here. At the same time, words like "firms," "knowledge" and "technology" are also frequently used. Moreover, the appearance of words like "performance," "value," "strategies" and "management" reinforce the observation that open innovation research commonly deals with (typically large) for-profit firms.

As a complement, Figure 1.4 presents a word cloud based on the most frequently occurring words in the articles' keywords (instead of the articles' abstracts in Figure 1.3).[7] Here we take out the words "open" and "innovation" in order to get a better understanding of the related themes within open innovation. Based on our analysis, we find that a large amount of research investigates the Outside-In (inbound) side of open innovation. This research deals with how firms can leverage external knowledge and technology to accelerate internal innovation. There is often a focus on R&D with a prominent role of absorptive capacity. There is generally less focus on the Inside-Out

Figure 1.3 Word cloud based on abstracts, top-30 words

Note: Generated with Wordle.net

Figure 1.4 Word cloud based on author keywords, top-50 words excluding "open" and "innovation"

Note: Generated with Wordle.net

(outbound) side of open innovation, although the growing interest in this area is reflected by the research on for example intellectual property (IP), licensing and selective revealing. This also relates to a growing interest in business models and business model innovation. Moreover, again only partly reflected in Figure 1.4, the broad literature on open innovation addresses several mechanisms, tools and processes to identify and leverage external innovation sources, such as networks, communities, and innovation contests, often with an interest in the role of users in innovation. More recently, the notion of broadcast search through crowdsourcing has attracted particular attention in the scholarly community. Finally, open innovation research is expanding to other domains, such as low-tech industries and services, while it is also increasingly being linked to other themes, such as outsourcing and (academic) entrepreneurship, although the exact linkages with open innovation are typically not yet fully developed.

1.3 WHAT OPEN INNOVATION IS AND IS NOT

We anchor open innovation in the prior economics and management of innovation literature, and accordingly further develop and clarify the conceptualization of open innovation, while also proposing a refined definition. Based on this, we then juxtapose open innovation to other related concepts, and discuss their differences and complementarities.

1.3.1 Clarifying the Open Innovation Concept

At the most fundamental level, open innovation is embedded in the notion that the sources of knowledge for innovation are widely distributed in the economy. As such, the idea that most smart people work for someone else, popularized as Joy's Law, harks back to Hayek's (1945) view of knowledge as being distributed across society. When Chesbrough (2003a) inaugurated the popular use of the term open innovation, it described a phenomenon of companies making greater use of external ideas and technologies in their own business, and letting unused internal ideas and technologies go outside for others to use in their business. The book proposed "erosion factors" that undercut the logic of the earlier "closed innovation" model of R&D and developed the logic of an open innovation model. These erosion factors, such as increased mobility of workers, more capable universities, declining US hegemony, and growing access of startup firms to venture capital, changed the conditions under which firms innovate. Here we propose yet another erosion factor that allows firms to leverage increasingly distributed knowledge sources, namely the rise of the Internet (and the related rise of social media), which has brought the knowledge access and sharing capabilities of previously firm-specific internal ICT networks to the World Wide Web. These erosion factors are at the core of why open innovation reflects a paradigm shift as they challenge the basic assumptions, problems, solutions and methods for the research and practice of twenty-first century industrial innovation (cf. Kuhn, 1962).

Following the introduction of the open innovation concept, Chesbrough's (2006b) "purposive inflows and outflows of knowledge" definition of open innovation was intended to make the definition more formal and to connect it to previous academic work. Accordingly, open innovation refers to an innovation model that emphasizes purposive inflows and outflows of knowledge across the boundary of a firm in order to leverage external sources of knowledge and commercialization paths, respectively. Here we provide further motivation for this conceptualization.

The definition of "purposive inflows and outflows of knowledge" hearkens back to a vibrant economic literature on spillovers that arise from the firm's investment in research and development. Because firms cannot fully specify the outcomes of this investment in advance, R&D inevitably produces outcomes that were not expected *ex ante*. These outcomes spill over beyond the ability of the investing firm to benefit from them, hence the term "spillovers." Richard Nelson observed back in 1959 that basic research generated many spillovers, and that firms who funded this research had only limited ability to appropriate value from these spillovers (Nelson, 1959). Kenneth Arrow (1962) also took note of this spillover problem, recognizing that these spillovers meant that the social return to R&D investment exceeded that of the private return to the firm undertaking the investment. Hence, he reasoned, private firms will underinvest in R&D from a social perspective. It is reasonable in this context for the public to provide a subsidy for R&D investment, in order to stimulate further R&D to move closer to the socially ideal level. Cohen and Levinthal (1990) in turn wrote about the importance of investing in internal research in order to be able to utilize external technology, an ability they termed "absorptive capacity." Nathan Rosenberg asked the related question, why do firms conduct basic research with their own money (Rosenberg, 1990), and answered that this research enhanced the firm's ability to use external knowledge. It is important to note, however, that the specific mechanisms to enable companies to absorb external knowledge were not identified by these scholars. Nor was there any consideration of companies opting to move unused internal knowledge out to the wider environment.

This prior research points out the presence of spillovers, and the benefits of being able to utilize them when they exist in one's surrounding environment. Throughout this literature, however, spillovers are deemed a cost to the focal firm of doing business in R&D, and are judged to be essentially unmanageable. This is the critical conceptual distinction made by the open innovation concept, which proposes that, in the open innovation model of R&D, spillovers are transformed into inflows and outflows of knowledge *that can be purposively managed*. Firms can develop processes to seek out and transfer external knowledge into their own innovation activities. Firms can also create channels to move unutilized internal knowledge from inside the firm out to other organizations in the surrounding environment. Specific mechanisms can be designed to direct these inflows and outflows of knowledge. Thus, what was unspecified and unmanageable before can now be specified and managed in the open innovation model.

These elements then give a basis to refine the definition of open innovation. Also following the original and more recent conceptualizations (Chesbrough, 2003a, 2006b; Gassmann and Enkel, 2004; Dahlander and Gann, 2010; West and Bogers, 2014), we define open innovation as a distributed innovation process based on purposively managed knowledge flows across organizational boundaries, using pecuniary and non-pecuniary mechanisms in line with the organization's business model. These flows of knowledge may involve knowledge inflows to the focal organization (leveraging external knowledge sources through internal processes), knowledge outflows from a focal organization (leveraging internal knowledge through external commercialization processes) or both (coupling external knowledge sources and commercialization activities), as we discuss further. In this definition, innovation refers to the development and commercialization of new or improved products, processes, or services, while the openness aspect is represented by the knowledge flows across the permeable organizational boundary, as also shown in Figure 1.5. As an organizational construct it is moreover the business model, which may be implicit or explicit, that puts the distributed innovation process into the organizational realm as it describes not only how value is created within the value network but also how it is captured by the involved organization(s).[8]

1.3.2 Types and Mechanisms of Open Innovation

Following the above conceptualization of open innovation, the purposive management of knowledge spillovers essentially implies two directions of knowledge flows across the boundary of the firm or organization more generally: Outside-In (or inbound); and Inside-Out (or outbound) open innovation. Following Gassmann and Enkel (2004), to these we add a third type, namely Coupled open innovation, which implies combined knowledge inflows and outflows between actors in the innovation process. Figure 1.5 provides an illustration of these different types of open innovation. The figure shows different paths that knowledge or technology may follow within and across the boundaries of the firm. It also extends the open innovation model from upstream R&D to manufacturing and marketing (as examples of functional areas) to thereby highlight the importance of more downstream activities in the overall innovation process (Bogers and Lhuillery, 2011) and to emphasize the importance of considering all activities from invention to commercialization in order to create *and* capture value from ideas and technologies (Chesbrough, 2006a; West and Bogers, 2014).

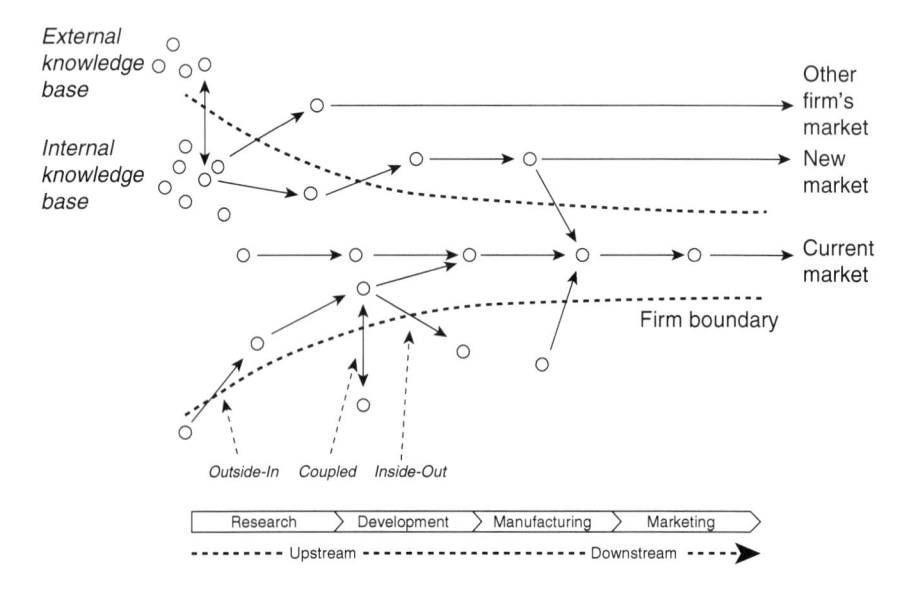

Figure 1.5 The open innovation model

As shown in the analysis above, extant academic research and industry practice has paid most attention to Outside-In open innovation, while the Inside-Out and Coupled types are less well understood. For example, in a review of 165 open innovation articles, West and Bogers (2014) find 118 addressing Outside-In open innovation, in contrast to 50 articles addressing the Inside-Out type. Moreover, while 70 articles in their sample address Coupled open innovation, there remains a lack of understanding of the interactive and reciprocal nature of such coupled innovation processes.

The Outside-In type of open innovation involves opening up a company's own innovation processes to many kinds of external inputs and contributions, for example through acquiring or sourcing, following Dahlander and Gann (2010), while West and Bogers (2014) identify obtaining, integrating and commercializing as phases of the inbound open innovation process. The company's business model, in turn, determines which external inputs and contributions will be taken forward into the market. In open innovation (Chesbrough, 2003a, 2006a), a variety of mechanisms are identified that help firms manage purposive inflows of knowledge: scouting, inlicensing IP, university research programs, funding startup companies in one's industry, or collaborating with intermediaries, suppliers and customers, and utilizing non-disclosure agreements. Subsequent research has identified additional mechanisms, including crowdsourcing, competitions and tournaments, communities, and spin-ins or spin-backs.

The Inside-Out type of open innovation requires organizations to allow unused and under-utilized ideas and assets to go outside the organization for others to use in their businesses and business models (cf. Arora, Fosfuri, and Gambardella, 2001a; Maarse and Bogers, 2012; Tranekjer and Knudsen, 2012). Following Dahlander and Gann (2010), interactions in this type of open innovation can involve either selling or revealing. The business model for the idea often will differ from that of the company from which it came, and often the business model must be discovered in order to take the idea to market. The mechanisms for managing outflows of knowledge from the firm as also identified by Chesbrough (2003a, 2006a; Chesbrough and Garman, 2009) include outlicensing IP and technology, donating IP and technology, spin-outs, corporate venture capital, corporate incubators, joint ventures and alliances (i.e., becoming a supplier to or a customer of a new initiative, vs. executing the initiative internally).

A third type of open innovation links Outside-In and Inside-Out open innovation processes (Bogers, 2012; Enkel et al., 2009; Gassmann and Enkel, 2004). This Coupled type of open innovation involves combining purposive inflows and outflows of knowledge to collaboratively develop and/or commercialize an innovation. Coupled open innovation involves two (or more) partners that purposively manage mutual knowledge flows across their organizational boundaries through joint invention and commercialization activities (Bogers, 2011; Bogers, Bekkers, and Granstrand, 2012). While

Coupled open innovation may in principle involve any combination of the respective mechanisms for Outside-In and Inside-Out open innovation, companies may implement specific mechanisms, such as strategic alliances, joint ventures, consortia, networks, ecosystems and platforms, all involving complementary partners.

1.3.3 Divergent Views on "Open" Innovation

There have emerged various perspectives that build on the distributed nature of innovation sources (cf. Bogers and West, 2012). Despite some common ground and complementarities, there are also stark differences between some of these perspectives. These definitional disputes matter to the further development of academic research in this area. Clay Christensen, a sympathetic outside observer of an Academy of Management symposium[9] from August 2012, reported the following observation about the confusion resulting from different ways open innovation was being defined:

> There is also a major downside to being *imprecise* in the definition of open innovation. An imprecise definition not only makes open innovation more difficult to understand, because little research is done on the actual phenomenon, but it also makes it more difficult to implement, because there are a lot of people who claim to talk about "open innovation" but are actually talking about something else. (Christensen, 2012)

The most notable complementary perspective to open innovation relates to the distributed social division of labor as pioneered by Eric von Hippel (1988, 2005). This perspective emphasizes the public good nature of much of the innovations that have been described by von Hippel and others, conceptualized through notions as "open, distributed innovation" (von Hippel, 2005), "open source innovation" (Raasch, Herstatt, and Balka, 2009) and "open collaborative innovation" (Baldwin and von Hippel, 2011). However, we observe that the divergent use of terminology for open innovation has contributed to some confusion in the literature. For example, the term "open" has been used as a synonym for "user-centric," as exemplified by von Hippel's use of the terms:

> In this book I explain in detail how the emerging process of user-centric, democratized innovation works.... Open, distributed innovation is "attacking" a major structure of the social division of labor. (von Hippel, 2005: 2)

His specific perspective on open innovation has more recently been described as follows:

> An innovation is "open" in our terminology when all information related to the innovation is a public good—nonrivalrous and nonexcludable.... It differs

fundamentally from the recent use of the term to refer to organizational permeability. (Baldwin and von Hippel, 2011: 1400).

While this schism in definitions may to some extent reflect a fundamental difference in the conceptualization of innovation, this difference may be more a matter of emphasis than reflecting incompatible perspectives. On the one hand, "open innovation" entails purposefully managing knowledge flows across the organizational boundary as well as the associated business model as defining features. On the other hand, "open collaborative innovation" and related notions refer to an innovation model that emphasizes low-cost or free production of public, non-rivalrous, non-excludable goods. There are indeed some signs of convergence in that these perspectives are becoming part of a larger holistic research domain that draws on the notion of distributed sources of knowledge for innovation (cf. Bogers and West, 2012). For example, Dahlander and Gann (2010) have developed a useful integration of the two definitions, yielding a framework based on inflows and outflows of knowledge, while also respecting both pecuniary and non-pecuniary motivations for participation. As such, firms may selectively reveal some of their knowledge, thus relying on "best of both worlds" (von Hippel and von Krogh, 2003). More fundamentally, some business models support open source development methods, and publishing one's discoveries has become an important part of IP management strategy (Chesbrough and Appleyard, 2007). The different perspectives moreover appear to reflect complementary perspectives on the innovation process given that users may be particularly important sources of innovation in the early phases of industry development, while firm involvement is typically needed for the further development and scaling up in the industry (cf. Baldwin, Hienerth, and von Hippel, 2006; Bogers, Afuah, and Bastian, 2010; Shah and Tripsas, 2007). In other words, after the initial inventions by users, business models help to further advance the relevant products and processes by capturing some of the public good knowledge, attracting capital, scaling the innovations, and thereby creating an economically sustainable business or industry.

1.4 RESPONDING TO CRITIQUES ON OPEN INNOVATION

One result of the growing academic attention that open innovation has received is the emergence of critiques of the concept. These academic critiques are an invaluable part of the scholarly process. They challenge weak or unsupported arguments. They highlight errors in logic or gaps in supporting evidence. It is only appropriate, therefore, to discuss these critiques, evaluate their main arguments, and provide a response that can guide future scholars in their own assessment of open innovation.

In general, the critiques are broadly characterized by two main arguments. First, some have argued that open innovation, as it has been introduced by Chesbrough (2003a), is not in fact a new phenomenon. This implies that open innovation does not rise to the standard of being a valid, novel insight into innovation. One critique summarizes its viewpoint with its pithy title, "Old wine in new bottles" (Trott and Hartmann, 2009), while the other states in its title, "Plus ca change…" (Mowery, 2009). A second, different contention is that open innovation is explaining something new, but these new phenomena can be explained by an already established concept. This implies that we don't need a new concept or theory to account for the phenomenon. This critique claims that open innovation may be a "communication barrier for theory development" (Groen and Linton, 2010).

1.4.1 Old Wine in New Bottles?

Trott and Hartmann (2009) engage the Chesbrough (2003a) book at some length, and subject it to a rhetorical analysis. They argue that the concept of Closed Innovation was a rhetorical strawman that no company ever followed. They claim that companies have always been open in their innovation processes, so that there was no real change in innovation paradigms from Closed to Open. There are, however, specific points in the 2003 book that do explicate a novel phenomenon. Chesbrough (2003a) identified erosion factors that influence the conditions under which innovation takes place (e.g., increased mobility of workers, more capable universities, declining U.S. hegemony, and growing access of startup firms to venture capital) have changed, giving rise to a new paradigm in which firms need to be and benefit more from being open for innovation. This does not imply that individual elements of open innovation were absent in the earlier paradigm, but rather that they now combine to form a new paradigm to manage innovation. These points, however, are not discussed in Trott and Hartmann's analysis.

Trott and Hartmann also criticize the book for taking insufficient account of prior academic literature in its development of open innovation. We note, however, that many of the 174 footnotes in Chesbrough (2003a) cite prior academic research. And we can observe that Trott and Hartmann adopt their own rhetorical strawman in their critique by considering only one work in a much larger stream of research. That is, their criticism is based entirely on the Chesbrough (2003a) book, which was aimed at managers as well as academics. In contrast to managers, who need frameworks to make sense of complex phenomena (and have only a limited tolerance for prior academic work), scholars are a different kind of audience, requiring limits and caveats to a new concept,

Explicating Open Innovation 23

as well as extensive references to prior literature. These are all present in the Chesbrough et al. (2006) treatment and the many academic papers that were subsequently inspired by this work, but none are thoroughly engaged in this 2009 critique.

Finally, Trott and Hartmann piece together prior literature from a wide variety of fields in their critique of open innovation. In our view, this actually shows one of the benefits of the open innovation concept, namely that it is a novel synthesis of many previously disparate points. Trott and Hartmann acknowledge this in their closing paragraphs, in which they themselves highlight the value of the open innovation concept in reaching "new audiences...that the innovation and R&D literatures failed to reach for so many years" and in creating "real-life laboratories" for studying the mechanisms of open innovation.

1.4.2 Plus Ca Change?

Mowery's (2009) article provides a different, and in many ways contrasting, critique of open innovation than that offered by Trott and Hartmann. As his title suggests, Mowery's thesis is that the developments of the past generation are returning innovation to a model not unlike that which prevailed during the latter part of the nineteenth and early twentieth century. Mowery's discussion of open innovation, regrettably, is incomplete, with little notice taken of any of the evidence or the analysis offered in the Chesbrough (2003a) book, such as the change in erosion factors noted above. Nor does he, like Trott and Hartmann, consider the subsequent open innovation research.

However, his detailed analysis of the U.S. innovation system over the past century is quite well done and actually provides further evidence for some of the erosion factors noted by Chesbrough (2003a). His analysis is based on secondary data sources, most of them from the level of the U.S. economy, while the Chesbrough (2003a) volume was supported primarily from qualitative studies of specific firms' R&D processes using primary data sources. Yet, though the evidence and methods of the Mowery article differ substantially from the Chesbrough (2003a) book, his article confirms that the process of industrial innovation has indeed changed in the past 40–50 years. This stands in stark contrast to Trott and Hartmann's critique, which asserts that little has changed.

Mowery's critique also overstates his argument. There is indeed some continuity between the innovation systems of a century ago and the systems of today. The large-scale industrial research laboratories that arose during the last century have receded. But to state that we have returned to innovation conditions of a century ago overlooks a great deal that is new. The roles of startup firms, of venture capital, the growth of federally funded university research, of Bayh-Dole IP rules for university ownership of that taxpayer-funded research, the strengthened IP protection arising from the creation of a dedicated federal circuit court, and the growth of R&D activity in SMEs relative to R&D activity today in large companies—all of these differ substantially from the period of the second industrial revolution (when some of the large businesses of that era were monopolies, such as railroad, oil, telegraph and telephone companies, etc.).

There is, however, a very good comment made in this otherwise incomplete critique. Institutional context is important to the operation of innovation processes in general and to open innovation in particular. Mowery is right to emphasize that this needs to be included in any analysis of innovation, including open innovation. This implies that open innovation will work differently in different institutional environments, and may not work effectively in at least some of them. This remains an important area for future research in open innovation.

1.4.3 Communication Barrier?

Finally, Groen and Linton (2010) critique open innovation from a very different perspective from either earlier critique. They acknowledge that much has changed about industrial innovation over the past generation, but question whether one need resort to a new concept like open innovation to explain it. The concept of the supply chain covers everything open innovation covers, in their view. They propose that the term "open innovation" may be "hindering growth in research and understanding [because it may create] false barriers that inhibit communication between different groups of academics." (Groen and Linton, 2010: 554).

This critique revolves again around definitions. Innovation is about the creation of new products, services, and processes, while the supply chain is about the management of existing products, services, and processes throughout the value chain. Open innovation also includes many more innovation actors than does the concept of the supply chain from raw material to the final consumer. Universities, research consortia, spin-offs, false negatives, business models, venture capital, IP management, ideation contests, innovation communities—to name a few of the open innovation stakeholders and processes—have no place in the supply chain literature.

While open innovation has some overlap with supply chain management (and also with product development models like Robert Cooper's Stage Gate model), we come to the opposite conclusion of Groen and Linton (2010), as open innovation (1) deals specifically with the creation of new products, services and processes and (2) covers a much broader range of possible stakeholders, which provides value in using the term open innovation as distinct from supply chain management.

1.5 ADDRESSING THE RESEARCH AGENDA FOR OPEN INNOVATION

In the earlier book, Chesbrough et al. (2006) devoted a final chapter to the topic of a research agenda for open innovation. It is useful to briefly review that chapter, and assess the research reported in this chapter according to the agenda articulated earlier.

1.5.1 Levels of Analysis

West, Vanhaverbeke, and Chesbrough (2006) outlined five different levels of analysis for future open innovation research. They were:

1. Individual and Groups
2. Firm/Organization
3. Inter-organizational Value Networks
4. Industry and Sector
5. National Institutions and Innovation Systems

The authors concluded that extensive work had been conducted at the organization level of analysis, and suggested that more research was needed in these other areas of analysis.

The articles examined in this chapter allow us to update our perspective on the research conducted to date under each level of analysis. Using the top-20 most cited articles in the past decade, Table 1.3 documented the primary level(s) of analysis for each paper.[10] As the table shows, the Firm/Organization level of analysis remains the most examined in these highly cited articles, while the Network level has received significant attention as well. The other levels of analysis have by comparison not received a similar amount of attention.

Since 2006, a few new possible levels of analysis have arisen that may offer a more fine-grained framework for open innovation research. For example, West and Lakhani (2008) argued that communities are a distinct level of analysis, given that communities are in essence an extra-organizational set of actors and thus different than the notion of inter-organizational networks as proposed by West et al. (2006). Moreover, there may be a number of intra-organizational units of analysis that show important heterogeneity with respect to firm's open innovation performance, such as business units, functional areas or innovation projects. For example, Bogers and Lhuillery (2011) showed how innovation in R&D, manufacturing, and marketing were each correlated with a distinct set of external sources of knowledge. On a higher level, we may also consider the society at large given the opportunities of open innovation in the public commons, through for example initiatives like open government and open data. As

Table 1.4 Possible units of analysis and research objects for open innovation research

Unit of Analysis	Possible Research Object
Intra-organizational	Individual
	Group/Team
	Project
	Functional area
	Business unit
Organizational	Firm
	Other (non-firm) organization
	Strategy
	Business model
Extra-organizational	External stakeholders: individual, community, organization
Inter-organizational	Alliance
	Network
	Ecosystem
Industry	Industry development
	Inter-industry differences
Regional innovation systems	Local region
	Nation
	Supra-national institution
Society	Citizens
	Public policy

such, the erosion factors noted above need to be supported by public policy, along with for example appropriate policies toward IP, and access to university research funded from public sources (Chesbrough and Vanhaverbeke, 2012; de Jong, Kalvet, and Vanhaverbeke, 2010).

Table 1.4 offers a larger list of possible units of analysis and research objects that could be further explored in future research. Some of the open innovation research, which we reviewed here, would embed multiple units of analysis or recombine multiple research objects, although a more comprehensive understanding of the interaction across various levels of analysis is generally still lacking (cf. Gupta, Tesluk, and Taylor, 2007).

1.5.2 Conclusion

Much has been written since the advent of open innovation a decade ago. Thousands of articles, which have received numerous citations, attest to the growing academic interest in open innovation. Some of this research is highly cited, suggesting that open innovation is not simply old wine in new bottles—there is a new paradigm being constructed for the conception of industrial innovation in the twenty-first century. Yet there have been disparate definitions employed during the past decade for open innovation, and the research that has been done is less coherent as a result.

We propose the following definition of open innovation, in hopes of unifying future work in this area: open innovation is a distributed innovation process based on purposively managed knowledge flows across organizational boundaries, using pecuniary and non-pecuniary mechanisms in line with each organization's business model. These flows of knowledge may involve knowledge inflows to the focal organization (leveraging external knowledge sources through internal processes), knowledge outflows from a focal organization (leveraging internal knowledge through external commercialization processes) or both (coupling external knowledge sources and commercialization activities).

We know from studies of the role of standards in innovation that establishing standards can help to direct further innovation in more efficient, productive ways (Shapiro and Varian, 1999). Adopting a consistent definition of open innovation, and building upon that shared definition, would help accelerate our understanding of this new approach to innovation. We hope that our compilation of open innovation research in this chapter, along with our discussion of definitions and critiques, will help us move in this direction as a community of scholars.

ACKNOWLEDGMENT

We thank Wim Vanhaverbeke and Joel West for useful feedback on earlier drafts.

NOTES

1. We retrieved the database that serves as a basis for our analysis on July 10, 2012. We cleaned the database by completing missing data (e.g., missing abstracts) by obtaining that information from the publisher website.
2. Specifically, the database we used was SCI-EXPANDED. We did not include conference proceedings, which are also indexed by the Web of Science.
3. Of these 941 articles, 731 cite Chesbrough (2003a) and 411 have "open innovation" in title, abstract or keywords.
4. Only looking at SSCI gives 679 articles that cite Chesbrough (2003a) and 337 with "open innovation" in title, abstract or keywords, which combined are 828 articles.

5. Notably absent from this list are the works of Eric von Hippel, a highly-cited innovation scholar who has written extensively on the sources of innovation. This absence can be attributed to the framing of his work, which represents a complementary perspective (more below) and which generally did not use the phrase "open innovation" or cite Chesbrough (2003a), thus not meeting the criteria to enter our sample—exceptions being von Hippel and von Krogh (2006) and the more recent Baldwin and von Hippel (2011).
6. For this purpose, we use Wordle.net, which relies on an algorithm to present word clouds that "give greater prominence to words that appear more frequently in the source text" (http://www.wordle.net; Accessed: July 9, 2012). As the relative size of the words is determined by the number of times it occurs, it is a useful tool to identify some of the recurrent themes in a body of research. Note that Wordle can automatically exclude stop words such as "the" and "and."
7. This concerns the keywords as suggested by the authors in the articles themselves. We have also looked at the keywords as proposed by the Web of Science, which are based on recurring words in for example the articles' references. Although there are some other dominant keywords/themes here (e.g., "alliances," "biotechnology," "science" and "systems"), the findings are roughly similar and reflected in our general analysis.
8. In our experience, public and non-profit organizations often resist the notion that they somehow operate with a "business model." Yet non-profits have to sustain their operations financially over time, and need to procure resources from the environment to do this. And public organizations need to maintain political support from resource providers in order to continue their activities as well. Thus, in this broader context, non-profit and public organizations need to create value and capture a portion of that value, in order to continue to operate.
9. Christensen reports on the symposium entitled "Open Innovation and the Theory of the Firm: (How) do Organizations and Boundaries (still) Matter?," organized by Marcel Bogers and Teppo Felin (see http://www.marcelbogers.com/?q=AOM2012-OpenInnovationTheory).
10. Our coding is based on our assessment of which level of analysis provided the key variable(s) in the articles. In most cases, we can tie this coding scheme to the open innovation concept as defined above, although some articles are considered based on their complementary and indirect relevance given that definition. For example, even though Rothaermel et al. (2007) consider openness from the university (i.e., organization) point of view, we code it as a Network as it relates to a company's innovation network and thereby potential of (dyadic) knowledge exchange.

2

Firms, Users, and Innovation

An Interactive Model of Coupled Open Innovation

Frank Piller and Joel West

2.1 INTRODUCTION

Researchers on open innovation (OI) and user innovation (UI) share certain assumptions and precepts. Perhaps most importantly, they agree that knowledge relevant for innovation is widely dispersed outside the firm (Bogers & West, 2012). Henry Chesbrough (the father of open innovation) writes that "useful knowledge is generally believed to be widely distributed, and of generally high quality" (Chesbrough, 2006b: 9), while Eric von Hippel (the father of user innovation) concludes that "the information needed to innovate in important ways is widely distributed" (von Hippel, 2005: 14).

However, OI and UI are at best partly overlapping perspectives on this distributed model of innovation. While the two differ in their values and assumptions, an important factor in their limited commensurability is their tendency to study different phenomena. Open innovation is a firm-centric paradigm that is primarily concerned with leveraging external knowledge to improve internal innovation and thus the firm's economic performance. User innovation is mainly about individuals using innovation to address their own (often unique) needs, without regard to firm success and often as part of a socially embedded community.

In this chapter, we focus on the overlap of these two perspectives: when individual users innovate in ways that improve the offerings of firms. In some cases, firms leverage existing innovations by users; in other cases, firms and users collaborate to create innovations that have both use or social value for users and commercial value for firms. We begin by reviewing the user and open innovation literature, and then contrast their overlapping and divergent assumptions. From this, we summarize and extend the research on the "coupled" process of open innovation suggested by Gassmann & Enkel (2004), identifying three distinct dimensions of coupled processes: the nature of the external actor (individual vs. organizational), the topology of collaboration (dyadic vs. network), and the locus of innovation (whether collaboration between separate efforts or in a joint process of interactive co-creation).

Combining these streams, we focus specifically on the joint production of innovation by firms and individuals. We propose a four-phase model of interactive coupled open innovation that combines earlier conceptions of inbound open innovation with the collaboration tools and processes that make such production possible (West & Bogers, 2014; Diener & Piller, 2013). Our model connects the open and user innovation literature with the perspective of co-creation, a school of research from the marketing literature that has been largely unconnected to open innovation. We then discuss each of the stages of this model from the perspective of both open innovation and user innovation. We conclude with specific suggestions for future research.

2.2 CONTRASTING USER AND OPEN INNOVATION

Researchers in user and open innovation have overlapping but not entirely congruent perspectives on the process of innovation outside the firm, including how firms can harness the innovations of external individuals, such as users or consumers. Table 2.1 summarizes some of the key attributes of these two large bodies of research.[1]

2.2.1 User Innovation: Learning from Lead Users

User innovation was proposed by von Hippel (1988, 2005, 2010) as an alternative model to the dominant view in management that innovation results from activities of producers and managers. In this model, users are not "consumers" of products created by "producer" firms, but instead are empowered (often as "self-manufacturers") to create their own products and services. The users may be individuals or firms—either focused solely on their own needs, or collaborating in communities to share their creations. User innovation thus has three key premises: users have unique ("sticky") information about their needs, when enabled they will create solutions to those needs, and they may freely reveal their results to others (von Hippel, 2010).

A large body of empirical work has shown that users have been the originators of many industrial and consumer products (Urban & von Hippel, 1988; von Hippel, Ogawa, & de Jong, 2012). Especially when markets are fast-paced or turbulent, so called lead users face specific needs ahead of the rest of the market. When they want something that is not available on the market, users have an incentive to innovate from the direct use benefits they obtain from their innovation effort. Users are defined in this regard as individuals (or firms) that expect to benefit from using a design, a product, or a service

Table 2.1 Contrasting open and user innovation

	Open Innovation	User Innovation
Core references	• Chesbrough (2003, 2006b)	• Von Hippel (1988, 2005)
Focal actor of study	• Firm (R&D Lab)	• Individual user
Key principles	• Knowledge is widely dispersed beyond any one firm • Innovations must be aligned to a firm's business model • Firms should embrace both internal and external alternatives	• Users have unique "sticky" information • When enabled, they will solve their own needs • Many will freely reveal to others
Focal object of transfer	• Technological knowledge in form of IP or technologies	• Information about needs and ideas how to transfer need into solution
Typical institutional arrangement for knowledge transfer	• Research contracts • In- and out-licensing; IP transfer agreements • Tournament-based crowdsourcing for technical solutions	• Lead user method • User communities
Representative IP practices	• Patents • Licensing contracts	• Free revealing • Open source or creative commons licenses
Governance of innovation process	• Private model	• Collective or private-collective model
Motivations of actors to engage in distributed innovation	• Monetary incentives • Innovation is seen as a "money market"	• Incentives of self-use • Social incentives • Innovation is seen as a "social market"
Key managerial decision	• Building absorptive capacity • Defining and defending IP • Internal organization for OI • Defining metrics for OI	• Identifying lead users • Establishing bridging strategies to lead user innovation • Defining fair regimes of coordination • Opening- up IP
Other streams of related research	• R&D networks / strategic alliances • University-firm research contracts • Absorptive capacity theory	• "Voice of the customer" methods of market research in innovation • Participatory design • Social production

(Baldwin & von Hippel, 2011). In contrast, producers expect to benefit from selling the innovation. These lead users are not "average customers," who are rarely innovative (cf. Christensen, 1997); instead, they are "extreme users" who (1) face needs that will become general in a marketplace much earlier than the bulk of that marketplace encounters them; and (2) who are positioned to benefit significantly by obtaining a solution for those needs (von Hippel, 1988).

The early user innovation literature clearly focused on the lead user as the focal actor who is innovating autonomously to solve his/her own need (von Hippel, 1988). However, later research also found them to engage in strong knowledge sharing and co-development in communities of other user communities (Franke & Shah 2003; Füller et al., 2008). Within these communities, users have been shown to frequently reveal innovative ideas freely towards firms and other users (Harhoff et al., 2003), i.e. they share their ideas, knowledge, and inventions with other users without request or even expectation of compensation. These communities may operate independent of firms or even deal with firms' products in an unauthorized manner (Flowers, 2008). For example, a study of four sports equipment communities found that one third of the community members improved or even designed their own equipment innovations, often driven by collaborations with other community members (Franke & Shah, 2003; see also Jeppesen & Frederiksen, 2006). Communities where innovating users collaborate to develop new products or services often build upon product-related discussion forums, where users exchange experiences and support each other in using a product (Sawhney & Prandelli, 2000; Füller et al., 2006).

Other recent research has studied collaboration processes between users[2] and producer firms who seek to commercialize the user innovations. First, such firms may engage in lead user research (Churchill, von Hippel, & Sonnack, 2009) or apply the lead user method (Lilien et al., 2001; Thomke & von Hippel, 2001), a systematic producer-driven search process to identify people with lead user characteristics, both from their own industry and from analogous markets, and engage in a collaborative problem-solving process using concept generation workshops. Second, some lead users create their own producer firms to commercialize their own innovations, a process Shah and Tripsas (2007) termed "user entrepreneurship."

Finally, in other instances firms facilitate user creativity by creating dedicated platforms to innovate with users (Piller & Walcher, 2006). One example is toolkits for user innovation which provide a convenient user interface so that users can create their own designs by utilizing a library of basic modules and functionalities (von Hippel, 2001; Franke & Piller, 2004). A related method is the implementation of ideation platforms for continuous user input, like Dell's Ideastorm (Bayus, 2013). Ideas generated on these platforms are often more radical (Poetz & Schreier, 2012) and also of higher commercial value (Nishikawa et al., 2012) compared to internally developed ideas, however also more difficult to realize. Hence, firms could profit from a deeper collaboration with innovating users to also get input on the technical implementation of these ideas. Concluding, user innovation literature has developed from its pure focus on innovating users in the original publications towards a notion of interaction among users and firms. But research that examines in-depth the process of collaboration between users and firm is still rather scarce.

2.2.2 Open Innovation: Accessing Purposive Inflows

The original conception of open innovation identified two modes of knowledge flow: the inbound (or "inside-out") and the outbound (or "outside-in") flow (Chesbrough, 2003a; see also West & Gallagher, 2006; and Chapter 1). The outbound mode is not directly related to user innovation, and so is not discussed further in this chapter. The inbound mode of open innovation involves, as Chesbrough (2006b: 1) put it, "the use of purposive inflows…of knowledge to accelerate internal innovation." Such a model combines externally and internally developed technologies to produce an offering that is commercialized by the focal firm. Key steps of this process include searching for external innovations, selecting and acquiring suitable innovations, integrating them into the firm's R&D efforts, and bringing them to market (West & Bogers, 2014).

As originally inspired by Chesbrough's (2003a) examination of large industrial firms such as IBM, Intel, and P&G, research on open innovation has tended to focus on organizational suppliers of such technology (Chesbrough, 2003b; West et al., 2006; see West & Bogers, 2014 for a review). The implicit (or sometimes explicit) assumption is that such organizational suppliers have economic motivations, whether firms seeking profit through outbound open innovation—such as the innovation merchants of Chesbrough (2003b)—or universities or non-profit research labs seeking to fund their R&D efforts (cf. Jensen & Thursby, 2001).

The majority of open innovation research has focused on corporations absorbing external knowledge stocks or Intellectual Property (IP) as an input for their innovation process, in exchange for monetary compensation. However, some researchers have gone beyond this focus and also investigated non-pecuniary exchanges and/or exchanges between individuals. A few studies have identified examples of why organizations might provide such innovations for non-pecuniary reasons (Chesbrough, 2003b; Dahlander & Gann, 2010). For example, the external partners may be individuals, or firms, they may be engaged individually or as part of broader communities, and their appropriation of the innovation may include personal use or rival commercialization (West et al., 2006; West & Lakhani, 2008).

A much smaller amount of work has identified the potential role of individuals as potential contributors to such firm efforts. These individuals may have economic, social, or some other combination of motives (West & Gallagher,

2006; Jeppesen & Lakhani, 2010; Dahlander & Gann, 2010). Although inspired and theoretically motivated by a different perspective, the research on firms commercializing user innovations or cooperating with user innovators is largely consistent with this open innovation perspective. However, inbound open innovation tends to focus on how firms benefit from such innovations without giving the motives of the suppliers too much attention, while user innovation tends to emphasize the utility motive of the individual inventor without paying much attention to the motives of individual contributors. An important exception is Dahlander & Wallin (2006), who contrast the motivations of individual and firm-sponsored contributors to an open source community

2.2.3 Contrasting User and Open Innovation

User innovation and open innovation have an overlapping interest in a distributed process of innovation. For example, when firms source innovative ideas from individuals—whether following the maxims of user innovation or open innovation—this requires firms to cooperate for the sourcing of technical knowledge across firm boundaries, in rejection of the traditional vertically integrated innovation model (cf. Bogers & West, 2012). However, there are essential differences. For example, open innovation continues the traditional view of the corporation as the locus of production, while user innovation anticipates (and often advocates) a decentralization of innovation from firms to individual users (cf. von Hippel, 2005; Baldwin & von Hippel, 2010; Füller, Schroll, & von Hippel, 2013). This parallels the thrust of the paradigm shifts respectively promulgated by Chesbrough (2003a) and von Hippel (2005)—one that advocates improved performance by making the boundaries of firms more permeable, while the other advocates supplanting firms by "democratizing" innovation.

In at least three ways the OI and UI perspectives are associated with a particular extreme on a continuum of alternatives: open innovation (consistent with its firm-centric outlook) continues the traditional role of the firm, while user innovation emphasizes independence from firm control. They include:

(1) For *intellectual property*, the open innovation model—particularly that of outbound commercialization of internally developed technologies—has tended to emphasize strong appropriability and aggressive IP enforcement as a precondition for OI success (e.g., Chesbrough, 2003c; West, 2006). Firms are certainly amenable to weak IP if it provides a cheap source of inbound innovations, i.e. if the innovators are willing to develop or government agencies are willing to fund innovations at no cost (Chesbrough, 2003b; West et al., 2006; Dahalnder & Gann, 2010).

User innovation research challenges this perspective with its emphasis on users "free revealing," i.e. voluntary surrender of appropriability for their innovations (Harhoff, Henkel, & von Hippel, 2003; Henkel, 2006). Users are interested in using the innovation; they benefit when a firm (or other users) take up their ideas—ideally into a fully supported commercial product. Users may also freely reveal if the cost of obtaining IP protection is too high. Finally, users freely reveal as a sign of reciprocity as they also use other freely revealed information within their own innovation endeavors (Harhoff et al., 2003; Jeppesen & Frederiksen, 2006). Together, the practice of free revealing helps both individual and societal welfare (von Hippel, 2005).

To emphasize his differences from the IP policies in Chesbrough's (2003a) definition of "open innovation," von Hippel has since referred this collaborative IP model as "open distributed innovation" (von Hippel, 2005; von Hippel & de Jong, 2010) and "open user innovation" (von Hippel 2010; Baldwin & von Hippel, 2011). As Baldwin & von Hippel (2011: 1400) write:

> An innovation is "open" in our terminology when all information related to the innovation is a public good—non-rivalrous and non-excludable.... It differs fundamentally from the recent use of the term to refer to organizational permeability—an organization's "openness" to the acquisition of new ideas, patents, products, etc., from outside its boundaries, often via licensing protected intellectual property (Chesbrough, 2003a).

(2) The differences in the conceptions of IP directly lead to a second pair of choices, between the *private vs. collective models* of how innovation is funded, organized, and controlled (von Hippel & von Krogh, 2003; Gassmann, Enkel, & Chesbrough, 2010). In the private model, the private control of the innovation and its returns provides an economic incentive for a private actor (typically a firm) to invest in developing and deploying an innovation; such a model is implicit in open innovation studies. The early user innovation literature implicitly followed an individualistic private model, when lead users serve the private interest of solving their own need (von Hippel, 1988). However, the user innovation model later expanded to include user communities which entail a cooperative process between multiple (often individual) actors who collaborate in both creating innovations and in sharing their benefits. While research has emphasized these extremes, a few hybrid private-public models have been identified, particularly in open source software (von Hippel & von Krogh, 2003; West, 2003).

(3) Finally, there is the distinction *between money markets and social markets* as incentives to organize participation (Piller, Vossen, & Ihl, 2012). Heyman & Ariely (2004) found that people expend more effort in exchange for no payment (a social market) than they expend when they

receive low payment (a monetary market). Much as Dahlander & Gann (2010) identified pecuniary and non-pecuniary motives for innovation sharing, here we apply the Heyman & Ariely typology of task motivation to suggest two types of markets for innovation:

- Money markets consist of markets for external innovation that are organized around economic (monetary) incentives exchanged for ideas and solutions (e.g. Terwisch & Xu, 2008; Jeppesen & Lakhani, 2010; Boudreau et al., 2011). This can lead to a Darwinian, zero-sum competition in which innovators compete among each other to get a maximum share of a limited award—as might be observed in an ideation contest where contributions are sought via "broadcast search." In general, open innovation follows this view of money markets as the regime to award external contributors to a firm's innovation process.

- Social markets rely on social-exchange relations, and are largely built upon the non-monetary incentives for participants such as enjoyment or task achievement (cf. von Hippel & von Krogh, 2003, 2006), for outcome expectations that enhance their own use experience or that of others (Harhoff et al., 2003), or through norms of mutual cooperation and reciprocity (Lakhani & von Hippel, 2003). This is the approach most often used in non-commercial distributed innovation, such as with user communities or lead user workshops. It dominates the original user innovation literature.

Of course, markets may be organized to combine both types of incentives, either across different members of the same community (Hars & Ou, 2002) or even with individuals who may have both economic and social motivations to contribute to innovation (Piller, Vossen & Ihl, 2012).[3] Together, these three distinctions between OI and UI identify areas of tensions between the interests of firms and those of individual users when they collaborate. While firms seek to collaborate to enhance their innovation, they tend to do so in the context of privately-controlled IP and the motivations of private economic returns. Conversely, by starting from personal utility rather than economic gain, users often seek to share their creations through a process of collective action and social exchange.

2.3 AN INTERACTIVE APPROACH TO COUPLED OPEN INNOVATION

Building on the open and user innovation literature, we now focus on collaborations where firms and individuals jointly create new knowledge or other inputs for an innovation process. In the OI literature, this understanding closely resembles the model of "coupled" OI, as identified by Gassmann & Enkel (2004). But our model differs from the formal interfirm alliances that were the primary interest of Gassmann and Enkel. We extend the coupled OI model by utilizing recent insights from open innovation, user innovation, and co-creation research. In particular, we propose an interactive model of coupled OI and offer a typology of different categories in this model. We then show how firms can manage interactive coupled open innovation with individual users.

2.3.1 Refining the Coupled Model of Open Innovation

Gassmann and Enkel (2004; Enkel et al., 2009) identified "coupled" as a third mode of open innovation, beyond the original inbound and outbound processes identified by Chesbrough (2003a). Defining this as "working in alliances with complementary partners," they elaborated:

> Companies that decide on the coupled process as a key process, combine the outside-in process (to gain external knowledge) with the inside-out process (to bring ideas to market). In order to do both, these companies co-operate with other companies in strategic networks (Gassmann & Enkel, 2004: 12).

As proposed by Gassmann and Enkel, the concept focused on the traditional perspective of firm alliances but has had limited theoretical development despite widespread potential application to open innovation research. In their review of 165 open innovation articles, West & Bogers (2014) found 70 articles (42%) that could be classified as relating to coupled open innovation (although many did not use that term). Here, we extend this broad conception of coupled processes by identifying four important dimensions (Table 2.2).

The first dimension is the *nature of the external actor*. The original emphasis for coupled open innovation was on firms as external partners; as with other open innovation collaborations, such firms may be suppliers, customers, complementors, or even rivals (West, 2006). However, the external partners for coupled processes may also include non-profit organizations (such as universities or research labs) or individuals; these potential partners differ in what and how they produce and commercialize innovation (West et al., 2006; Chapter 9). Differences in actors may also lead to differences in incentives, coordination and governance of the collaboration between the two parties.

The second dimension is the *topology* of the relationship with the external actors. Dyadic collaboration with a single external partner corresponds to the long literature on strategic alliances (e.g. Gomes-Casseres, 1996). It is the most common form, accounting for about half of the research on the coupled process studied by West & Bogers (2014), who identify two additional topologies for coupled open innovation: networks of collaborators (e.g., Vanhaverbeke, 2006) and collaboration with voluntary communities (West & Lakhani, 2008).[4]

Table 2.2 Multiple dimensions of coupled open innovation processes

Dimension	Alternatives
External actor	• Firms: customer, supplier, complementor, rival • Other organizations: university, research lab, government, other non profit • Individual: customer, user, inventor, citizen
Coupling topology	• Dyadic: single partner • Network: multiple partners • Community: a new interorganizational entity
Impetus for Collaboration	• Top-down: initiated by upper management • Bottom-up: developed through employee or customer collaborations
Locus of innovation	• Bidirectional: innovation created within each organization • Interactive: innovation jointly created outside the organizations

The third dimension is the *impetus for collaboration*. Most of the open innovation literature—including that for networked alliances and other approaches to coupled OI—emphasizes the strategic intent of the firm to achieve certain objectives. For example, Lee et al. (2010) subdivided OI collaborations of small firms into three strategies: customer–provider, dyadic strategic alliances, and networked inter-firm alliances. Meanwhile, research on open source software (e.g. Henkel, 2009) has documented the role of individual employees in initiating, directing, and implementing collaboration with external communities. In this regard, the top-down vs. bottom-up distinction roughly parallels the Mintzberg (1978) distinction between intentional and emergent strategies.

The fourth dimension refers to the *locus of the innovation process*. Here we draw a distinction between two approaches, which we term "bidirectional" and "interactive." In the bidirectional case, two actors (typically organizations) continue their separate efforts at creating innovation and other useful knowledge, but then share that knowledge; this most closely matches the Gassmann & Enkel (2004) definition of combining inbound and outbound flows. In some cases, the knowledge sharing parallels the reciprocity of user innovation sharing, but formalized through contracts or other legal instruments. In other cases, the reciprocal flows are explicitly monetized through licenses or other payment terms, as common, for example, for mobile phone standards (West, 2006; Bekkers & West, 2009).

However, we believe that there also is an interactive collaboration between two actors that is qualitatively and quantitatively different from the bidirectional form. Instead of using knowledge flows to augment the firm's internal innovation creation (and commercialization) efforts, in the interactive approach the knowledge creation takes place outside one particular firm. Instead, innovative outputs are being created in a collaborative activity of all parties together. Such external joint creation of innovation differs from the bidirectional forms in where the innovation (or innovative knowledge) is created, by whom, in how the process is governed and how the returns can be appropriated (Chesbrough, 2011). It is this latter understanding of coupled open innovation as an *interactive, collaborative process* of joint value creation that is the focus of the remainder of this chapter. We see this as a second form of coupled open innovation, distinct from the original bidirectional conception by Gassmann & Enkel (see Figure 2.1).

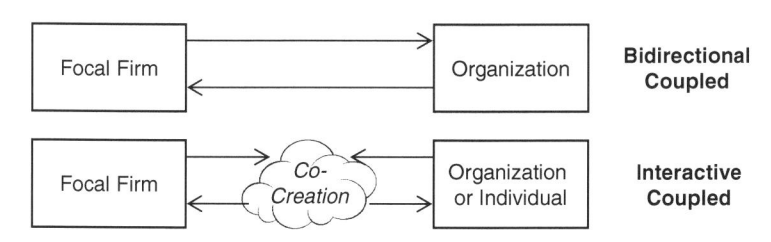

Figure 2.1 Two forms of coupled open innovation

This interactive process is similar to "co-creation," a term that has been popularized in a series of books and papers by Venkat Ramaswamy and colleagues (Prahalad & Ramaswamy, 2004b; Ramaswamy & Gouillart, 2010), who define co-creation as "the practice of developing systems, products, or services [by a firm] through collaboration with customers, managers, employees, and other company stakeholders" (Ramaswamy & Gouillart, 2010: 5). Their starting point is the question how firms can leverage the input from external entities to create value across the value chain. Co-creation originated in an earlier debate in the strategic marketing literature by Normann & Ramirez (1993), Wikström (1996), or Vargo & Lusch (2004). In the innovation management literature, "co-creation" has almost exclusively been used for firms collaborating with their customers or other users—although not always for product innovation. Building on Roser et al. (2009: 9), we define co-creation as an active, creative, and collaborative process between a firm and individuals during a new product/service development process in which participants contribute to a task initiated and facilitated by the firm.

2.3.2 A Model of the Collaboration Process

Based on earlier research, we develop a process model for interactive coupled OI between firms and users. Our model combines the OI interaction model by Diener & Piller (2008, 2013) and the inbound OI models of West & Gallagher (2006) and West & Bogers (2014). In addition, we consider the recent literature on organizing contest-based crowdsourcing for ideation and technical problem solving (Spradlin, 2012; von Krogh et al., 2012). Together, these literatures suggest that firm-initiated co-creation efforts entail four major steps (Table 2.3):

Table 2.3 A process model for coupled open innovation projects

Process Stage	Key Activities
Defining	• Problem formulation • Institutions and rules: including contract terms, IP • Resource allocation and strategic commitment
Finding Participants	• Identifying participants with right characteristics • Motivating and retaining a critical mass of collaborators • Selecting the right participants
Collaborating	• Governance of the collaboration process: organizing, monitoring, policing • Interaction platform and other tools • Openness of firm attitudes, structure and processes
Leveraging	• Integrating external knowledge • Commercializing the knowledge through products and services

Adapted from West & Gallagher (2006), Diener & Piller (2008), West & Bogers (2014).

1. **Defining.** The firm needs to define the problem that it is seeking to address via engaging external partners in the co-creation effort (cf. von Krogh et al., 2012). It depends on institutions and rules of engagement, whether the rules of communities that it creates or might join (West & O'Mahony, 2008), or broader appropriability rules of the society or economy (cf. Teece, 1986; West, 2006). Finally, the firm needs to determine the resources that it is willing to provide and, more broadly, its level of strategic commitment to the collaboration process (cf. Lazzarotti & Manzini, 2009).

2. **Finding Participants.** A major theme of open innovation research has been on searching for suitable external partners with the right knowledge relevant for the firm's needs (see West & Bogers, 2014 for a summary). Both the search for and the acquisition of such knowledge will depend on understanding and strengthening the motivations of external partners to create and share their knowledge (West & Gallagher, 2006; Antikainen et al., 2010).

3. **Collaborating.** The key value creation process in our model is the interactive collaboration process that creates new innovations. Even after a decade, open innovation has a lot to learn from research on co-creation that has focused on how firms collaborate with external partners in a collaborative exchange of knowledge and benefit. This includes creating and implementing the processes for collaboration (Prahalad & Ramaswamy, 2004b) as well as providing suitable tools (such as IT-enabled platforms) that facilitate the collaboration process (Diener & Piller, 2013). Finally, firms face the daunting challenge of selecting the most promising ideas from dozens or thousands of potential contributors (Terwiesch & Xu, 2008). Such external

interactions assume that the firm is willing to open itself to the external partners: the risk of leakage of internal firm insights must be weighted against the new insights gained by empowering external collaborators (cf. Prahalad & Ramaswamy, 2004; Enkel et al., 2009).

4. **Exploiting.** Even if these collaborations are successful in creating new knowledge or innovations, there is no guarantee of firm success from such efforts. Internal co-creation advocates must overcome suspicion and other resistance to externally sourced ideas by their colleagues, whether an overt culture of "Not Invented Here," or structural barriers impair collaboration (Chesbrough & Crowther, 2006; Dodgson et al., 2006; Schiele, 2010). In general, we know little about how (or how much) firms ultimately benefit from externally sourced innovations: do they use the same commercialization process as the internal ones, and are they more or less valuable than their traditional counterparts? (West & Bogers, 2014)

Here we apply the general model for the interactive process of coupled open innovation to the specific challenges of firms working with customers, users, and other external individuals. While our focus is on collaboration with external individuals, we believe that the model is also applicable to collaboration with firms or other organizations.

2.3.3 Defining Collaboration Tasks and Rules

To launch an interactive process of coupled OI, firms seeking external collaborators must define the tasks and rules for this collaboration and allocate sufficient internal resources to this endeavor.

Problem formulation: An interactive process of coupled open innovation starts with crafting the problem statement (Jeppesen & Lakhani, 2010; Sieg et al., 2010). The objective is to create a task description that can be used to attract external contributors, and also to think about the characteristics of such contributors. This signals external individuals about the opportunity for collaboration and asks interested individuals to submit either a solution proposal or just to indicate their interest in further collaboration. The formulation is aided by problem modularity that allows partitioning tasks between internal and external contributors (Langlois & Garzarelli, 2008).

The process of task formulation has been described well in the literature with regard to tournament-based crowdsourcing. Firms broadcast their problems, performance criteria, and contracting terms to an audience of potential solvers, usually in the form of a "request for proposals." Writing such specifications entails many challenges, including defining the problem (and scope) precisely, using terminology that will be clear to potential solvers with knowledge from other fields, and preserving the confidentiality of the firm's current and future technology needs (Afuah & Tucci, 2012; Spradlin, 2012; Lüttgens et al., 2014).

While a few researchers have started to study this activity of task formulation for contests seeking technical information (von Krogh et al., 2012; Lüttgens et al., 2014), we are unaware of research on problem formulation for other forms of coupled open innovation. User innovation research has only briefly identified the definition of a "search field" as the beginning of a lead user search process (Churchill, von Hippel, & Sonnack, 2009). Similarly, the co-creation literature has not covered this aspect beyond very brief references to its importance (in, e.g. Ramaswamy & Gouillart, 2010; O'Hern & Rindfleisch, 2010). But in either user or open innovation, defining the initial scope is crucial to initiate a coupled innovation process and avoid "garbage in, garbage out."

Rules of cooperation: In dyadic open innovation—whether inbound or coupled—firms typically acquire rights to knowledge via a contract that assigns all necessary rights to the firm (e.g. Frenz & Ietto-Gillies, 2009; Jeppesen & Lakhani, 2010). However, in more complex collaboration contexts, other arrangements are necessary: the most-studied example is that of open source software communities. If firms tightly control the output of a community then they discourage participation by individual contributors; thus, firms use a variety of selective openness strategies—controlling the IP rights, creation process, and community governance—to maximize the alignment to firm goals while attracting outside participants (West, 2003; Shah, 2006; West & O'Mahony, 2008).

Resource allocation: A firm initiating a sustained process of collaborative innovation must also commit the organization and dedicated resources for this process, particularly for ongoing interactions with external participants as their contributions are developed and evaluated. An often neglected activity is providing feedback to contributors, which is crucial to motivate future contributions and (particularly with customers) avoid developing a negative firm reputation.[5] Research has shown that companies often underestimate the effort required for these activities (Diener & Piller, 2008; Lüttgens et al., 2014). These resources must be supported by an internal structure that supports such external collaboration (Bianchi et al., 2011; Dahlander & Gann, 2010). Firms may be more successfully integrating external input if they have norms and explicit procedures for open innovation (Foss, Laursen, & Pedersen, 2011). At the same time, the firm needs internal guidelines for communication and exchange with external parties to improve cooperation by internal employees and units (Cordón-Pozo et al., 2006).

2.3.4 Finding Participants

The first step of any collaboration is *identifying participants* who have the relevant skills and interest to contribute to the firm's goals for the collaboration. Some participants may initiate this identification by publicly proclaiming their expertise (Droge et al., 2010) or even actively "pushing" their ideas to firms (Spaeth et al., 2010). In general, firms will actively engage in recruiting participants. We distinguish between three approaches for finding qualified participants (Diener & Piller, 2008, 2013):

- *Open Call:* In these cases, firms enable a broad range of participants and then select their ideas after they've been contributed (Piller, Ihl, &Vossen, 2011). This resembles the original understanding of "crowdsourcing" as an open call for participation to an undefined, large network of external actors, as defined by Howe (2006).
- *Selective Open Call*: Other firms identify the characteristics of suitable participants *a priori* (e.g. market segment, field of expertise, revenue potential by customers), and then limit their call for collaboration to that select list (Diener & Piller, 2008, 2013).
- *Open Search:* In other cases, firms engage in own search efforts to identify suitable actors within a large set of possible partners, and then explicitly invite then to join the co-creation activity. This is the typical approach in a lead user project (Poetz & Pruegel, 2010).

The *nature of participants* in a coupled OI initiative—identified by any of these three approaches—can span across a broad range of actors. Füller et al. (2009: 93) found that "potential task involvement of participants, their creativity, and experience in generating new product ideas" influenced their ability and willingness to participate in co-creation efforts. The typical individual participant is an expert for a specific domain or task, either because of her profession (i.e., industrial designers participating in ideation contests; a lab scientist participating in a technical contest), or her prior use knowledge in a similar situation. Experts are often motivated extrinsically, as discussed below. In the case of consumer products, participants may be either customers with lead user characteristics or "average" users with a high level of product involvement or a sense of belonging to a brand community.

Motivating external participants to engage in collaboration with the firm is an important task in coupled OI, emphasizing different participant incentives (Dahlander & Gann, 2010). Not surprisingly, theories of open innovation are explicitly about firm success, hence addressing the pecuniary monetary incentives. Because the supply of innovations to other firms corresponds to the outbound mode of OI—which also assumes that firms are seeking to maximize economic returns to innovation—the initial work on open innovation assumed that firms would be selling, licensing, or otherwise providing innovations in exchange for payment. Open innovation research on the motives of individual collaborators is less common. External innovation contests are often organized around financial incentives to attract and engage external collaborators to address a firm's needs (Jeppesen & Lakhani, 2010). At the same time, firms collaborating with external communities may find that individuals are more effectively motivated through non-economic (or indirect) incentives such as ego and career visibility (West & Gallagher, 2006; Boudreau & Lakhani, 2009).

Conversely, user innovation tends to consider non-pecuniary motivations. As noted earlier, the original user innovation work emphasized individuals utilizing their own unique "sticky information" to address their own unique and unsolved needs (von Hippel, 1988, 1994). However, more recent research has examined the social motives of users participating in collaborative communities (see von Hippel, 2005 for a summary). In his review of "social production," Benkler (2006) suggests that monetary incentives tend to crowd out intrinsic motives for contributing to communities, and thus social motives are most effective in motivating individuals to contribute when "pricing and contracting are difficult to achieve, or because the payment that can be offered is relatively low" (Benkler, 2006: 95).

However, this distinction between monetary and social motives is perhaps more sharp in theory than in real life. Later research on hybrid models of participation in communities like von Hippel and von Krogh's (2003, 2006) model of collective–private innovation suggest that some communities are driven by both social and monetary (private) motives. At the same time, users are increasingly able and willing to monetize their contributions when they create a new firm to commercialize their innovations; such innovation is often created through collaboration with a user community, and that collaboration continues after the formation of the new organizational entity (Shah & Tripsas, 2007). Also in many ideation contests with users, monetary incentives are clearly positioned as a complementary incentive next to social incentives ("helping others") or intrinsic motives ("having fun by co-creating"). In result, former "money markets" characterizing open innovation are becoming more "social," while former "social markets" characterizing user innovation are becoming more "monetary" (Piller, Vossen, & Ihl, 2012).

2.3.5 Collaborating with Participants

The heart of our model is the joint co-creation process of innovation by the firm and external actors. While open innovation has emphasized finding and obtaining external knowledge (West & Bogers, 2014), research on the joint creation process of such knowledge has been comparatively rare in the open innovation literature. Much of the prior research on this topic has focused on formal (contractual), long-term collaborations such as R&D alliances (e.g., Hoang & Rothaermel, 2005). Some open innovation research has considered collaboration *within* a firm as an enabler to connect with innovative knowledge from its periphery (Dahlander & Gann, 2010; Van de Vrande et al., 2010).

But little research has examined the structures and processes supporting collaborative knowledge creation with external actors (Blazevic & Lievens, 2008).

Similarly, the original UI literature on lead users did not look upon the collaboration stage, except (as noted earlier) for collaborations within communities of innovating users (e.g. Franke & Shah 2003; von Krogh, Spaeth, & Lakhani 2003). These studies largely ignored collaboration between users and firms. Here, we consider the gap of research on the collaboration stage of coupled open innovation in three important areas: governance of the collaboration process, tools and dedicated infrastructures facilitating this stage, and internal attitudes and capabilities of the focal firm supporting the collaboration.

Governance of the collaboration process. Unlike OI and UI, the co-creation literature covers more explicitly the activity of joint collaboration between firms and individuals, suggesting structures and processes that allow the firm to stir, monitor, and police its value creation through collaborative efforts with external partners (e.g. Prahalad & Ramaswamy, 2004b; Ramaswamy & Gouillart, 2010). A central point from the perspective of the firm is to define the span of control that the firm provides to the external co-creators (Diener & Piller, 2008; O'Hern & Rindfleisch, 2010; West & O'Mahony, 2008). Different regimes of co-creation provide different degrees of influence to the participants (Doan et al., 2011). Participants are engaged when given more control, freedom to operate and responsibility (Koch & Gates, 2010). Defining the span of control is a key firm decision when setting up interactive coupled OI. For example, in ideation contests a key decision is the extent participants can evaluate and rank contributions by other participants. If firms allow the final decision on the "best" contribution to be made by participants, then such empowerment may motivate contributors—but the firm gives up important control on the outcome of the contest (Gatzweiler et al., 2013).

Tools and collaboration infrastructures: Software tools play an important role enabling a broad collaboration with customers and other individuals at low transaction cost. For example, in ideation contests tools facilitate the search for participants, the collection and evaluation of ideas, user feedback, and clustering the submitted ideas (Piller & Walcher, 2006; Adamczyk et al., 2012). Social enterprise software facilitates a more general exchange within a participant community and between participants and the firm in form of web-forums, blogs, tweets, and the like (Sawhney, Verona, & Prandelli, 2005). This software can be seen as the backbone of modern co-creation activities (Piller, Vossen, & Ihl, 2012). Finally, toolkits for user innovation provide users a design space to create products meeting individual requirements, based upon libraries of modular or parametric components that can be modified and freely combined by users (von Hippel & Katz, 2002; Franke & Piller, 2004). These tools have been discussed extensively from a technology perspective in the information systems literature and, to a smaller extent, in the co-creation literature.

Rather than use the tools directly, some firms will utilize the services of specialized intermediaries and brokers for open innovation; these open innovation accelerators help clients by providing proprietary tools and methods, access to an established community of solvers or participants, and education and process consulting (Chesbrough, 2006a; Diener & Piller, 2008; Lopez-Vega, 2009; Mortara, 2010a). Intermediaries differ with regard to their task specialization, their software platform, and the characteristics of their participant community (Diener & Piller, 2013). Selecting the right intermediary to meet the contingencies of an innovation project is a key decision for firms who want to engage in an interactive model of coupled open innovation.

2.3.6 Leveraging the Collaboration Results

Once the firm has completed a collaborative effort, the challenge remains to realize the benefits of such efforts by *integrating the innovation* into the firm and then *commercializing the innovation at market*. In many cases, open innovation research assumes that commercializing external knowledge through products and services happens exactly the same as from knowledge created via internal sources of innovation (West & Bogers, 2014). But reality is more complex. Integration of the results will depend on the nature of the contribution and what part of the R&D pipeline is informed by that contribution. Some contributions will be in generating ideas for further internal development; some will come in the design of a product or service, while others will come in evaluating new offerings being tested prior to market (Füller & Matzler, 2007). One of the most common outcomes of co-creation is the incremental improvement of existing products that customers use and understand (Piller, Ihl & Vossen, 2011). It is more difficult to utilize co-creation to create radical innovations—whether new to the world or new to the firm—but it can be done if firms are able to use appropriate tools to help users to surface their unmet needs (Füller & Matzler, 2007).

Each type of integration may require interaction with different parts of the organization as well as different tools and processes. But all types of integration share a similar challenge: overcoming "not invented here" (Chesbrough, 2006c; Chesbrough & Crowther, 2006; Gassmann et al., 2010). Such an attitude is symptomatic of the cultural barriers that firms—particularly successful innovators—face in collaborating with external partners (West & Bogers, 2014). Additional challenges exist for integrating the results from co-creation, including the need to maintain transparency with partners, to adapt external ideas (of variable quality) to a firm's high quality standards, and the additional time required for an interactive process (Prahalad & Ramaswamy, 2004b).

When considering the extent to which firms can profit from external distributed knowledge, one aspect frequently studied is absorptive capacity, i.e. a firm's ability to recognize, assimilate, and apply external knowledge for innovation (e.g. Laursen & Salter, 2006; Foss et al., 2011). Incentive and reward systems have been shown to be instrumental for successful exploitation, reinforcing the use of external learning (Quigley et al., 2007). In addition, firms that emphasize external learning and absorption behavior as bases for appraisals and rewards will be more likely to acquire and utilize external knowledge via co-creation (van Wijk et al., 2008).

Another antecedent to integration is open-mindedness. Research has shown that mental models evolve with past successes and failures. They manifest expectations about action-outcome relationships in organizational routines, assumptions, and beliefs (Ringberg & Reihlen, 2008; Lin & McDonough, 2011). Hence, when employees are encouraged to use input from interactions with external users and other experts to think in new ways, knowledge generated in co-creation is more likely to be acquired and assimilated. Moreover, when engaging external experts and reflecting on own mental models at the same time, previously undetected technical connections can be recognized.

2.4 CONCLUSIONS AND OUTLOOK

Focusing on how firms collaborate with individuals, this chapter makes three contributions. First, it reviews and contrasts how such collaboration has been covered by user and open innovation. It identifies three important differences between these literatures: over the role of IP, the private vs. collective model of innovation, and the distinction between social and money markets to incentivize individual participation. Future research should examine further examples of hybrid models that combine the best of both approaches on these dimensions.

Second, the chapter expands the conception of "coupled" open innovation as proposed by Gassmann and Enkel (2004) with a multidimensional typology of different forms of coupled collaboration. The first dimension considers the nature of the external partner—whether individual, firm or non-profit organization—while the second identifies the topology of the collaboration process—either the dyadic collaboration commonly found in open inbound innovation research or the various forms of collaboration. Finally, the typology draws the distinction between the original coupled model bidirectional collaboration (where each actor pursues its own innovation) and a new, interactive coupled model, where the two parties jointly produce new innovation.

From this understanding we developed a four-phase process model of interactive coupled innovation: defining collaboration tasks and rules, identifying and engaging external partners, collaborating in the joint innovation process, and leveraging the results of that collaboration. This model links the user innovation, co-creation, and other literatures to an area of emerging interest in open innovation. Demands for future research exist in each stage of our processes model.

With regard to defining a collaboration task we need more large scale (quantitative) research on the influence of task formulation on OI performance. In which situation, for example, is a broader-formulated task superior to a highly specific task? Are special incentive schemes for participants more effective for specific tasks? How can a firm balance the trade-off between revealing too much information in a task on the one hand side and providing the right detail of input for productive contributions on the other?

Similar research is required with regard to finding and selecting the right external individuals for participation (Hoffman, Kopalle, & Novak, 2010). While many firms prefer to carefully control participation, this may exclude individuals who can offer valuable input for the task at hand. Research is needed on the optimal selection and recruitment strategies of external participants for a given innovation task. Research is also needed on whether the increasing use of coupled OI is creating a scarcity of capable and willing collaborators, i.e. the "crowd". Modeling the scarcity of "innovative external actors" could become a fascinating topic for future research. Related to this is a more nuanced examination of the interdependence of cooperative and competitive interactions within a crowd, as demonstrated by Boudreau & Lakhani (2013) in their recent study of 733 contributors to a TopCoder competition.

While research on the collaboration stage has focused on the tools and platforms, much less research exists on the rules and conditions that set the governance structure of using these tools. Which "optimal" rules and conditions of ideation contests both form a legal and an ethical point of view. When is an IP arrangement seen as "fair" from the perspective of potential participants? How do these conditions influence the willingness to participate by individuals? What is the right mix of monetary and social or intrinsic incentives for a given task?

We also still have little knowledge on what happens inside the firm that helps or hurts a firm's ability to profit from coupled OI. Beyond the limited existing research at the firm level, we need studies at the group and individual level on the open-mindedness and willingness of employees to engage with and leverage the contributions of external individuals—and how that translates into successful commercialization outcomes.

Finally, we acknowledge that the reality of competitive versus collaborative modes of open and user innovation is more nuanced than our discussion may indicate. We purposefully focused on the extremes of dimensions which in reality are continuums with many shades of collaboration. Understanding these nuances and the contingencies that make one particular configuration more successful than another for a given innovation task can drive plenty of fascinating new research in the field.

NOTES

We thank Morgane Benade, Vera Blazevic, Johann Füller, Alexander Vossen, and participants at the 2013 Open and User Innovation Workshop and especially editor Henry Chesbrough for their helpful suggestions on earlier drafts of this chapter.

1. A simple table cannot capture the depth and complexity of hundreds of articles in these two major streams of innovation research. For more in-depth summaries of user innovation, see von Hippel (2005) and Bogers et al. (2010); for open innovation, see West & Bogers (2014) and Chapter 1 of this volume.
2. Users may be organizations (aka "user firms"), and in fact process innovation is often driven by user firms (Lettl, Hienerth, & Gemuenden, 2008). However, most user innovation research focuses on individual users (Bogers et al., 2010).
3. Boudreau & Lakhani (2009) make a related distinction between competitive and cooperative communities as sources for technical solutions. These forms of communities differ, like social and money markets, in the form of incentives and norms that drive the interactions between community members.
4. Chapter 4 summarizes the similarities and differences between various network forms of organizing external open innovation collaboration, including communities, ecosystems and platforms.
5. While there are both coupled and non-coupled ideation contests, our focus is on how firms collaborate with individuals. We thus excluded those contests following the "inbound OI" mode where firms intentionally set up a one-way knowledge flow from participants, typical of intermediaries such as Innocentive and Nine Sigma (see Chesbrough, 2006a; Diener & Piller, 2013).

3

A Classification of Open Innovation and Open Business Models

Wim Vanhaverbeke and Henry Chesbrough

3.1 INTRODUCTION

Open innovation and open business models have received a lot of attention during the last decade both from practitioners and academia since Henry Chesbrough launched the two concepts respectively in 2003 and 2006 (Chesbrough, 2003a and 2006a). Careful observation of publications about these two concepts reveals that most researchers as well as practitioners do not make a proper distinction between them. In this chapter, we intend to clarify the distinction between the two concepts and, more importantly, to extend the range of potential innovation strategies by combining open innovation and open business models in different ways. This should lead to a comprehensive classification of possible innovation strategies in which open/closed innovations are combined with open/closed business models.

The classification starts with simple combinations of open or closed innovations and traditional vs. open business models in their new product development. Combinations of open innovation and open business models generate interesting models to create and capture value, which to our knowledge are not specified before in the open innovation literature. The classification will also illustrate that (open) innovation targeting new product or new business development is just one possible strategy how firms can create a competitive advantage. Product innovation may not be an option for companies producing commodities (e.g. crude oil) but the competitive drivers in these industries (e.g. finding the best oil wells) may be affected by the product innovations of their (technology) partners (e.g. new technologies to explore oil wells more effectively). Reframing open innovation in this way allows us to shed light on innovation networks in which the instigators of the network are not the innovators themselves, but they nevertheless form the hubs in a broader innovation ecosystem in which they benefit from the innovations of their technology partners.

In sum, this approach in which we define open innovation and open business models more carefully, in combination with a change in focus from new product development to (other) competitive drivers will result in a rich classification of different types of open innovation research. The clarity of the structure should make it attractive as a starting point for several new developments in open innovation research.

The rest of this chapter is structured as follows. The next section clarifies open innovation and open business models. The third section develops our proposed categorization combining the two concepts. The fourth section illustrates each of the resulting cells in the classification with numerous examples. The fifth section summarizes some thoughts about how this classification may lead to new developments in research about open innovation.

3.2 A CLARIFICATION OF OPEN INNOVATION AND OPEN BUSINESS MODELS

Open innovation and open business models are the titles of the two first books of Henry Chesbrough (2003a; 2006a). Although both concepts have been clearly defined, there is no explicit analysis of the difference between open innovation and open business models in the second book. Chesbrough (2006a) states that the first book treated the business model as static, and utilized open innovation to create more ways to create and capture value *within the given business model*. In the second book, the business model itself could be innovated, enabling new ways to obtain more value from the company's innovation activities. However, Chesbrough did not explicitly combine choices of open or closed innovation with choices of alternative closed or open business models as we do in this chapter.

Nowadays, practitioners and researchers tend to use both concepts interchangeably. Scholars (and managers) need to be careful in defining open innovation and open business models. Making the difference between the two concepts explicit allows us study a broad range of phenomena that have not been related to open innovation or open business models so far. Using a precise and specific definition for both concepts will lead to a more concise study of the two concepts and when we should utilize them.

3.2.1 What is Open Innovation?

There are many definitions of open innovation available. This issue has already been discussed at length in Chapter 1 of this volume. For the purpose of this chapter, we prefer to stick to the original definition. Chesbrough (2003a: XXIV) defines open innovation as "a paradigm that assumes that firms can and should use external ideas as well as internal ideas, and internal and external paths to market, as firms look to advance their technology". This is the most common definition used in the literature, and it underscores that valuable ideas emerge and can be commercialized from inside or outside the firm. This definition is echoed in Chesbrough, Vanhaverbeke, West (2006: 1): "Open innovation is the use of purposive inflows and outflows of knowledge to accelerate internal innovation, and expand the markets for external use of innovation, respectively." The business model concept is not included in the definition of open innovation but is tightly related to it. Chesbrough (2003a, XXIV) writes for instance: "Open innovation combines internal and external ideas into architecture and systems whose requirements are defined by a business model. The business model utilizes both external and internal ideas to create value, while defining internal mechanisms to claim some portion of that value." There is no way to conceive open innovation without business models: The value of an idea or technology depends upon the business model. There is no inherent value in technology per se. The value is determined instead by the business model used to bring it to market. The same technology taken to market through two different business models will yield a different return. This link between technology and business model is further strengthened by the intensive use of the open innovation funnel where business models are prominently represented at the right-hand side of the funnel. They determine which external technologies have to be sourced because they are indispensible for the business model and which technologies have to be monetized externally because they are not aligned with a firm's business model.[1]

3.2.2 What is an Open Business Model?

A business model is a framework to link ideas and technologies to valuable economic outcomes. At its heart, a business model performs two key functions: (1) it creates value, and (2) it captures a portion of that value. Organizations can create value by defining a range of activities that will yield a new product or service valued by a (target) customer group. Organizations also capture value by establishing a unique resource, asset or position within that series of activities where the firm enjoys a competitive advantage. Business models can be analyzed in great detail and different frameworks have been generated to develop new business models or change existing ones. As other authors have examined business models extensively,[2] we choose to focus directly on the specific characteristics of open (and closed) business models.

In explaining open business models, Chesbrough (2006a) starts with the current trend towards the "division of innovation labor". In this type of division of labor one party develops a novel idea but does not carry this idea to the market itself. Instead, it sells it to other parties, who carry the idea to the market. The division of labor is a new and powerful way to speed up innovation and improve R&D productivity. An open business model uses the division of labor *to create greater value* by leveraging more ideas (external ideas) and *to capture greater value* by using key assets, resources, or positions not only in the company's own business but also in other companies' businesses (Chesbrough, 2006a: 2-3). An open business model is thus a powerful organizational model of innovation. Open business models may lead to better financial performance by reducing the costs of innovation on the one hand and generating extra revenues on the other hand by monetizing technologies through licensing agreements and spin-off activities when the technology cannot be adopted profitably in the product markets of the company. In this way, open business models are still tightly linked to innovation activities of a firm or its external innovation partners. It is not by accident that the open business model is also called "open innovation business model" or "new business model of open innovation."

This interpretation of the concept "open business models" has proven to be a valuable extension of the original open innovation idea, which was launched in 2003. In the following sections we will combine closed/open innovation with two different types of business models—the stand-alone and the linked business models. Stand-alone business models reflect the idea of closed business models and linked or networked business models have a lot in common with open business models. In our view, combining different ways of innovation with different business models *in a systematic way*, will broaden our understanding about the strategic value of open business models and will extend the range of business contexts where "openness" can be applied as viable strategies.

3.3 CLASSIFYING COMBINATIONS OF OPEN INNOVATION AND OPEN BUSINESS MODELS

Open innovation and open business models can be considered separately. As we will see, it is a viable strategy to engage in open innovation and stick to a closed business model. A company can also "open" its business model, but rely on a closed innovation strategy. Yet, in our view more interesting strategies emerge when companies combine an open innovation strategy with open business model thinking. A systemic analysis of the possible combinations will lead to an unexpected variety of possibilities, linking "open innovation" and "open business model" strategies to phenomena such as innovation ecosystems.

Table 3.1 A classification of combinations of open innovation and open business models

	Closed/Stand Alone Business Model	Open/Linked Business Models
Outside-in Open Innovation	3. Use others' knowledge to develop a new offering Early iPod—Apple Swiffer—P&G	6. Use others' knowledge to develop a new BM iPod/iTunes Store—Apple SkyNRG—KLM Better Place
Inside-out Open Innovation	2. Unused knowledge used by others Food ingredients—P&G/ConAgra Foods Nodax—P&G Glad—P&G	5. Internal knowledge accessible to others to develop a new BM Amazon WS—Facebook Salesforce.com IBM-Linux
Closed Innovation	1. Closed innovation model Tide—P&G Nylon—Du Pont	4. Search for assets owned by others to develop a new BM iPhone—Apple

In Table 3.1 we offer a simple model to combine different types of innovation with open and closed business models. Innovation is here defined as the activities during the innovation process (or in different stages in the innovation funnel). Accordingly, *open innovation* can be defined as before: firms can and should use external ideas as well as internal ideas, and internal and external paths to market, as they look to advance their technology. Open Innovation combines internal and external ideas to develop products, services or processes whose requirements are defined by firms' business models. Open innovation requires the knowledge input from external sources, but it does not necessarily imply that external partners help create value—that is, innovation partners are involved in the development of the product not in the commercialization of the new offering.

In contrast, *open business models* are intrinsically related to the value creation and value capture and delivery through the introduction of new products and services in the market. Once research and development of an R&D project are successfully finished a company still has to launch the new product and grow sales over the next years. Companies can produce and distribute offerings on their own or they can rely on existing transactions with value chain partners. Most companies indeed take products to market without the help of strategic partners. It is of course always necessary to involve channel partners, but these relationships can be handled by (standard) transactional agreements. Therefore, *closed or stand alone business models* refer to the situation where companies market a product using their own assets and relying on other value chain partners through market transactions. However, the introduction of a new offering in the market can be a complex process where companies need critical inputs from strategic partners. In these cases, value is created *jointly* with strategic partners and they have to come to an agreement how to share the pie (as the distribution of the value among the partners has to be managed guaranteeing that all partners stay on board). Many innovations have not made it to market because the innovating company did not manage the ecosystems of market partners.[3] *Open or linked business models* thus refer to the situations where the innovating company relies on its partners' competencies to jointly create value for customers and share that value according to agreements they have negotiated prior to the collaboration.

There is a major difference between open innovation and open or linked business models. Tapping into external technologies and setting up collaborative deals in open innovation is usually temporary: Collaboration with partners comes to an end once the research project is finished.[4] This is not the case with linked business models: because value is jointly created, partners usually team up during the whole product lifecycle according to agreements set up at the start of the cooperation.

In Table 3.1 we combine open and closed innovations with stand alone (closed) and linked (open) business models. This results in a classification scheme with six different situations. First, business models can be stand-alone or linked. Second, innovations can be categorized as closed and open, but we prefer in line with existing definitions of open innovation to make a distinction between inside-out and outside-in open innovation.

(1) In the outside-in process open innovation activities enrich the company's own knowledge base through the integration of knowledge of external partners.
(2) In the inside-out process a company earns extra profits by bringing internal ideas to market, selling and licensing IP, and spinning off ventures whose business model is not aligned to that of the company (Gassmann & Enkel, 2004).[5]

We discuss each of the six cells in Table 3.1 in the following section.

3.4 A CLASSIFICATION OF DIFFERENT TYPES OF OPEN AND CLOSED STRATEGIES

3.4.1 Closed Innovation Model

The closed innovation model represents the classical case of the closed innovation paradigm described by Chesbrough (2003a). In large companies, the adage was for a long time that successful innovation requires control. In this type of strategy companies generate their own ideas, develop them, build them, market them, and service them on their own. This is exactly what this cell describes. Closed innovation implies that a firm looks for ideas in its own R&D labs since the brightest people in the industry are working for the company. The ideas are also developed internally. The company has state of the art research infrastructure and test facilities which, in turn, guarantee control, secrecy, and if necessary an acceleration of the development. The closed business model implies that the company markets it through its own (global) sales channels. Developing new products is a routine in large companies leveraged by their leading edge R&D-infrastructure, massive manufacturing system, and global sales reach.

The development and commercialization of Nylon at DuPont illustrates this situation.[6] In 1928, the DuPont chemical company opened a research laboratory for the development of artificial materials, deciding that basic research was the way to achieve and sustain competitive advantage. In the lead of Wallace Carothers a team of scientists investigated the acetylene family of chemicals. In 1931, DuPont started to manufacture neoprene, a synthetic rubber. The research team then turned their efforts towards a synthetic fiber that could replace silk. Japan was the United States' main source of silk, and trade relations between the two countries were breaking apart at that time. By 1934, they developed a synthetic silk by producing fibers formed by a polymerizing process. In 1935, DuPont patented the new fiber known as nylon.

Nylon was introduced to the world in 1938.[7] It was first used for fishing line, surgical sutures, and toothbrush bristles. DuPont first announced and demonstrated nylon stockings to the American public at the 1939 New York World's Fair and began commercial production in late 1939. From the time nylon went on sale to the general public in May 1940, nylon hosiery was a huge success. In 1942, nylon went to war in the form of parachutes and tents. Today, it is still used in all types of apparel and is the second most used synthetic fiber in the United States.

Nylon is a typical case of closed innovation: research, development, and commercialization have been realized within and financed by DuPont. Everything was done within the corporate boundaries of the company, from the first detection of the silky fibers till the highly successful B2C applications. It is also a typical example of a closed or stand alone business model. Value is created and captured by DuPont by leveraging its manufacturing infrastructure and global sales apparatus. DuPont was not relying on any strategic partners to develop, manufacture or sell nylon.

3.4.2 Unused Knowledge Used by Others

The inside-out mode of open innovation has been well documented in several case studies. What most case descriptions do not mention is the business model behind these agreements. The underlying business model of most licensing agreements and spin-offs is a stand-alone business model: the recipient of the technology (licensees or corporate spin-offs) will further develop the technology, manufacture the product and launch it in the market. The firm who originally developed the technology is no longer involved in the commercialization of the technology. The recipient uses the insourced technology to bring an offering to the market without relying strategically on the innovator (or other organizations) to market the product. This situation illustrates cell 2 in Table 3.1.

Take for example the licensing and capabilities agreement between P&G and ConAgra Foods. The latter was established in 1919 and has been growing in the last decades through purchasing over one hundred prepared food brands. It moved heavily into the frozen food business and the packaged meat industry, and picked up a selection of other brands from firms like RJR Nabisco and Beatrice Foods among others. ConAgra Foods is selling different types of food products than P&G but can use the unique nutrition-enhancing food ingredients and packaging capabilities of P&G, creating unique competitive advantages for their business. However, the licensing agreement with P&G is not interfering with the business model of ConAgra foods. The company is producing and distributing the products on its own, creating and capturing value without relying on P&G.[8]

The classification scheme in Table 3.1 is not a straight jacket and some situations can be labeled as hybrids. We illustrate this with the Glad case.[9] P&G developed a promising plastic film technology through diaper research. It was proved successful in test markets but it was not strategic for P&G to become a new player in such a well-established market where market leaders are protected by strong brands. Clorox is one of P&G's competitors, who already had a leading plastic wrap business—Glad. In 2002, both companies established a joint venture. P&G brought in marketing expertise and IP behind Press'n Seal as well as future innovations such as ForceFlex. Clorox brought in brand equity, focused R&D in plastics and resins, and was responsible for the manufacturing and distribution of the new plastic film products. This set up allows continuing collaboration on additional initiatives in the plastic film business and proves competitors can work together successfully. The continuing collaboration turns the Glad example into a hybrid between cell 2 and cell 5 in Table 3.1. It is an obvious case of inside-out open innovation but the business model is not entirely closed as both companies continue to work together as strategic partners to deliver value to the market. Joint ventures tend to be acquired over time by the dominating partner—here Clorox. Therefore, it is not excluded that Clorox acquires the joint venture when the innovation dynamics and the market sales run out of steam. In that scenario, the Glad case would move back to cell 1.

3.4.3 Use Other's Knowledge to Develop a New Offering

Companies frequently combine outside-in open innovation with a stand-alone business model. The strategy is very similar to the one discussed in the previous section. The only difference is that the company is now looking for external knowledge which can be used in developing its new products or services within its own business model. A company is systematically searching for proven technologies or ideas that can improve existing or introduce new products. In this strategy it is important to know exactly what the company wants, based on carefully defined targets. It is crucial that the company focuses on ideas and technologies that can create value through the application of internal knowledge, marketing and distribution skills, or other capabilities because the company who is sourcing external knowledge has to manufacture and sell the new product on its own.

If we return to P&G the most successful products developed by external partners but brought to market by P&G are Clean Magic Erasers, TidePods, Olay Regenerist, Swiffer Dusters, and the Crest SpinBrush.

Take Swiffer Duster as an illustration.[10] Procter & Gamble wanted to produce a duster as a follow-up to its successful Swiffer mop but the internally developed prototype was not appealing. The Japanese company UniCharm had developed an attractive duster which it sold only in Japan. UniCharm did not have the manufacturing, distribution, or marketing strength to take the innovative product into other markets. P&G's research team recognized the superiority of UniCharm's duster and saw an opportunity to work together. P&G signed a licensing deal with UniCharm to distribute the duster under the P&G name everywhere in the world except Japan. The Duster hit the market in 2003 and has made millions of dollars for both P&G and its Japanese partner. This case illustrates how a large company can insource external knowledge and innovative products to drive sales growth. It leads to a win for both partners: UniCharm had the right innovation but not the strength to market it globally. P&G is globally operating but did not have the same innovative product as UniCharm. Note that once the innovation was insourced P&G relied completely on its own manufacturing and distribution strengths to launch the product and grow the market. P&G uses a stand-alone business model.

Another example can be found in P&G's skin care products organization.[11] This organization was looking both internally and externally for antiwrinkle technology options for next-generation Olay products. At a technical conference in Europe, P&G first learned of a new peptide technology developed by a small cosmetics company in France. After they shared some of their work at a conference attended by P&G's skin-care researchers, they accepted an invitation for their technologists to visit P&G and present their data on the anti-wrinkle effects of the new peptide. The peptide became a key component used in the blockbuster product, Olay Regenerist. This company now continues to collaborate with P&G on new technology identification and further upstream research projects. Again, the technology is sourced from an outside developer and the deal connects the technology strengths of the French firm with the global brand power of P&G in skin products. The technology provider—although it continues to be a strategic technology partner—is not involved in the commercialization of this product.

3.4.4 Search for Assets Owned by Others to Develop a New Business Model

Companies can develop new technologies mainly in-house. Still, the business model requires an open approach to create and capture value. Chesbrough (2006a) describes different cases of this strategy. We illustrate it with the development of the iPhone.[12]

The iPhone can be considered as an internal innovation and design project of Apple. Apple was already working on the iPhone in 2002. After several years of keeping the details of the phone under wraps, Apple announced the launch of the iPhone in June 2007. This new, internally developed innovation had 200 patents attached to it. The iPhone stood out from its competitors because of its touch screen, advanced features, and elegant hardware design.

With hindsight, however, the real value for the iPhone resides in the apps, rather than in the device itself. With each new application that is developed customers can use their mobile phone in new and unprecedented ways. Remarkably enough, unlike smartphones enabled by Symbian, Palm, and Microsoft, the original iPhone did not provide native support for third-party applications. It was only after there was evidence of tremendous demand for third-party applications (further propelled by successful efforts by third parties to install software without Apple's cooperation) that Steve Jobs reversed his initial position. He announced that a software development kit (SDK) would be made available to third-party developers in 2008, and the iPhone SDK was released on March 6, 2008.

Even here, though, Apple retained control of the ecosystem. Loading a third-party application onto the iPhone is only possible after the developer pays an Apple Developer Connection membership fee and the app is approved by Apple for distribution through the App Store (run by Apple). Developers are free to set any price for their applications to be distributed through the App Store, of which they will receive a 70% share. Developers can also opt to release the application for free and will not pay any costs to release or distribute the application beyond the membership fee. The App Store was launched with the release of iOS 2.0, on July 11, 2008.

While the ecosystem remains under Apple supervision, the uptake of third-party apps has been impressive. As of March 2012, Apple has passed 25 billion app downloads. Opening the iPhone to third-party developers turned the device into a platform where Apple and the external developers jointly create value for customers by leveraging the ideas of a fast growing set of apps. Apple captures value by using the iPhone and the (access to the) App Store as key assets to appropriate value in a systematic way.

3.4.5 Internal Knowledge Accessible to Others to Develop a New Business Model

Companies can outsource internally developed knowledge not to monetize on it (see cell 2 in Table 3.1), but to create a platform similar to the one we have been discussing in section 3.4. In this case, companies with considerable knowledge in specific technological fields farm out technologies and know-how in order to reinforce a platform from which they may profit indirectly. Strong platforms are necessary to induce third parties to develop proprietary applications, which, in turn, generate more profits for the platform owner. In other cases, opening the platform technology allows application developers to customize products and extend the range of applications far beyond the original set of applications offered by the firm that developed the platform—similar to the iPhone case. We illustrate this in cell 5 of Table 3.1 by IBM's support for Linux.

IBM has supported Linux for more than a decade. The company has donated hundreds of patents and it invested $100 million a year to support the Linux OS.[13] Furthermore, IBM is working closely together with Linux to generate and accelerate new applications based on the Linux OS. IBM's support increased the chances that Linux succeeds in its competition with Microsoft in OS markets (Henkel, 2006).

Linux was the first operating system that could challenge the dominance of the Microsoft's Windows OS. In 1999, when IBM adopted Linux, it already had a large installed base of customers and a large community of developers committed to contributing to its development. Through the investments in Linux, IBM could be independent from Microsoft's licensing terms and it could reveal interface specifications for its platforms. IBM profited from an open source platform such as Linux because open source software is less expensive than proprietary software. This allowed IBM to charge its customers less for applications and services. One of the advantages of developing an open source OS is that the risks and costs of designing and developing software can be distributed among many contributors. Although IBM was spending a lot of money on the development of Linux, other firms such as Nokia, Intel, and Hitachi made substantial investments as well. Commercial investments in Linux are estimated to exceed $1 billion a year. Sharing costs and efforts developing a core infrastructure reduced the costs of having a widely accepted OS that could compete with Microsoft's Windows.

The Linux platform also provided a common platform on top of which IBM could build special applications and services. IBM was increasingly focusing on selling high-end hardware, proprietary software running on top of Linux, and integration and other customized services to enterprise customers. By helping to establish Linux, IBM was strengthening its own business model in selling proprietary software solutions for its clients running on top of Linux. The openness of Linux also gave IBM more freedom to co-develop products together with its customers.

3.4.6 Use Others' Knowledge to Create Your Own Business Model

The upper right cell in Table 3.1 combines a linked business model with outside-in open innovation. Here, a company sources knowledge from other organizations to set up a business model that is linked to business models of other (partner) organizations. As we will see, this combination enables companies to come up with new strategies that go beyond the classical open innovation examples. We illustrate this with a few examples.

3.4.6.1 Better Place

Better Place was a venture-backed start-up company, established in 2007 and based in Palo Alto. The start-up's mission was to make electric vehicles (EVs) appealing to the mainstream buyers. It developed and sold battery-charging and battery-switching services for electric vehicles. Better Place could thus be considered as an *electric-car infrastructure* firm: its approach was not to innovate the electric car but rather to innovate the ecosystem around the car. Alongside its smart-grid network of charging stations, battery packs, and cars, Better Place was best known for its concept of battery swap stations, where cars fitted with appropriate battery packs can exchange a discharged pack for a fully-charged one in less than 5 minutes. The company got into financial difficulties by the high investment required to develop the charging and swapping infrastructure and the market penetration which was significantly lower than originally, which, in turn, was caused by the resistance of car manufacturers and other players to switch to a battery swap system.

Better Place intended to make EVs a mainstream product in the car industry by a unique business model. The model started identifying the current problems with the use of EVs: Adner (2012) identified a number of hurdles to overcome: First, EVs were more expensive to purchase than comparable gas-powered cars, mainly because batteries are very expensive. Second, electric cars have a limited driving range which is primarily determined by its battery. Third, the battery charging infrastructure is scant compared to the network of gasoline service stations. Moreover, charging takes a long time.[14] Fourth, since batteries are expensive and the technology is evolving fast, the resale value of the battery is extremely low. Finally, the limited driving range precludes drivers from reaping the full benefits of an EV thanks to the lower power costs per kilometer compared to gas-powered cars.

Better Place's business model tried to resolve the problems mentioned above. First, consumers should not own batteries in order to reduce the price of EVs. Second, EVs should offer the same driving range and convenience as the gas-powered cars. Better Place invited customers to enter into subscriptions to purchase driving distances similar to the mobile telephone industry where customers contract for minutes of airtime. The initial cost of an electric vehicle may also be subsidized by the ongoing per-distance revenue contract just as mobile handset purchases are subsidized by per-minute mobile service contracts. In this way, electric cars could be sold cheaper than the average gasoline car. The Better Place approach enabled manufacturing and sales of different electric cars separately from the batteries. The monthly payment covered electric "fuel" costs including battery, daily charging, and battery swaps. Better Place allowed customers to pay incrementally for battery costs including electric power, battery life, degradation, warranty issues, maintenance, capital cost, quality, technology advancement, and anything else related to the battery. An extra advantage for the customer was that when the car owner sells the car, he doesn't have to sell an outdated or degraded battery. The resale value was in this way tied to the car and not the battery. The Better Place electric car charging infrastructure network tackled the problem of the limited driving range. The company invested in national or region-wide battery switching stations.[15] Better Place has already rolled out a nation-wide network of battery switching stations in Israel and was working with partners to build standards-based networks in Denmark, Australia, California, and Hawaii.

Better Place is an illustration of cell 6 in Table 3.1. Its business model could only be successful if different partners in the ecosystem aligned their innovations to the business model of Better Place. The success relied on the innovations of others. Moreover, Better Place's success hinged on the willingness of the partners to subscribe to the business model. Car manufacturers were very important in that respect: They had to change the car design so that battery packs could be switched according to Better Place standards. Renault-Nissan adapted its Laguna, eRogue and Luence car models, but there was not enough backing from other car manufacturers to turn Better Place into a success. Also battery manufacturers had to adapt their technology to switchable packs. Moreover, they would no longer sell their batteries to car manufacturers but to Better Place itself. Similarly, relations with electricity producers/distributors could be important for Better Place. Finally, governments played a crucial role in the adoption process of the business model. They could set or adopt standards, set prices of oil and electricity, and they had the power to decide about the development of the network and could help in setting the price of new EVs through tax regulations and subsidies.

Although Better Place filed for bankruptcy, the case illustrates that making EVs attractive to the mainstream car driver requires an organization whose business model tackles the different hurdles in the existing ecosystem. Better Place could only succeed if other players in the ecosystem innovate (outside-in open innovation) and adopt or co-align their business model to that of Better Place. Lack of support by car manufacturers was one of the reasons why the company never turned into the success it promisd to be in its first years. It is nevertheless an interesting example of so-called linked business models.

3.4.6.2 SkyNRG

A major concern for the airline industry is greenhouse gas emissions from aviation and their impact on climate change. One interesting initiative of the aviation industry in that respect is SkyNRG, which was launched in November 2009.[16] Founding partners are Air France KLM Group, North Sea Group (a group of companies delivering products and services to the oil market) and Spring Associates (a strategy consulting firm helping companies in becoming more competitive through sustainable products and solutions). The mission of SkyNRG was to help create and accelerate the development of a market for sustainable jet fuel, yet safe and affordable.

SkyNRG is not established to develop new biofuel technologies. Many of the technical hurdles facing aviation in its move towards sustainable aviation biofuels have now been overcome and much of this work has been achieved within the industry. However, to reduce greenhouse gas emissions biofuel has to become mainstream in the aviation industry. Therefore, commercialization and scaling up of the supply of aviation biofuels are crucial. SkyNRG's mission is to set up an ecosystem of strategic partners to introduce biofuels as an alternative source of energy. SkyNRG has spent two years on research and development to create a *one-stop shop* for airlines, airports, military, and other kerosene end-users integrating the complete supply chain for sustainable jet fuels. From feedstock-to-flight, the supply chain covers elements such as strict selection of sustainable feedstock, refining contracts, distribution to any airport in the world, quality assurance, plane fuel service, insurance, marketing and project (co) funding with airports and end customers. In this way, SkyNRG can be considered as a global market maker in sustainable jet fuel.

SkyNRG is an example of how an organization can unfold an innovation ecosystem including organizations from different industries that need to work together for a new value proposition to succeed. The instigator, KLM, is one of the (potentially many) airline companies that can profit from a new source of kerosene also helping them to achieve the target to reduce the carbon footprint of the industry. KLM has no knowledge or internal capabilities to develop and produce biofuel, but it has a vested interest as a customer of aviation biofuel to guarantee a stable and competitively priced supply of bio-kerosene. With the establishment of SkyNRG, KLM was setting up a large ecosystem including the different types of partners that are necessary to convince airline companies around the globe to switch gradually from petro-based fuel to biofuel.[17]

If we look at SkyNRG from the perspective of KLM, the initiative illustrates how a company can tackle a major strategic challenge (dependence on petro-based fuel) by setting up a hub company that unfolds an ecosystem to guarantee a growing supply of aviation biofuel. KLM and its network of airline partners in Skyteam are (potential) customers of the aviation biofuel. The required technical innovations (such as the development of second generation feedstock for fuel production) are developed by the biofuel industry and specialized technology centers. However, technology is only one determinant of the mainstream adoption of biofuel in the airline industry. A breakthrough can only be achieved by combining essential expertise and experience in the fields of regulations (standard setting), effective sustainability criteria, product knowledge, and air transport. There are considerable adoption risks, as several players may choose not to switch to aviation biofuel. KLM is interested as a customer in a steady supply of competitively priced biofuel which is a major driver of profitability. Outside-in open innovation plays a role but in a different way than in cell 3 of Table 3.1: in that cell the knowledge sourced from external partners empowers the focal firm to develop a new product or service. Here (in cell 6), KLM is not developing a new product or service using external knowledge, but it established SkyNRG to accelerate the technological progress in the biofuel value chain, guaranteeing in this way the supply of aviation biofuel. KLM adopts an open business model: value is created through the concerted action of several partners in the ecosystem and is shared according to predetermined agreements. The ecosystem is orchestrated by a dedicated company (SkyNRG) who crafts inter-organizational ties to create an ecosystem between organizations that have never been in contact with each other before. While in the previous case, Better Place took on the role of a hub firm, KLM tries to realize its strategic objective through the establishment of an ecosystem orchestrated by another organization.[18]

3.5 EXTENDING OPEN INNOVATION

The classification scheme presented in Table 3.1 is a simple framework to examine different combination of innovation strategies and business modeling. The outcomes of the left and right columns in Table 3.1 are quite different from each other. The stand-alone business models are focusing on the new product or service development. Most open innovation examples that have been described in the literature can be classified in the left-hand column. In contrast, linked or networked business models have received less attention.[19] With linked business models a company can use inside-out or outside-in open innovation beyond new product development. In the left-hand column of Table 3.1, external knowledge is sourced to develop a new product or business, or internal knowledge is marketed to another firm that will use it for its own product development. In the right-hand column, a combination of linked business models and open innovation can be used to leverage *any strategic driver* as in the case of SkyNRG. We discuss this point in greater detail below using a hypothetical example of the crude oil business in BP.

Assume you are a manager in the crude oil business at BP. The product you are selling is inevitably a commodity and product innovation is by definition excluded. Competitiveness in the crude oil business depends on various technologies that increase the productivity of exploration and extraction. Oil companies have to detect the richest oil wells earlier than competitors and drill them more effectively through new technologies that allow them to extract oil at larger depths. Although the oil industry is dominated by large companies with strong R&D capabilities, they rely on specialized oil-services companies such as Schlumberger and others to develop new technologies for oil exploration and extraction: the oil-services sector is a beacon of innovation within the energy industry. Oil-services firms typically receive more patents each year than most of the large integrated oil companies. BP can gain a competitive advantage if it partners with Schlumberger (usually in combination with other specialized services companies) with leading-edge exploration and drilling technology. BP can set up a research program with these firms and (co-) finance the research and development of new exploration and drilling technology. They become strategic partners in advancing this technology. BP will typically require exclusive use of the technology for several years before Schlumberger can sell the technology to other oil companies.

This example illustrates how a combination of open innovation can be applied in commodity businesses where product innovation is not a competitive driver. Open innovation applied to new product or business development should be considered as a specific competitive driver relevant in particular situations but not in others. Open innovation can be applied in a wider range of situations if we start with the strategy of a business, identify the key value drivers that should be acted upon, spot and select the potential innovation partners and set up a joint project to develop technologies or solutions that will strengthen the firm's competitive position. Thus, even in the absence of any product or service innovation in the business, firms can still "nurture" their network of innovation and value chain partners to become more competitive. Nambisan and Sawhney (2010) have shown how such a network has to be managed.

This shift away from product innovation also shows that the competitive position of firms may rely on a broad set of value drivers, going from process innovations, an increase in the productivity, or enhancing the quality of products. Increasing throughput time, reducing operational complexity and costs, or integrating processes are other examples. Which one to focus on depends on the business context, but in each case the focal firm can set up a joint research initiative and encourage (technology) partners to join forces to accelerate the required technological innovation to augment the competitive advantage of the former.

Finally, extending open innovation in this way makes it more relevant for companies and organizations who are recipients of technologies/innovations—service industries, low-tech manufacturing industries, governments, etc. Recipient organizations can initiate and orchestrate the collaborative initiative while technology providers are implementers within this framework.

3.6 CONCLUSIONS

Open innovation from its inception has examined the connection between the firm's R&D processes and the firm's innovation model. As the concept has been explicated and expanded, that connection has itself evolved. In the initial 2003 book, the business model was treated as static (Chesbrough, 2003a). The canonical experience of Xerox PARC with its many technologies that did not fit the copier/printer business model of the parent company showed the importance of linking innovation and the business model. In Chesbrough (2006a) the evolution of the business model was considered, and a six-stage maturity model was propounded to represent different levels of development. The platform business model was the highest, most valuable type of business model. In the 2011 volume that directly considered open innovation in services, the shift of many companies toward services required their business models to change as well.

With this chapter, we think that we have extended the evolution of these concepts further. We invite readers to engage in this research themselves, and help us further understand the novel ways in which more open innovation processes can combine with more open, connected business models to create and then capture value.

NOTES

1. Grönlund, Sjödn, and Frishammar (2010) is an interesting article about the stage-gate process in an open innovation context.

Classifying Open Innovation and Open Business Models 67

2. One of the most influential books on business model development is Osterwalder and Pigneur (2009). Amit and Zott (2001), Zott and Amit (2007; 2008) Afuah and Tucci (2001), and Chesbrough and Rosenbloom (2002) provide great insights on the different dimensions of business models.
3. A good example is the Flavr Savr genetically modified tomato of Calgene (Vanhaverbeke & Cloodt, 2006) or the examples provided by Adner (2012) such as e-books, digital cameras, inhalable insulin, electronic health records, etc.
4. In a webinar on November 28, 2012, the Manager of Open Innovation at Procter & Gamble, Nick Nicholides, stated that P&G had done more than 1,300 collaboration contracts in its Connect and Develop program since 1999, and that 40% of P&G's partners had done multiple deals with P&G. So, some of the open innovation transactions in this instance had turned into ongoing relationships.
5. We also could add the "coupled process" type of open innovation as argued by Gassmann and Enkel (2004). To keep the number of combinations tractable within a short chapter, we chose to limit our attention to the outside-in and inside-out type of open innovation.
6. The development of Tide at P&G is another great example. Tide was the first heavy duty synthetic laundry detergent which was developed at P&G by a small team long after the company had officially given up the project to develop a synthetic detergent. It is a perfect example of skunkworks. For more information, we refer to Dyer, D., Dalzell, F. & Olegario, R. (2004) *Rising Tide: Lessons from 165 Years of Brand Building at Procter & Gamble*, Boston: Harvard Business School Press. An overview of the Tide development can be found at http://laundry.about.com/od/laundrydetergents/ss/Tide-Laundry-Detergent-Through-The-Decades.htm.
7. See http://www2.dupont.com/Phoenix_Heritage/en_US/1939_c_detail.html and http://inventors.about.com/od/nstartinventions/a/nylon.htm.
8. P&G also developed Nodax, a technology that can be used to manufacture new biopolymers using renewable resources. However, P&G is not a producer of plastics and it sold the invention to Meredian, Inc., a privately-held corporation. Meredian uses the acquired technology to manufacture polymers that will biodegrade. Similar to the ConAgra Foods case, P&G's technology helps Meredian to develop and produce a new product category, but the recipient launches the product and grows the market without the help of the Cincinnati based multinational. For more information see http://www.pgconnectdevelop.com/home.
9. This example is taken from Chesbrough, 2006: 200–201. See also: http://www.pgconnectdevelop-la.com/espanhol/anexos/cd_brochureWEB.pdf.
10. See http://www.enablingideas.com/?p=1418.
11. See http://www.pg.com/connect_develop/cd_success_stories.shtml.
12. This account is synthesized from two key sources. One is http://apple-history.com/ipod; the other is an article from *Wired Magazine* in 2004, http://www.wired.com/gadgets/mac/news/2004/07/64286?currentPage=all; see also West & Mace (2010).
13. See Chesbrough, 2006, chapter 8, for more discussion of IBM and Linux.
14. DC fast charging is presently considerably slower than Better Place's 59.1 seconds battery-switchover. The same DC fast chargers can recharge the battery of the Nissan Leaf to 50% in 30 minutes because the car's software controls the rate of charge and not the fast charger (LaMonica, 2010).

15. www.betterplace.com.
16. For more information, consult the following websites: http://skynrg.com, http://argosenergies.com/en/innovation, http://www.klm.com/csr/en/climate/footprint/biofuels/index.html.
17. The industry has been forging ahead with pilot projects in a number of countries worldwide. But airline companies and the biofuel producing value-chain cannot do it alone. Political support and financial investment will have to come from a number of stakeholders. Global sourcing and marketing of sustainable jet fuel require promoting R&D throughout the entire supply chain, advancing the technical certification and economic viability of next-generation aviation fuels, pushing for mechanisms that help to create a level playing field for sustainable jet fuel, and finding ways to finance the premium to be paid for sustainable jet fuel until the bio-fuel industry is big enough to guarantee competitive prices.
18. This, in turn, raises interesting questions how large stakeholders such as KLM keep control of the decision making in hub firms such as SkyNRG.
19. Linked business models are, however, not new: Chesbrough (2006) provided interesting examples of open business models and showed how companies can create and capture value through the development of a platform.

Part II

Analyzing Open Innovation at Different Levels of Analysis

4

Challenges of Funding Open Innovation Platforms

Lessons from Symbian Ltd.

Joel West

4.1 INTRODUCTION

Two core concepts of the open innovation paradigm are the centrality of the business model—particularly creating and capturing value—and the need for multiple firms to cooperate in creating value (Chesbrough & Rosenbloom, 2002; Chesbrough, 2003; Chesbrough, 2006a, 2006b). Research in open innovation has examined how firms have used open innovation strategies to create value using external networks, communities, and ecosystems (Vanhaverbeke & Cloodt, 2006; West & Lakhani, 2008; Rohrbeck et al., 2009). This builds on a broader body of research about how firms utilize alliances, networks, communities, consortia, ecosystems, and platforms to support their innovation strategies (Gomes-Casseres, 1993; Powell, 1990; West & Sims, 2012; Pisano & Verganti, 2008; Adner, 2012; Gawer, 2009). Such cooperation is particularly important for sponsors of general purpose computing platforms, who have for more than 30 years run formal ecosystem management programs to obtain third-party complements that complete the value proposition of their platforms (Kawasaki, 1990; Gawer, 2010).

This chapter considers an example of an innovation ecosystem that created value but faced major challenges with the allocation of the value capture—using a case study of Symbian Ltd., a London startup company that created the most successful smartphone platform of 2003–2010. At its peak in 2007, the Symbian platform accounted for 63% all smartphones sold; two years later, the sponsoring company had ceased to exist. By 2011, the platform was orphaned when, in a once-unthinkable move, its largest remaining customer announced plans to discontinue Symbian smartphone sales in favor of Windows.

This study uses a combination of primary and secondary data, internal and public sources to analyze the transitory success of Symbian Ltd. and its Symbian OS platform. It discusses the company's ecosystem strategy during its entire decade of existence, and the internal stresses within the ecosystem over efforts at value capture.

I argue that many of Symbian's difficulties reflect the inherent difficulties of its open innovation approach to platform leadership. As a cash-starved startup, the corporate venture investments by Symbian's customer-shareholders (handset makers) both sustained its R&D efforts during its initial six years of losses while heavily constraining its strategic options. Symbian also faced conflicting goals between managing its own survival and that of its ecosystem members. To contrast with Symbian's failed strategy, the chapter identifies the commonly used strategy of "platform chaining" from a previously successful platform that has provided other sponsors the resources necessary to launch a new platform and ecosystem.

4.2 PRIOR RESEARCH

The goal of this chapter is to explain the challenges that one firm faced in creating and managing an ecosystem of external partners to support its platform. Here I review the degree to which open innovation research—particularly the research on firm use of external innovations—has considered the role of ecosystems and related concepts, and how such research can be informed by other bodies of research that consider firm interactions with networks, ecosystems, platforms, and related constructs.

4.2.1 Networks, Ecosystems, and Platforms

To support their innovation efforts, firms have engaged in a range of strategies for managing relationships with external counterparts, including alliances, networks, communities, consortia, ecosystems, and platforms. In each case, the investments by the exchange partners in assets, capabilities, and strategies reflect a pattern of recurrent relationships rather than a single market transaction, demonstrating an interdependency of reciprocity and repeated interactions that helps mitigate the risks of opportunism (Powell, 1990; Jones et al., 1997).

Research on *alliances* generally focuses on the relationship with one partner. These dyadic partnerships tend to be long-lived (multi-year) and created through formal (but incomplete) contracting to manage opportunism. The success of the alliances often depends on the complementarity of the partners, whether through differing technologies, between innovation creation and commercialization (as in biotech), or through strengths in different parts of the value proposition or value chain (Hagedoorn, 1993; Gomes-Casseres, 1996; Rothaermel & Deeds, 2004).

When firms have a pattern of building multiple alliances, it may be more appropriate to consider these alliances as *networks* of interfirm interactions (Powell, 1990; Gomes-Casseres, 1996). Research on such networks has focused on the complementarity and reciprocal interactions of multiple independent actors, such as the supplier and customer relationships within a given industry, industrial trading group or regional economy (Powell, 1990). Firms in a network may work together to create value through coordinated innovation efforts, particularly in the presence of network effects, increasing product modularity, and when enabled by communications technology (Staudenmayer, Tripsas, & Tucci, 2000; Nambisan & Sawhney, 2011).

Research has identified specific patterns of networks that share common characteristics and theoretical mechanisms. For example, firms work with external self-governing *communities* organized for a common purpose to produce a shared common good; these communities may be composed of firms, individuals, or both (West & Lakhani, 2008; O'Mahony & Lakhani, 2011). Such communities differ from networks both in terms of governance and in their sense of shared social identity (Markus, 2007; von Hippel, 2007). The communities vary markedly in terms of their degree of innovativeness and their alignment to firm innovation goals (West & Sims, 2012). Two of the most frequently studied types of communities are those that produce product compatibility standards (Rosenkopf et al., 2001; Simcoe, 2012) and open source software (Dahlander & Magnusson, 2008; West & O'Mahony, 2008).

Consortia are also a specialized network of organizational members,[1] whose members jointly provide resources to fund research efforts guided through a form of centralized control or governance. Such consortia are driven both by common goals and a desire to share in the outputs of the collaboration (Sakakibara, 1997; Doz et al., 2000). However, they differ from communities (particularly open source communities) in their ability to exclude others from the benefits of joint production (West & Gallagher, 2006b; Pisano & Verganti, 2008). Unlike networks organized for the benefit of a single firm, such consortia tend to be organized as heterarchical networks with no single dominant actor or beneficiary (Müller-Seitz and Sydow, 2012).

An important extension of the network perspective came with the metaphor of the business *ecosystem* of firms that sell complementary goods and services. The success of the member firms both contribute to and depend on the health of the ecosystem, although (as in environmental ecosystems) these ecosystems are marked by constant competition for overall leadership and dominance of specific niches (Moore, 1993; Iansiti & Levien, 2004a).

In some industries, distribution of innovation between ecosystem members is often the direct consequence of technical modularity (Baldwin, 2012). The success of the ecosystems in jointly creating value through innovation depends not just on the ecosystem leader, but also the efforts of the member firms in overcoming their own technical challenges (Iansiti & Levien, 2004b; Adner & Kapoor, 2010). While some ecosystems lack formal governance, others may be associated with a community or consortium and its governance mechanisms; a successful example of such an ecosystem is the Eclipse open source community (Fitzgerald, 2006; West & O'Mahony, 2008). Sharing control of an ecosystem encourages third party participation and a greater provision of complementary goods (West & O'Mahony, 2008; Boudreau, 2010).

Finally, firms work with external third parties to create a *platform* in which the joint value creation and integration of complementary products is mediated by compatibility standards that define a systems product (Gawer, 2002; Gawer & Cusumano, 2008). For a "proprietary" or "closed" platform, a single sponsoring firm controls the platform and its standardized interfaces to assure its own value capture, while sharing enough of the returns from the ecosystem to attract third-party complements (Gawer & Cusumano, 2002; West, 2003). An "open" platform is one where control (including defining the interfaces) is shared across a self-governing community; with the assurance of greater access to the benefits of the ecosystem, such shared control is often more successful in attracting external participation and complement production (West & O'Mahony, 2008; Baldwin & Woodard, 2010; Simcoe, 2012). However, in the real world, there is a wide range of intermediate points between these two extremes, as measured by the degree to which one or more central firms can control access to use and benefit from platform innovation—which in turn determines the cost paid by customers and complementors to use the platform (West, 2003, 2007a). For example, some platforms are tightly controlled by multiple sponsors, with the proprietary benefits accruing to multiple firms (Eisenmann, 2008).

Here I am particularly interested in sponsored platforms: the management of the ecosystem of a supply of complementary products by a firm that defines the interfaces. The practice of ecosystem management to support a platform predates formal academic theory on either topic. From the birth of the mainframe application software business with IBM's 1969 unbundling decision through the sale of retail packaged software such as Visicalc for personal computers a decade later, system vendors have increasingly recognized the importance of third-party complements for the success of their products (Campbell-Kelly, 2003). Beginning in 1983, Apple Computer even created a new job category called "evangelist" to attract new ecosystem member companies and to coordinate interactions with the ecosystem sponsor (Kawasaki, 1990).

4.2.2 Research in Open Innovation

To date, open innovation has been less complete in its coverage of networks, ecosystems, and platforms—perhaps because of its origins as a normative theory for profit-maximizing firm managers. Research on firm use of open innovation has tended to emphasize the dyadic exchange (usually market exchange) between the focal firm and external sources of innovation (West et al., 2006; see also Chapters 1 and 2).

For example, in the earliest and most-cited study of open innovation using Europe's Community Innovation Survey, Laursen & Salter (2006) studied firms and their potential collaboration ties with eight different external sources, including suppliers, customers and non-profit research labs. In their review of research on the inbound and coupled modes of open innovation, West & Bogers (2014) found a dyadic emphasis for inbound modes, but a nearly equal split between dyadic and network interactions in research on coupled modes that consider bi-directional flows of knowledge and innovation creation.

In the coupled category there is a limited amount of research that has examined how firms utilize networks to support their open innovation strategies, beginning with the 2006 book *Open Innovation: Researching a New Paradigm* (Chesbrough et al., 2006). In addition to the challenges of sourcing (or selling) innovations with external partners found in the dyadic perspective, firms must also coordinate the activities of the networks "both to develop new technologies... and to exploit technology-based business opportunities" (Vanhaverbeke & Cloodt, 2006: 277). Firms externally source technology, components, and products from suppliers and third parties, particularly to create and complement complex assembled systems (West, 2006).

Much of the work of open innovation has examined firms creating a complex integrated product in the information-communications technology (ICT) sector. West and Gallagher (2006b) considered how firms leverage open source communities to support their computing or software products. Meanwhile, two studies examined the collaboration of Nokia in mobile telecommunications: Maula and his colleagues (2006) explained how Nokia planned its own long-term innovation efforts and those of its complementors. Dittrich & Duysters (2007) showed how Nokia shifted how it used external partners— from exploitation to exploration of knowledge—as its traditional radio-based competencies became less valuable.

4.2.3 Research Questions

As Chesbrough (2006b: 1–2) noted in our earlier book, "open innovation explicitly incorporates the business model as the source of both value creation and value capture" (cf. Chesbrough & Rosenbloom, 2002; Simcoe, 2006). Such a perspective is one of eight ways that the open innovation paradigm differs from earlier innovation studies (Chesbrough, 2006b).

The business model perspective is also essential for explaining the success of an open innovation ecosystem. As with other innovation ecosystems, firms leverage an OI ecosystem for the joint value creation that makes its products more valuable to the prospective customer (Maula et al., 2006). At the same time, an open innovation ecosystems strategy must consider the allocation of value capture that both allows the focal firm to succeed, and also motivates the external partners to continue to participate (Vanhaverbeke & Cloodt, 2006; West & Gallagher, 2006b; see also Iansiti & Levien, 2004a). Firms that manage open innovation ecosystems must confront the inherent tension they face in maximizing both their value creation and value capture (Simcoe, 2006; Henkel et al., 2014), and also the degree to which business models within the ecosystem are aligned or are in conflict in the value capture (Vanhaverbeke & Cloodt, 2006).

I am particularly interested in how ecosystem management is handled by new companies such as Symbian, which by their nature must access, mobilize and then generate resources if they hope to grow to large, successful companies (Garnsey, 1998). Nascent high-tech entrepreneurs are particularly concerned with gaining legitimacy and access to external resources (Liao & Welsch, 2008).

4.3 CASE STUDY: SYMBIAN'S SMARTPHONE PLATFORM

This chapter examines ecosystem strategy in the first decade of a new type of computing platform: the smartphone, which combined the computing capabilities of a personal digital assistant (PDA) with a mobile phone to eventually create an Internet-aware mobile computing device. From 1997–2002, manufacturers Nokia, Qualcomm, Ericsson, Handspring, and Research in Motion released a series of first-generation devices, experimenting with size, form factor, application software, and input modes. Eventually a dominant design emerged that included a color screen, email, and a web browser (West & Mace, 2010).

The focus is on the Symbian smartphone platform during the entire life of its sponsoring company, Symbian Ltd. (1998–2008). From 2007–2013, I compiled data regarding the firm's platform and financing strategies from a wide range of primary data, including information on its website, earlier web pages at the Internet Archive, and interviews with ecosystem managers; this was supplemented by news accounts published during and after the firm's existence (see West & Wood, 2013 for additional information on the data sources).

4.3.1 Ecosystem Strategy

Symbian Ltd. was a privately held London-based software developer created in June 1998 as a spinoff of Psion PLC, a successful British maker of keyboard-based personal digital assistants. Psion held licensing negotiations with the world's leading handset makers to adapt its operating system for use to create what later would be called a "smartphone". Symbian was founded as an independent company by approximately 160 employees transferred from Psion's software subsidiary, and it began to adapt the PDA software to support mobile phones with PDA features that would run on GSM (and later W-CDMA) mobile telephone networks.[2]

Symbian proclaimed itself as an open platform, because it was not controlled by any single firm (as with Microsoft). Unlike the vertically-integrated strategies used for the earliest mobile phone production, Symbian's business model of selling its software to a wide range of manufacturers anchored it explicitly in what later would be called the open innovation paradigm. However, Symbian OS was not open in that the platform's interfaces were controlled by an independent committee such as the POSIX committee that standardized Unix (cf. West, 2007a).

4.3.1.1 Symbian's Partners

Internally and in public, Symbian normally used the term "ecosystem" to refer to its network of customers and complementors (e.g. Northam, 2006). Symbian sometimes used the word "community," but "ecosystem" was generally preferred since "ecosystem" recognized that companies have competitive relationships as well as the "friendly" relations implied by the word "community" (West & Wood, 2013).

The Symbian ecosystem concept was modeled after earlier computing ecosystems, particularly Psion's PDAs. Because Psion was vertically integrated, Symbian's open licensing of its eponymous operating system more closely resembled that of Microsoft Windows, with the operating system sold to system integrators (in this case handset makers) who combined the CPU (and other hardware) with the operating system (and other software) to create value for end-users. Even so, the Symbian ecosystem was far more complex than that for Windows, with nine distinct categories of ecosystem partners (Table 4.1). Some of the increased complexity was a matter of degree, as with the number of CPU vendors (five major suppliers) and the early importance of in-house software development by large enterprises.

Symbian's ecosystem had two types of stakeholders not found in the Windows counterpart. One was network operators that ran the telephone networks to which Symbian smartphones would be connected. Handset makers depended on them for distribution of more than 90% of mobile phone handsets, and they were highly fragmented, with more than 500 networks across

Table 4.1 Categories of Symbian ecosystem members ca. 2002

Partner Category	Product	Customer	Examples
Handset makers ("Licensee")	Handset	Network operator, end user	Nokia, Sony Ericsson, Motorola, LG, Matsushita
CPU vendors	CPU	Handset maker	Infineon, Intel, Motorola, Renesas, Samsung, ST Micro, TI, Toshiba
Other hardware suppliers	Hardware	Handset maker	ATI, CSR, Wacom
User Interface companies	Software	Handset maker	Nokia, NTT DoCoMo, UIQ
"Licensee Suppliers" (pre-loaded software)	Software	Handset maker	Access Systems, Hantro, Macromedia, Opera, RealNetworks, PacketVideo
Independent Software Vendors (ISVs)	Software	End user	AppForge, Borland, Psion, Symantec
Enterprise software developers	Software	End user (self)	
Consulting and training	Service	Handset maker	Atelier, Digia, K3, Omron, Wipro
Network operators	Service	End user	Vodafone, T-Mobile, Orange, Telecom Italia

Source: Categories taken from West & Wood (2013); partner examples adapted from 2008 Symbian internal list of partners

200 countries.[3] Although the operators did not make products, they imposed requirements upon Symbian and the handset makers in key areas, such as pre-loaded software and security.

Another new stakeholder was the user interface supplier. To allow customization of the look-and-feel by handset makers, the Symbian OS relied on separate UI software developed by its handset or operator partners. Because Symbian did not control all of the APIs or user experience, over the long term this created major difficulties in evolving both the ecosystem and the underlying technical architecture (West & Wood, 2013).

While the formal ecosystem program evolved in three phases from 1998 to 2008, the most significant changes in the conception of how the ecosystem created value came from 1998–2002 (West & Wood, 2013). This came due to two under-appreciated differences between developing smartphones and the earlier PDA and PC antecedents.

The first was that the creation of a smartphone was far more difficult than a standard PDA device or the (already mature) PC. The smartphone category was brand new, with a technical complexity that no company had previously mastered. Smartphones were not only computing devices, but also had to control voice, messaging, and data access to the telephone networks. The first devices were released at a time when the 3G network standards were being developed and then the first 3G networks were deployed. Finally, all these functions had to be delivered within the weight, power, and battery constraints of a portable, pocket-sized device.

A second difference was that (unlike with a PC), the operating system software had to be finalized before the handset was manufactured.[4] This meant that the availability of the operating system and key pre-installed software (such as a Java interpreter or web browser) could become the key bottleneck in the availability of a new handset. In recognition of this bottleneck, in 2000 Symbian revised its ecosystem categories for software developers to distinguish between those who developed preloaded software (licensee suppliers) and those that created software that was downloaded later (independent software vendors).

This final constraint also limited Symbian's leadership of its open innovation ecosystem. Unlike Psion, it did not sell devices directly to consumers, and unlike Microsoft it could not sell its operating system or upgrades directly to end users. This meant that adoption of its latest technology depended on new adoption of smartphones and replacement purchases by existing owners. And because it effectively had no direct relationship with customers, Symbian (unlike Microsoft or Apple) focused its branding efforts on system integrators and made little effort to increase public awareness of the Symbian operating system.

4.3.2 Funding Platform Development

4.3.2.1 Customers as Strategic Investors

A unique[5] strategy of Symbian's ecosystem was to use its most important partners—its handset licensees—as its investors and shareholders (Table 4.2). Symbian was launched in June 1998 with a joint announcement by the three largest handset makers—Nokia, Motorola,[6] and Ericsson—and won investments by the Matsushita Electric (owner of Panasonic, then the fourth largest maker) in May 1999 and Samsung (by then the third largest) in February 2003. These corporate venture investments provided Symbian instant legitimacy, ties to prospective licensees and funding to develop its new platform.

The handset makers shared a common need to make smartphones possible. They also had a common desire to block Microsoft from repeating its PC rent-seeking in the mobile phone world, which is why Bill Gates later cited Symbian as "serious competition" (West & Wood, 2013). However, from the start there were tensions among the investors, reflecting both the divergence of interests between these competing handset makers and between Symbian and

Table 4.2 Shareholders of Symbian Ltd., 1998–2008

Company	HQ	Peak Capital Investment	Equity 1998 Q2	1998 Q4	2002	2003	2004[§§]
Psion	UK	£5.8 mil.[§]	40%	30.7%	26.6%	31.1%	
Nokia	Finland	£67.4 mil.	30%	23.1%	20.0%	32.2%	47.9%
Ericsson[†]	Sweden	£44.4 mil.	30%	23.1%	20.0%	17.5%	15.6%
Sony Ericsson	UK	£17.0 mil.				1.5%	13.1%
Motorola	US	£33.2 mil.		23.1%	20.0%		
Matsushita	Japan	£23.8 mil			8.4%	10.5%	10.5%
Siemens[††]	Germany	£24.3 mil.			5.0%	4.8%	8.4%
Samsung	Korea	£17.0 mil.				5.0%	4.5%

Source: News coverage, Symbian press releases, Symbian website
Notes:
[†] In 2001, Ericsson transferred its handset business (but not its Symbian investment) to the Sony Ericsson joint venture
[††] Siemens sold its handset business to BenQ in 2005 but remained a Symbian shareholder
[§] Does not include the value of technology transferred at time of Symbian's founding
[§§] Shareholding unchanged from July 2004 until Nokia acquired 100% ownership in late 2008

its customers. This divergence was magnified by Nokia's growing influence as the largest Symbian shareholder, developer, and customer.

From Symbian's standpoint, the major role that the investor-manufacturers played was as a source of working capital to help support more than £200 million in Symbian R&D from 1998–2004 until it became profitable (West & Wood, 2013).[7] The company raised a total of £233 million ($370 million) from 1998–2004: £154 million from the initial equity purchases of seven handset manufacturers through 2003, plus £79 million from additional shares sold to existing investors in 2000 and 2004 (Table 4.3). Three transactions highlight the tensions between the shareholders.

Near the peak of the dot-com boom, in August 2000 Psion announced plans to cash in its Symbian stake (worth as much as £1.7 billion) by spinning off its shares via IPO that could have valued Symbian Ltd. at £2–6 billion (Daniel, 2000).[8] The IPO plan was also popular with Symbian employees holding stock options, and would have provided the company future sources of capital. The manufacturer–shareholders (notably Nokia) forced Psion to cancel its spinoff plans (Lettice, 2004), although employees continued to hope that the IPO might be revived.

In 2004, seeking liquidity for its core operations, Psion announced plans to sell its 31.1% share to Nokia—giving it 63.3% of the company—but was swiftly opposed by Ericsson's CEO. After elaborate negotiations, Nokia's holdings

Challenges of Funding Open Innovation Platforms 81

Table 4.3 Changes in Symbian Ltd. capital structure, 1998–2008

Date	Transaction	Proceeds to Symbian	Transaction Amount	Implied Valuation[†]
June 1998	Nokia, Ericsson each buy 30% share in new company	£80 million		£133 million
Oct. 1998	Motorola buys 23.1% share	£28.75 million		£124 million
May 1999	Matsushita buys 8.9% share	£22 million		£244 million
Aug 2000	Psion announces intended IPO of its 28.1% stake			£2-6 billion (est.)
Jan. 2002	*Pro rata* capital infusion by existing shareholders	£20.75 million		£265 million
Apr 2002	Siemens buys 5% share	£14.25 million		£285 million
Feb 2003	Samsung buys 5% share	£17 million		£340 million
Oct 2003	Nokia, Psion buy Motorola's 19% share	-	£57 million	£300 million
July 2004	Nokia, Sony Ericsson, Panasonic and Siemens buy Psion's 31.1% share	-	£137.7 million	£480 million
	Nokia, Sony Ericsson and Siemens buy new shares from Symbian	£50 million		
June 2008	Nokia proposes to buy 52.1% of shares held by other manufacturers	-	£264 million	£401 million
	Total	£232.75 million		

[†] Post-money valuation implied by transaction amount and share
Source: Symbian press releases, news coverage

were limited to 47.9% when existing investors shared in buying Psion's holdings and new shares from Symbian.

Shareholdings remain unchanged until June 2008 when Nokia announced its plans to buy out rivals for £209 million. Although some shareholders initially objected to the price, Nokia completed its purchase in November 2008, integrating the company into Nokia and launching an (ultimately unsuccessful) effort to establish Symbian as an open source platform (West & Wood, 2013).

4.3.2.2 *Formal Control*

Openness to customers was a central principle in the creation of Symbian: both the ownership and management control were carefully structured to prevent proprietary control by any one firm. However, both directly and indirectly, the shareholder-customers made crucial decisions that affected the company's financial viability.

Internally, Symbian had a two-board structure. Its senior managers governed the company through an "Operational Board." The shareholders were represented on the company's "Supervisory Board," whose "role is to set the standard licensing terms and conditions for Symbian OS" (Symbian 2006), such as the company's royalty rates. The investor–manufacturers had a strong influence over the technical direction of the Symbian platform, not only through their role in governing Symbian and its allocation of development resources, but also through their own R&D investments in the various user interfaces and their own handsets (West & Wood, 2013).

In theory, a shared platform would allow each handset maker to leverage common R&D expenses, but would limit the opportunity for differentiation between the vendors (a problem for both PC and handset makers licensing Microsoft's software). As a compromise, the Symbian platform allowed for separate user interfaces—funded directly or indirectly by the handset makers—that allowed makers to offer a distinct "look and feel." Five such interfaces were shipped, with three accounting for more than 99% of the unit sales: Series 60 by Nokia (83%), MOAP by NTT DoCoMo (14.6%) and UIQ by Ericsson (2.2%). Each interface was in effect a sub-platform of the Symbian platform, with its own UI-specific APIs and thus third-party applications (West & Wood, 2013). For example, each UI had its own preferred web browser which later proved a major problem when competing with the browser-centric iPhone.

4.3.2.3 End Users

Although it first relied heavily on consulting income from helping licensees develop their handsets, Symbian's path to profitability depended on royalties from sales of Symbian-equipped handsets. From 2002–2010, the Symbian platform reported record unit sales for every year except 2008 (Figure 4.1).

Symbian Ltd. initially hoped to receive $10 royalty per unit, but dropped its price to $5 with surcharges for new releases. Still, its royalty income had an annual growth rate of more than 100% from 2002 until 2006, until its shareholder-customers pressed Symbian to adopt a graduated royalty of $2.50–$5.00 per unit; Nokia was the only manufacturer with enough volume to receive the lower royalty rate (West & Wood, 2013).

Although equity was carefully balanced among manufacturers, the share of Symbian handsets (and royalty payments to Symbian) was highly skewed. From 1998–2008, Nokia accounted for about 80% of Symbian Ltd.'s unit sales, and never less than 75% from 2004 onwards, selling about 350 million Symbian handsets overall. At the peak of Symbian's success, Nokia sold N-series handsets for premium prices that provided its largest handset margins.

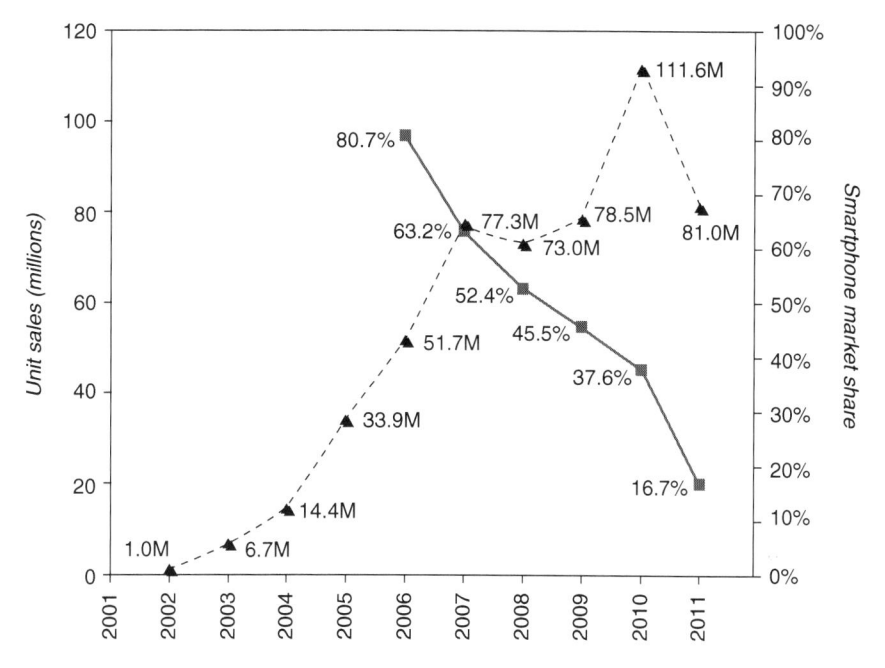

Challenges of Funding Open Innovation Platforms 83

Figure 4.1 Symbian global smartphone unit sales and market share, 2002–2012
Source: Analyst reports (Canalys, Gartner, Tomi Ahonen), Symbian press releases, author's estimates

After Nokia, the next most important customer was NTT DoCoMo and MOAP (Mobile Oriented Applications Platform), whose handsets accounted for 10–20% of Symbian sales each year from 2004–2010. Fujitsu produced 61 MOAP handsets from 2003–2012, Sharp made 37 from 2005–2012, while Sony (later Sony Ericsson) and Mitsubishi together created 30.

From the beginning, Ericsson (later Sony Ericsson) had invested heavily, but reaped few financial rewards. Overall, I estimate that it sold about 15–20 million handsets—about half UIQ handsets (for which it bore nearly all of the UI development expense) and half MOAP handsets in Japan. While Samsung made 15 handset models, the remaining investors shipped even fewer: Motorola (7), Panasonic (3) and Siemens (1) (West & Wood, 2013).

4.3.3 New Rivals and Paradigms

With the iPhone, Symbian faced its first serious market challenge. With its June 2007 launch, the iPhone created the new dominant design for mobile phones: a large touch screen display that provided access to standard web pages. When combined with integration to the Apple music store, the iPhone was an instant PR success that became the best-selling single phone model (West & Mace,

2010). The following year brought the first of a series of phones using Google's (Linux-derived) Android operating system, which provided iPhone-like features with a wide range of vendors, products, and price points (Kenney & Pon, 2011).

In addition to product features, Symbian and Nokia also faced a challenge to their fundamental ecosystem strategy—first from iPhone on openness to complementors, and then from Android on openness to handset vendors. As discussed below, both posed challenges that Symbian was unable to meet.

In July 2008, Apple launched the iPhone App Store, which provided a convenient and inexpensive way for ISVs to sell their software directly to handset owners. While Symbian had taken 7½ years to acquire nearly 10,000 applications, the iPhone app store offered 15,000 apps after six months and 100,000 after 16 months (West & Mace, 2010). In response, other platforms launched their own app stores, but Symbian was blocked by Nokia and its operator partners from creating its own direct-to-consumer store. Symbian had considered launching its own app store in 2005, but dropped the plan due to internal opposition and likely opposition from handset makers and operators (West & Wood, 2013).

The other challenge came from the Android platform, which shipped its first smartphone in 2008. Symbian's "open platform" was a consortium in which source code developed by Symbian and its licensees was available only under non-disclosure and a royalty-bearing license. Meanwhile, Google offered a royalty-free Android license and source code to any external partner. The promise of openness and Google's backing attracted a wide range of handset makers: when the Android sponsoring organization (Open Handset Alliance) launched in 2007, founding members included two Symbian shareholders and licensees—Motorola and Samsung—as well as NTT DoCoMo, Symbian's main sponsor in Japan. Symbian shareholders Ericsson and Sony Ericsson joined 13 months later (Table 4.4). By 2009, Android had achieved what the Symbian platform ultimately failed to do: provide an open innovation platform shared by a wide range of handset makers and controlled by none of them.

Compared to the iPhone and Symbian, Android was widely seen as "open" because its source code was released under an open source license. However, unlike independent open source projects such as Apache or Linux, Google tightly controlled the development process; providing open access to IP but not sharing governance is a common way that firms control open source for their direct benefit (West & O'Mahony, 2008). In an independent analysis of open source community governance in the mobile phone industry, Android was judged to be the least open of eight projects, after Eclipse, Linux, WebKit, Mozilla, MeeGo, Symbian, and Qt (Laffan, 2011). In May 2012, Google completed its purchase of Motorola's handset business, but abandoned vertical integration when it announced plans to sell Motorola in early 2014.

Challenged by the iPhone and the threat of Android, Nokia bought out the other Symbian shareholders and integrated Symbian with the S60 development team to create a single platform. It created a non-profit open source

Table 4.4 Licensees of smartphone operating systems

Handset Maker	Home Country	Symbian Ltd Shareholder	First Handset	Symbian Foundation Member	Open Handset Alliance Member	First Handset	First Handset (Windows)
Nokia	Finland	*1998*	2001	*2008*		-	2011
Huawei	China		-		2008	2009	2013
ZTE	China		-		2010	2010	2012
Fujitsu	Japan		2003	2009	ca. 2012	2010	2007
Mitsubishi	Japan		2005			-	-
Matsushita (Panasonic)	Japan	1999	2005			-	2003
Sharp	Japan		2005		2008	2010	2007
Toshiba	Japan		-		2008	2008	2007
LG	Korea		2007	*2008*	*2007*	2010	2008
Samsung	Korea	2003	2004	*2008*	*2007*	2009	1998
HTC	Taiwan		-		*2007*	2008	2002
Sony Ericsson[†]	UK	*1998*	2000	*2008*	2008	2010	2008
Motorola	US	1998	2003	*2008*	*2007*	2009	2003

Founding members shown in *italics*
[†] 50/50 joint venture from 2001–2012; Ericsson (Sweden) before that and Sony (Japan) afterwards

foundation to own the Symbian source code (cf. O'Mahony, 2003) and in February 2010 released 40 million lines of Symbian OS source code in what was then the largest single open source release in history. However, all but Nokia and the DoCoMo manufacturers abandoned the platform, and in February 2011 Nokia announced it would be phasing out its Symbian phones in favor of Microsoft's Windows Phone (West & Wood, 2013). The switch failed to stem Nokia's falling market share, and in September 2013 it decided to sell its entire handset division to Microsoft for €5.44bn (including patent royalties).

4.3.4 Conflicting Ecosystem Interests

As a startup, Symbian leveraged an open innovation strategy both to raise funds and bring its technology to market. In retrospect, these brought two fundamental problems: the conflicting interests of the investor-manufacturers and a scarcity of resources to support the platform.

In conceiving and implementing a new approach to smartphone design, the vertically integrated Apple had a huge advantage. Decisions about software, hardware, APIs—even distribution of third-party applications—could all be made within one firm. Symbian did not control key aspects of its platform and had to work closely with (or through) handset makers and UI companies to implement other crucial changes and support third-party software.

4.3.4.1 Difficulties Aligning Interests

Many of the difficulties facing Symbian came from the inherent tensions of aligning the conflicting interests of the competing investor–manufacturers, and the processes put into place to manage those tensions. The employees of Symbian Ltd. spent a decade trying to provide a platform that would serve the need of its shareholders who were also direct competitors. Symbian's CTO from 2003–2008, Charles Davies, said that this tension was unresolved from the very beginning:

> There was no understanding or discussion of how the owner-licensees would compete. I don't think that was ever discussed or resolved. I don't remember people saying, "OK, so how are our devices going to be different from each other?" (Orlowski, 2011)

Meanwhile, the handset makers fought for market share (Figure 4.2) and thus the interests of the shareholders continued to diverge:

- *Nokia* enjoyed great success with Symbian, having the most successful UI, the most phones, the greatest unit sales and profits. But the sizable resources it applied to smartphones benefitted mainly S60 and its handsets, not the Symbian platform. As Davies recalled, "Nokia were, understandably, more concerned about making a success of their devices rather than making a success of Symbian" (Orlowski, 2011).
- *Ericsson*[9] and *Motorola* were in decline, as their strategically valuable competencies of the 1980s—deep radio expertise—became largely irrelevant in an era where value was defined by software and not analog radio reception.
- *Samsung* and *LG* were indifferent to platforms: they used Symbian for access to the European smartphone market, Windows for the U.S. market, but switched to Android as the latter gained features, application, and market share.
- The four main Japanese licensees (*Fujitsu, Sharp, Sony,* and *Mitsubishi*) made phones for DoCoMo using a subplatform and handset designs that were not exported; except for Sony, none was ever a major player in the global handset market.

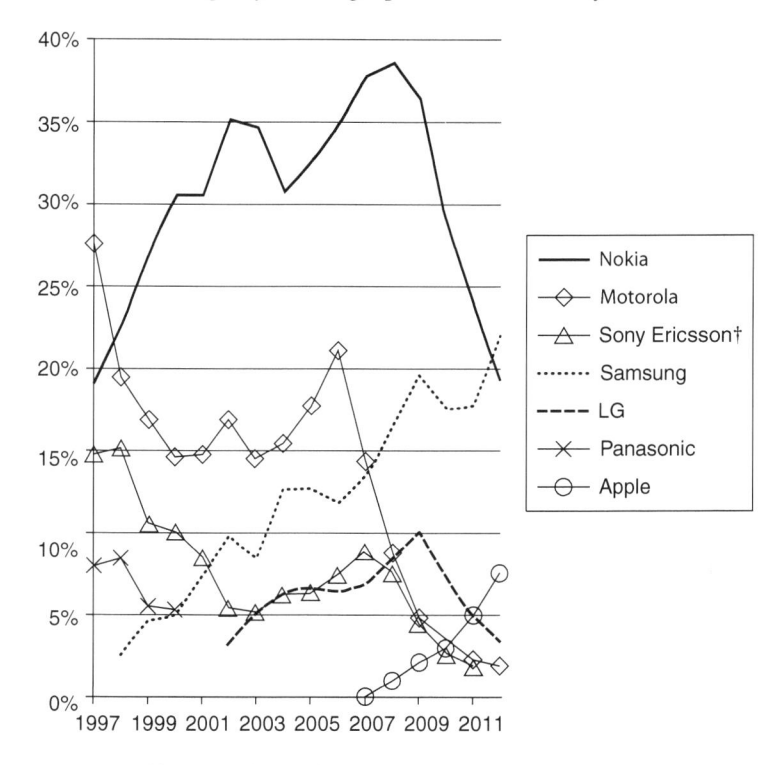

Figure 4.2 Overall market share of leading handset makers, 1997–2012
Source: Global smartphone and other mobile phone sales as reported publicly by Gartner
† Ericsson only market share from 1997–2000

- *Psion* wanted Symbian to be a market and financial success as an independent company, but over time its interests diverged from those of the remaining shareholders and their desires to keep Symbian captive.

4.3.4.2 Health of the Ecosystem

Symbian's success also depended heavily on the success of its ecosystem: as with handset makers, some had more stakes in the success of Symbian than others. Some partners—notably chip makers and network operators—had a strong stake in the success of the smartphone category, but not the health of Symbian *per se*. Instead, each sought to align itself with the most popular handsets and platforms, which meant they were loyal to Symbian when its unit sales were rising, but quickly moved to back the iPhone and Android as they gained momentum.

Some software vendors were similarly platform neutral, particularly those with a two-sided revenue model based on free mobile clients and expensive server-side software: Macromedia (maker of Flash software for multimedia web pages) was happy to work with any provider, as was Oracle (which made mobile phone clients for its mainframe databases). Other smaller software companies—that relied on revenues from selling a mobile-based application—tended to be loyal to a single platform and thus invested in the stake of that platform. Among these, Symbian won the early loyalty of smartphone app makers—particularly in Europe.[10]

Overall, as a startup Symbian lacked the resources to do everything it might have liked to support its ecosystem. Because of its open innovation strategy, it depended on partners to bring its core product (Symbian OS) to market—partners that (due to other alternatives) were not fully committed to platform success. And because its primary funding came from its customers, its strategic choices were heavily constrained. The decision to unify the platform under a single set of APIs—making it easier for third-party suppliers, but reducing the differentiation between handset makers—came only after Symbian was no longer an independent company, but a wholly-owned subsidiary of its largest customer and investor.

4.4 DISCUSSION

4.4.1 Contrasting Open Innovation Platform Strategies

While competing vertically integrated platforms were the norm in the early computer era, Intel CEO Andy Grove (1996) argued that it was more cost-effective if systems providers shared a common component supplier (such as Intel) and thus amortized R&D cost across a broader customer base. When combined with network effects and other demand-side economies of scale, this would encourage the use of open innovation in platform industries (West, 2006).

In the twentieth century, there were three notable exceptions to the vertically integrated pattern: Windows, Unix, and Linux. Unix was licensed by AT&T to IBM's mainframe competitors who sought to create a rival to IBM. The IBM PC platform evolved into the "Wintel" platform as Windows replaced MS-DOS and IBM lost control of its platform. Meanwhile, Linux became a low-cost server alternative[11] both to Unix and Windows by combining an open source knock-off of Unix with the Wintel hardware (Bresnahan & Greenstein, 1999; Gawer & Cusumano, 2008; West, 2003; West & Dedrick, 2001). In response to the success of Windows and theories of platform strategies (such as Grove's), the twenty-first century smartphone platform wars brought three open innovation software platforms: Windows, Symbian, and Android.[12] Most of these devices shared the same CPU architecture, based on an open innovation licensing strategy by ARM Ltd. (Chesbrough, 2006a).

Challenges of Funding Open Innovation Platforms 89

Table 4.5 Successful open innovation platforms

Category	Mainframe	PC	PC Servers	Smart-phone	Smart-phone	Smart-phone
Platform (product dates)	Unix (1972–)	MS-DOS (1981–2000) Windows (1991-)	Linux (1994-)	Windows CE, PocketPC, Windows Mobile, Windows Phone (2002–)	Symbian (2000–2012)	Android (2008–)
Sponsor	AT&T, Unix Systems Laboratories, Open Group	Microsoft, Intel	Intel, IBM, Linux Foundation	Microsoft	Symbian	Google
Rival platforms	IBM S/360	Mac OS	Windows NT, Windows Server	Palm, iPhone OS, BlackBerry, (various) Linux		
Market share	10–90+%, depending on segment	90+%	0–50%	4–15%	40–67%	75+%

Together, these comprise six major open innovation computing platforms of the past 40 years (Table 4.5). All six platforms registered notable successes. Unix was too late to displace IBM's mainframe lead, but (thanks to buyer demands) dominated the final years of the minicomputer era and had nearly 100% share in engineering workstations, while enjoying a disproportionate impact on computing tools and computer science education. Leveraging IBM's legitimacy, MS-DOS and then Windows attracted the widest range of desktop (later laptop) PC systems and complements, garnering the largest market share of any major platform category. More recently, Linux has garnered nearly a 50% market share against Windows in the market for PC servers. In all cases, the platforms benefitted from a wider range of complements, lower switching costs, and bandwagon effects from multiple supporters.

Discerning a pattern in smartphones is less clear-cut: while Symbian defeated Windows and Android defeated Symbian, the cause and effect are not clear. Handset manufacturers both feared Microsoft's monopoly rents (as in PCs) and high royalties, discouraging adoption. Compared to Symbian, Android offered lower direct royalties,[13] an improved Internet-centric user experience, and also a newer, more Unix-like architecture for software developers. Meanwhile, the second most popular smartphone platform in 2013 (the iPhone) was vertically integrated, but offered the best user experience and easiest distribution for third party software.

This pattern suggests that open innovation as a platform strategy is here to stay, but that further research is needed to discern how important openness is to platform success when compared to other factors such as platform capabilities and attractiveness to ISVs (cf. Gallagher & West, 2009; Gawer, 2010; West, 2003).

4.4.2 Ecosystem Challenges of Startup Companies

Entrepreneurs have been long advised that to succeed, they must focus their attention and limited resources (Bird, 1988). Or, as various experts have advised entrepreneurs, "If you have more than three priorities, you don't have any."

In some cases, limited resources can be an advantage in that young firms come up with new ways of creating value that transcend existing conceptions of the market (Baker & Nelson, 2005). However, I believe that in orchestrating the development of a complex ecosystem, this entrepreneurial focus is a major disadvantage.

In Symbian's case, it was focused on trying to ship new revisions of its operating system, keep its investor-customers happy and (for the first six years) find cash to pay for R&D as it continued to lose money. It had attracted a large number of handset makers and independent software vendors—as well as the most end-user adopters of any smartphone platform—and so by many measures it had a vibrant smartphone ecosystem.

However, the partners most dependent on Symbian—the small ISVs—were not generating enough revenues to become successful companies. Symbian considered creating its own online app store, but it would be difficult without cooperation from Symbian's downstream partners—the manufacturers and network operators—and so Symbian management concluded this was not its priority. Several years later, Apple (a *Fortune* 500 company with decades of ISV support) created its iPhone App Store that attracted 100 times as many applications and became a key differentiator for its platform.

4.4.3 Funding New Platforms

Finally, I consider the challenges of a startup firm finding the resources to launch a new platform. Needing more capital than the European VC market could support, from 1998–2004, Symbian raised some $370 million in outside funds. All of that came from corporate venture capital (CVC), and all but Psion's 2.5% (in 2002) came from handset manufacturing customers. Symbian had greater difficulty launching and funding its platform than did Facebook (and its numerous social media imitators) because of more complex software and the dependence on manufacturing and distribution partners to reach customers, and was also forced to share the value capture with these downstream partners throughout its existence.

As noted earlier, Symbian and 3DO are the only examples I could find where a startup platform company was funded by its customers (which for 3DO included not just hardware companies, but content creators and distributors). Symbian's strategies and access to funds were constrained by the same partners who were part of its business model, which contributed to (but did not determine) its strategic challenges.

Symbian's challenges highlight the need for an ongoing stream of resources to support platform development. Unlike the one-time standardization as in VHS vs. Betamax, computing platforms reflect a series of linked contests (Gallagher & West, 2009). These point to an underappreciated aspect of the Gawer and Cusumano (2002) conception: platform leadership requires an ongoing investment in both the architecture and ecosystem.

Prior platform research has also emphasized sustained competitive advantage (and thus profits) as the outcome of successful platform leadership. However, such profits are not just a consequence, *but a necessary antecedent of platform success*. In particular, I believe that Symbian demonstrates that a key role of a platform leader is to extract profits from the value chain and reinvest those profits into expanding the technical and organizational reach of the platform. Instead, while one CVC investor (Psion) sought the greatest possible financial success for Symbian—as would an independent VC—the remaining CVC investors favored their interests as customers, explicitly structuring Psion to prevent Wintel-type profits.

In fact, I suggest an empirical regularity in the role of cross-subsidies in launching and sustaining a new platform. Table 4.6 lists various examples where the "cash cow" profits from an earlier platform were successfully used by companies diversifying into a related industry segment, a process I term "platform chaining."[14] This resembles "platform envelopment" (Eisenmann, et. al 2011) in that existing technology and customers help a firm lever its way into an adjacent platform market—but differs because (unlike with envelopment) the new platform remains distinct from the earlier platform. Conversely, the case of Real Networks cited by Eisenmann and colleagues (2011) is another example of a firm that lacked either an existing cash cow or a sizable revenue stream from existing customers to maintain its early platform lead.

More generally, I believe this has important implications for a broader class of challenges facing companies seeking to profit from their innovations. In the Teece (1986) profiting from innovation framework, "the implicit assumption was made that risk capital was available" to fund a firm's commercialization efforts (Teece, 2006: 1140). Symbian's use of strategic corporate venture capital should have been a way to obtain such capital, but its investors' control (including blocking its hopes for an IPO) ultimately limited the success of its business

Table 4.6 Examples of successful chaining from a cash cow platform to a new platform

Company	Cash Cow Platform	New Platform	Reference
IBM	System/360	IBM PC (1981)	Moschella (1997)
Apple	Apple II	Macintosh (1984)	Malone (1999)
Intel[†]	Wintel-compatible PCs	Lintel servers (1995)	West & Dedrick (2001)
Sun	Solaris	Java (1995)	Southwick (1999)
Microsoft	Windows	Xbox (2001)	Takahashi (2002)
Apple	Macintosh	iPhone (2007)	West & Mace (2010)
Google	Google search engine	Android (2007)	Kenney & Pon (2011)

[†] Wintel: Windows on Intel; Lintel: Linux on Intel

model. While the CVC literature has acknowledged risks to startups from receiving such investments, it has mostly focused on risks due to misappropriation (e.g. Katila et al., 2008; Maula et al., 2009) rather than the divergence of interests between partner–investors and startups. As Chesbrough (2000) posits, corporate venture capitalists are far more likely to constrain the business model choices of startups than are independent venture capitalists solely focused on financial returns; as suggested here, they may also constrain their exit strategies as well. This suggests opportunities for future research on both constraints.

NOTES

Thanks to David Wood for all the help and insights during this research, and to Annabelle Gawer, Markku Maula, and editor Wim Vanhaverbeke for helpful feedback on earlier drafts.

1. In addition to corporate members, consortia often include university or non-profit research labs (cf. Dimancescu & Botkin, 1986).
2. For a summary of the transition from the 2nd generation GSM into the 3rd generation 3GSM (WCDMA) network standards, see Bekkers (2001) and Bekkers and West (2009).
3. For example, in December 2008, the trade association for GSM mobile phone network operators reported that it represented "more than 750 mobile networks across 219 countries and territories," according to an Archive.org copy of the GSMA.com website. Many of these networks were owned completely or in part by larger holding companies such as Vodafone, Orange, Deutsche Telekom, and Telefónica, but handset procurement decisions were influenced (and sometimes determined) by the local subsidiaries.

4. Eventually handset makers developed technology and processes to update the operating system software over 3G or WiFi networks, as with the first update to the iPhone OS in 2008.
5. Nearly all computing platforms since the 1960s have been vertically integrated, with only a handful (such as AT&T's Unix or Microsoft Windows) being licensed to external parties (cf. West, 2003, 2007a). The only other example I can identify of a platform funded by investor-customers is 3DO, an unsuccessful videogame console (1993-1995) funded by Matshushita (Panasonic), Goldstar (later LG), AT&T and various content providers.
6. Although Motorola was not mentioned in the June 24, 1998 press release, it was prominently featured in news coverage of the announcement. It signed a shareholder agreement August 28 and announced its investment on October 28.
7. Symbian's first profitable year was 2005, with a £15.3 million net income vs. a £23 million loss the year before (West & Wood, 2013). The company's audited financials distributed to shareholders showed that it had a positive cash flow from operations of £2.6 million, vs. an outflow of £32 million the year before.
8. Although that valuation was far greater than ever used by Symbian's shareholders, the £6 billion ($9 billion) was much less than the $42 billion public valuation of Palm Computing, its most direct competitor.
9. In response to increased competition, Ericsson combined its handset business with Sony in 2001, and then exited the handset business in February 2012 by selling its half of the joint venture to Sony. In May 2012, Motorola's handset business was acquired by Google, more for its patent holdings than its product revenues.
10. In the US, ISVs were focused on Palm and Windows from 1998–2005, and on the iPhone and Android after 2008. While Symbian was popular in Japan, DoCoMo's closed Symbian-based MOAP platform prevented the creation and installation of native downloadable applications.
11. Linux has since been used for other applications, such as embedded computing (Henkel, 2006)). Android itself was a derivative of Linux that was incompatible with Linux until the latter developers merged the two code bases in early 2012 (Kennedy, 2012).
12. For a summary of open innovation and vertically integrated smartphone platform strategies, see Kenney and Pon (2011).
13. Since 2010, the actual cost of smartphone platforms has included patent royalties paid to competing platform owners such as Apple and Microsoft, but the specific rates have not been publicly disclosed. New entrants in the smartphone market—using Android and without large patent portfolios of their own—have been particularly vulnerable to such patent litigation.
14. This is not meant to minimize the examples of failed attempts at platform chaining—such as Apple into PDAs (with the Newton) or Intel into mobile phone CPUs. As with any other strategy, adequate resources are necessary but not sufficient for success.

5

Open Innovation and Industrial Dynamics—Towards a Framework of Business Convergence

Jens Frøslev Christensen

5.1 INTRODUCTION

The central proposition of this chapter is two-fold: First, industrial dynamics must increasingly be conceived in terms of convergence and divergence rather than industry-bounded trajectories. Second, these dynamics represent central drivers for more open and industry-transcending patterns of innovation.

Three widely used key frameworks assume stable industry boundaries as context for strategic management and innovation. These are the Product Life Cycle (PLC), the Five Forces, and the Innovation Life Cycle. The PLC model, grounded in industrial economics, presents a systematic sequence of stages in the evolution of the economic structure of industries (Gort & Klepper, 1982; Klepper, 1997). In the embryonic stage, volume is low, product design is experimental, and many firms enter. In the growth or "shake-out" stage, volume growth is high, product design stabilizes, and large-scale operations lead to lower entry, intensified exit, and an oligopolistic market structure. This stage is followed by stages of lower growth and decline. Porter's Five Forces framework is likewise grounded in industrial economics (Porter, 1980). It identifies five forces framing the competition and strategic attractiveness of industries (rivalry, supplier power, buyer power, substitutes, and entry barriers). The Innovation Life Cycle (ILC) is rooted in Schumpeterian economics and innovation studies.[1] It is complementary to the PLC in that it points to changing patterns of innovations as key drivers in the evolution of industries. The early stage of industries is characterized by technological discontinuity, radical product innovation and rivalry among a diversity of product designs. This stage ends when a dominant design is selected by the market. Radical innovation in production processes leads to scale economies and high growth followed by incremental product and process innovation within the constraints of the dominant design.

These frameworks have been particularly effective in explaining evolving market structures and innovation dynamics in many manufacturing industries (e.g. the automobile, bicycle, or television industry) characterized by well-defined and stable boundaries making the notions of life cycles meaningful. However, three generic tendencies have weakened their scope of validity:

- First, business dynamics are to a *decreasing* extent confined to single-product industries that behave according to the PLC and ILC.[2] Instead, business dynamics are characterized by increasingly open boundaries associated with dynamics of industry and product market *convergence*,[3] and *divergence* into disaggregated (sub-) markets.
- Second, the increasing prevalence of convergence/divergence makes the *collaborative* nature of the firm-environment relationship more central to strategic management than what is reflected in the Porterian "industry analysis" with its exclusive focus on *competitive bargaining* relationships. While the competitive game remains vital in business dynamics, the collaborative game and its interplay with competition has become equally decisive (Brandenburger & Nalebuff, 1996). This means that the Five Forces is insufficient in assisting firms' strategies.
- Third, convergence and divergence dynamics imply other types of innovation than those addressed in the ILC (radical versus incremental innovation, and product versus process innovation). These categories cannot grasp the particular dimensions of innovation that are embedded in convergence and divergence processes, in which innovation must open up for ideas, technologies, designs, and market features residing in other— often unfamiliar—industrial contexts.

This chapter presents a framework for understanding the business dynamics of convergence and divergence. It serves four purposes. First, it offers a systematic way of analyzing business and innovation contexts that are not well explicated by the Five Forces framework. Second, it proposes a theory of the generative mechanisms underlying convergence and divergence, and a new "life cycle" model, the Convergence Life Cycle. Third, the framework contributes to explain the increasing prevalence and particular patterns of open innovation, and fourth, it enlightens the relations between open innovation and the notion of dynamic capabilities in strategic management (Teece et al., 1997; Teece, 2007).

These theoretical and analytical endeavors take their point of departure in the business dynamics of the IT security sector.

5.2 THE CASE OF IT SECURITY

One dark side of the Internet and related information and computer technologies has been an explosion of different types of security problems such as viruses, spyware, pitching, and hacking. These, again, have given rise to swarms of firms exploring the commercial opportunities of offering safeguarding measures. We have witnessed the emergence of a large range of new security products and services and the creation and hyper-growth of a new complex and volatile sector. IDC estimated global IT security revenue to reach $35 billion in 2003, reflecting a doubling of revenues since 2001 (IDC, 2003) and comparable to the size of the global market for recorded music.[4] Just 15 years before, this sector did not exist. IT security entails software- and hardware-based products and systems on the one hand, and large variants of services on the other. These services assist users in selecting, implementing, and managing security products, and in adopting organizational procedures for safeguarding. IT security products and services are dedicated to alleviate IT security problems of different kinds. But despite this commonality, the enormous diversity of IT security doesn't lend itself fruitfully to being analyzed as one industry. Rather, it must be conceived as an ecosystem of evolving and interrelated product markets or submarkets.

For the period 1988–2004, we have identified three partly overlapping stages of business dynamics. The late 1980s and early 1990s saw the formation of numerous product markets based on specialized firms' development of autonomous security products, each responding to specific types of upcoming security problems. While product market formation continued after the mid-1990s, even if with decreasing intensity, two trajectories of convergence-based dynamics came to dominate. One associated with the bundling of two or more products, the other with the integration of products into broader platforms and systems. In the following, we provide an overview of these trajectories. In the context of open innovation, we maintain that these strategic leaps from one trajectory to another impose particular requirements on innovation management to reach out for integrating technologies and business models from unfamiliar business contexts. The case is based on our study of the evolution of IT security (for details, see Christensen, 2011) and other research on the sector.

5.3 TRAJECTORIES OF PRODUCT MARKET FORMATION AND THE ROLE OF TECHNOLOGY LICENSING

Early security technologies were not associated with commercial products but with university and government-sponsored research and development especially in the U.S. (Giarratana, 2004). But from the late 1980s to the late 1990s, hundreds of security specialist firms were established based on product innovations or the invention of security technologies. Many have later exited or been subject to mergers and acquisitions. During the early years, in particular large IT enterprises and government institutions in the U.S. were important lead-users of innovative products and licensees of technologies provided by technology specialists (Giarratana, 2004).

Among the early examples of security innovations are antivirus, firewalls, Virtual Private Networks (VPN), vulnerability assessments scanners and services, authentication, and Intrusion Detection Systems (IDS), all invented and initially commercialized between 1989 and 1996. Each of these products (and numerous others) was subject to specialized product market formation, with many rival firms engaged in innovation to expand their capacity and performance.[5] From the early start, numerous firms also developed a market for security technology. In 2002, technology licensing accounted for 17.4% of overall revenues in the software part of security, while revenues from products and services accounted for, respectively, 52.3% and 30.3% (Gambardella & Giarratana, 2007). One important security technology, encryption technology, is patent-intensive and applied in product markets dealing with encryption, network security, and authentication.[6] These markets were also the most licensing-intensive. Sixty-five percent of firms entering these markets during the period 1989–2004 were based on in-licensing of encryption technology. This was only the case for 13% of new entrants in other security product markets such as firewalls, antivirus and anti-spam (Arora & Nandkumar, 2007). Technologies with more product market applications tended to be out-licensed to firms in markets different from those in which the licensors were operating implying that the latter avoided direct competition from the licensees (Gambardella & Giarratana, 2007). Thus, the technology market in IT security has contributed to enhance entry into numerous product markets and stimulate market fragmentation. At the same time the economic functioning of the technology market in IT security seems to have benefitted from the strong market fragmentation. This pattern of open innovation in the early stage of business formation in IT security reflects a dynamic division of labor between product/services companies and technology specialists in a context of high market fragmentation.

The ubiquitous tendency since the late 1990s to move ever-increasing volumes and types of data into ever-more distributed IT networks, have given rise to new forms of security breaches, which again have paved the way for new waves of security innovations (including anti-phishing, anti-spam, anti-spyware and various forms of biometrics). As employees and consumers increasingly operate from mobile devices (notebooks, tablets, smartphones) that were initially without the same safeguards as desktops, innovative endeavors have increasingly addressed security measures for these devices (Morgan Keegan, 2006).

5.4 TRAJECTORIES OF CONVERGENCE THROUGH PRODUCT-BUNDLING INNOVATION

The many and diverse products associated with product market formation were more or less complementary—sometimes with overlapping, sometimes with potentially synergistic features. In the early years, linking these products to form interoperable systems was either not attempted, or left to the IT departments of the customers, and assisted by security services firms that appeared in large numbers. For a time, competition was confined within narrow product markets. Only gradually did the specialized security vendors, together with their corporate customers and assisting services firms, accumulate precise knowledge of the complementary and synergistic prospects of the different products hence of opportunities for *systemic* innovation that would require a different pattern of open innovation than the one prevailing in the early stage. The trajectory of product-bundling innovation was initiated by a small group of security specialists, who over a few years transformed themselves into large security integrators.

The first steps were taken by Networks Associates (now McAfee) and AXENT Technology, which through numerous acquisitions developed a broad portfolio of security products. Both companies launched the first product suites (bundles) around 1998/99, but they were not well received by the market (Wall Street Transcript, 2001). Like the two first-movers, Symantec had acquired numerous security specialist firms during the late 1990s. In 2000 the company also acquired AXENT Technology enabling it to make a profound relaunch of product suites. Over the few years of acquisition-based growth and early trials in systemic innovation, Symantec expanded its core capabilities in numerous security technologies and products and also built powerful distribution assets and brand recognition. By 2004 Symantec was recognized as the leading product suite vendor in IT security. Several other specialized security companies, including Internet Security Systems (ISS), likewise embarked on product-bundling strategies (SG Cowen & Co., 2004: 24; Christensen, 2011).

Product suites evolved to include all or most of the following products: firewalls, VPN, Intrusion Detection, antivirus, content filtering, anti-spam and privacy control systems (Wall Street Transcript, 2004c). Also, the emergence of managed security services during these years reflects a move towards integrated solutions. Security managers like Counterpane and Cybertrust delivered unified operations, maintenance and updating of their customers' security systems.

Such converged solutions responded to the rapidly expanding complexity for the customer in dealing with the ongoing add-on budding of new products and services and associated increases in security costs.[7] The value proposition of the new solutions was clear: overall reduced costs and simplification for the user in terms of implementation, multi-functionality, management, and (more or less) one-stop shopping. The professional customer, however, would face the potential disadvantage of being locked into a single vendor offering one or more weak functionalities in the package. The alternative would be to shop for autonomous products to get the best products for each specific function, in industry jargon "best-of-breed strategy." However, the disadvantage of this strategy was the high costs incurred by the need to develop in-house expertise in searching, reviewing, selecting, negotiating, and contracting with the "best" vendors in different product categories, as well as the costs of integrating and managing the products to operate as one system.

By 2004 the best-of-breed strategy had primarily survived in some high-end segments such as in the military and financial sectors with a high-risk profile (Wall Street Transcript, 2004a, and 2004b). Companies like Counterpane (managed security services) or CipherTrust (messaging security), for example, had positioned themselves as among the best in their respective product markets. By far the largest share of the overall market—the mid- and low-end segments—was controlled by security integrators providing product suites, while specialized providers of autonomous products maintained strong positions in parts of the high-end segment.

The pattern of open innovation took a different course during this transformation. While technology in-licensing was an important way for new entrants to gain a foothold in specialized product markets, acquisitions of firms became a dominant lever for incumbents engaging in product-bundling. They not only needed access to technologies but also to the particular market knowledge and business models associated with the unfamiliar product markets to be integrated. Therefore firm acquisitions became a key means for realizing product-bundling opportunities.

5.5 TRAJECTORIES OF CONVERGENCE THROUGH CONTEXT-EMBEDDING INNOVATION

Convergence through product-bundling contributed to drive the highly fragmented security sector or ecosystem in the direction of a unified (even if still fragmented) industry in which leading firms began to compete against each other with rivalling product suites. By contrast, the trajectory of convergence through context-embedding innovation contributed to undermine this "unification" process because security in this trajectory became embedded features of *other* offerings (IT networks and platforms) rather than dedicated security products with business opportunities in their own right (Morgan Stanley, January 5, 2005). Both convergence tracks responded to user needs for reduced complexity associated with the surge of autonomous products.

Obviously private users want simple solutions that are practically invisible. But also professional users prefer reduced complexity. "They are looking for plug and play appliances that require little installation expertise and hence are less prone to configuration errors—especially for offices with little to no on-site security staff" (Morgan Keegan, 2006). Context-embedding innovation gained momentum around 2000 and subsequent years. It signified a tendency for security to become built into every layer of the IT infrastructure, from the network up through the application. The agents driving this trajectory were incumbents in established industries, especially network and systems vendors and Internet service providers. Two of the most prominent examples were Cisco and Microsoft, whose security strategies are briefly outlined below.[8]

Cisco, with its strength in network equipment (especially routers, switches, and dial up access servers), early on took an offensive stance in providing security at the networks level. From 1995 to January 2005, Cisco acquired ten security firms, primarily specialists with autonomous products across a broad spectrum of IT security. This made it possible for Cisco to introduce an increasing number of network-embedded security offerings (e.g. firewall, VPN, and Intrusion Prevention functionalities).[9]

Microsoft began actively to engage in IT security around 2002. This happened against the background of increasing vulnerabilities in operative systems. Microsoft's response was to implement more secure coding standards and design more inherently secure software using threat-modelling, filtering, and penetration testing. In addition, Microsoft began to integrate security functionalities into its systems.[10] Key to this endeavor was the acquisitions of three security firms, GeCAD (2003), a small Romanian anti-virus company, Giant Company Software (2004), an anti-spyware firm, and Sybari (2005), an anti-virus and data protection specialist. Specialized security knowledge from these firms was applied in the OneCare subscription service, aligning anti-virus (based on the GeCAD technology), anti-spyware (based on technology from Giant Company Software), firewall protection and PC cleanup tools to Windows private users. Microsoft's security strategy during this period was to offer users basic security and to enhance user convenience by providing security systems that were simple to use and with automated updating. By 2004, Microsoft also began to emphasize the creation of partnerships with security vendors, networking companies, and Internet service providers (Morgan Stanley, 2005; Morgan Keegan, 2006; SG Cowen & Co, 2004). Thus, several of Microsoft's competitors (i.e. in firewalls, anti-virus, and anti-spyware) also became partners, invited to build extended security solutions on Microsoft platforms and services for users with more specialized security needs.[11]

Even if the two convergence trajectories identified in this case reflects rival strategies (e.g. bringing Symantec into competition with Microsoft), they also represented a division of labor in terms of product specialization and market orientation, making it possible for both trajectories to run in parallel, rather than having one fully replace the other (e.g. Symantec also became a Microsoft partner).

Like the case in product-bundling convergence, the acquisition strategy in context-embedding convergence signified that the pioneering firms not only needed to have access to unfamiliar technologies but also to the broader market knowledge of the acquired firms.

5.6 TOWARDS A FRAMEWORK OF CONVERGENCE-BASED BUSINESS DYNAMICS

IT security is not an industry, hence cannot be explained by industry-bounded frameworks. IT security must instead be understood as a collection of complementary product markets and adjacent industries with changing boundaries shaping the context for convergence and divergence dynamics.

During the pioneering phase of the IT security in the 1980s and into the 1990s, business dynamics were driven by innovative start-ups targeted at autonomous products to cover an expanding array of security holes of IT systems (here and in the following we use the term product to also refer to specialized services). Competition was segmented within narrow product markets. According to the PLC and ILC models, one would predict that each product market (e.g. firewalls, anti-virus) would undergo a stage of design rivalry, resulting in a dominant design, paving the way for an era of incremental innovation and the build-up of operations, distribution, and marketing systems necessary to establish and sustain mass markets. Indeed, we can trace such dynamics in the trajectories of product market formation. Within each product market, different solutions were tried out, one or a few designs came to prevail in the market, their underlying technologies were stabilized, and a few successful companies managed to bring these designs into leading positions for mainstream markets. Product market formation was dominated by small entrants often using technology in-licensing as a key means of innovating their products. Other firms specialized in developing security technologies and gained their revenues from out-licensing these to product firms.

However, when the products approached the stage of dominant design and shakeout, the rules of the game changed in the direction of convergence. This first materialized in product-bundling innovation, and subsequently, with a time-lag of a few years, in context-embedding innovation—proliferations of cycles that are *not* accounted for in the PLC and ILC models. During that transformation, many specialized firms exited or were acquired by a smaller group of "integrators" taking the driver's seat in shaping these trajectories. During the period 1993–2005, we registered 291 acquisitions of IT security firms (Christensen, 2011). By far the largest share was specialized security firms taken over by product-bundling security integrators, or, especially after the turn of the millennium, by large network and systems vendors. By gaining control over requirements for adaptations at the specialized product level of expertise, these acquiring firms became better able to address the particular challenges of boundary-crossing innovation.

Although the early stages of segmentation and stabilization around dominant designs along the PLC can be identified in early security product markets, the subsequent dynamics became dominated by convergence across product markets. Likewise, while a stage of radical product innovation and experimentation along the ILC can be traced in each of the product markets, the subsequent dynamics after the settlement of a dominant design became driven by *systemic innovation*, reflecting convergence across rather than *process innovation* within product markets. In order to conduct such systemic innovation, a radical form of in-bound open innovation was needed, namely acquisitions of specialized firms—sometimes supplemented by technology licensing.

Despite the dominance of convergence, the specialized product markets tended to live on. But rather than continuing into the mature stages of the PLC/ILC implying scale-intensive provision of commodities, they positioned themselves towards high-end niche markets while the vendors of converged products came to lead mainstream markets.

In order to understand business dynamics characterized by convergence and radical requirements for open innovation, we need a conception of the business environment that goes beyond the classical industry conception, and we need a theory that specifies the contingencies under which convergence and divergence come to reside.

5.7 A THREE-PRONGED CONCEPTION OF THE BUSINESS ENVIRONMENT: ECOSYSTEMS, PRODUCT MARKETS, AND INDUSTRIES

In recent strategy literature the notion of "industry" as the organizing context for business strategies and innovation practices has increasingly been replaced by the more open-ended concept of business or innovation "ecosystems" (Adner, 2006; Iansiti & Levien, 2004a; Moore, 1996; Teece, 2007). Teece (2007: 1325) defines business ecosystems as "...the community of organizations, institutions, and individuals that impact the enterprise and the enterprise's customers and suppliers." Adner (2006: 98) defines innovation ecosystems as "...the collaborative arrangements through which firms combine their individual offerings into a coherent, customer-facing solution." Compared to the classical industry concept, the ecosystem implies a more rich and flexible understanding of the business environment and two analytical "turnarounds." First, while the industry view tends to emphasize the exogenous state of the environment that firms should take for given and adapt to, the ecosystems view underscores the symbiotic and co-evolving relationship between the strategies and innovations of firms and their business environment. This view sometimes leans towards the "endogenous" position that proactive firms contribute to shape the business environment, rather than reactively being shaped by it.[12] Secondly, while the industry view in Five Forces exclusively addresses the competitive bargaining game, the "ecosystems" view addresses complementarities and associated collaborative arrangements.

In order to more fully understand the environmental side of the dynamics of convergence or divergence, we must consider both the competitive and the collaborative arenas. Thus, we cannot simply *replace* the industry concept with the ecosystem concept. However, the competitive arena needs a differentiated typology of concepts (see Figure 5.1) that, unlike the industry concept, does *not* assume robust boundaries and long life cycles. Therefore we use the term "product market" to signify the emergent context for competition among firms with identical or similar categories of specialized products. A product market may evolve along the PLC/ILC, hence eventually prove to become a "real" industry, *or* it may undergo more or less frequent and concurrent dynamics of convergence or divergence. Or it may eventually vanish. Thus, while "product market" here denotes the *emergent* context for product-specific rivalry, the term "industry" refers to *mature* product markets reflecting well-established patterns of competition and typically dominated by large incumbents.[13] Like product markets, industries may also be subject to convergence (e.g. the convergence of the cell phone and the camera industry) and divergence (e.g. the widespread outsourcing of services and components in many industries). Finally, we use the concept of "ecosystem" to address the evolving collaborative context for product market or industry convergence or divergence.[14] In this sense IT security constitutes an ecosystem of evolving clusters of complementary product and technology markets among which convergence may take place.

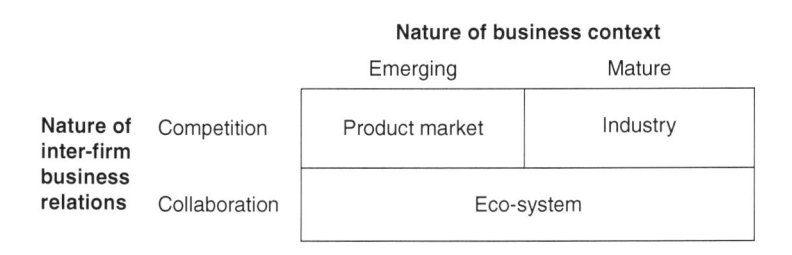

		Nature of business context	
		Emerging	Mature
Nature of inter-firm business relations	Competition	Product market	Industry
	Collaboration	Eco-system	

Figure 5.1 Conceptions of business environment

From both a theoretical and managerial perspective, there are advantages gained from using this three-pronged conception of the business environment. Theoretically, the combined use of these concepts provides complementary lenses for analyzing business dynamics associated with, respectively, rivalry (the product market, the industry) and coordination (the ecosystem). The realization of convergence prospects within an ecosystem implies the creation of integrated product markets forming new bases for competition. The realization of divergence prospects within a product market or an industry implies the creation of specialized (sub-) markets and new coordination requirements between these and the residual players in the original product market or industry.

From a management perspective, this conceptualization can contribute to a more adequate understanding of the context for what we may term Open Business Dynamics. It provides a systematic way of specifying the arena for both competition and coordination, whether addressing firms in emergent product (sub-) markets or in mature industries. It offers an analytical perspective on strategically thinking ahead of *current* competition and collaboration and out-of-the-box, that is, across the boundaries of current product markets and industries. Finally, it offers an industrial dynamics perspective for explaining changing patterns of open (and less open) innovation.

5.7.1 The Concepts of Convergence and Divergence

Convergence is here defined as the process of full or partial *integration* of two or more product markets or industries that were previously not interconnected through competitor or supplier relations. This means that some firms begin to integrate product functionalities from their own product market or industry with those of others through systemic product innovation and subsequent production and sales of converged products. Such "producer-based" convergence gives rise to new forms of "user convergence," that is, opportunities for joint use of the constituent product functionalities.[15] The prototypical case of convergence is that taking place over the last couple of decades across the broad array of computer and telecom product markets and industries (Yoffie, 1996).

Divergence refers to the reverse process of full or partial *disintegration* of one product market or industry into one or more new specialized product markets or submarkets. This means that some firms carve out a market by specializing in the provision of functionalities that were previously developed and produced as an integrated part of a product or system. The evolution of the PC industry provides a widely studied example of divergence.[16] Another example is the emergence of a specialized submarket providing class D digital amplifiers (or components) replacing class A/B amplifiers that were (and to some extent still are) integrated by incumbents in the consumer electronics industry

(Christensen et al., 2005; Christensen, 2006). Dynamics of divergence has been widely research under the headings of outsourcing and vertical disintegration of large companies (Langlois, 2003).

In recent years, convergence and divergence have in particular been studied in regard to software-based dynamics under themes such as platform strategy and the integration of "complementor" products or applications.[17] Such applications may originally constitute end-product markets comprising specialist vendors with no concern for user convergence as we have witnessed in the IT security case. Alternatively, application vendors may from the beginning be "complementors" providing their products as dedicated to and sold via a platform controlled by a platform leader. For users of the particular platform, this assures a basic level of user convergence *without* producer-based convergence. One example is TomTom's early provision of applications (e.g. personal financial planners, cooking guides, and eventually route navigation) to become embedded in and sold via a PDA (Personal Digital Assistant) platform like Psion or Palm Pilot.[18] A higher level of user convergence may be obtained through innovation strategies that pursue synergistic features between particular applications. This requires the co-specialization of these applications so that they can "speak together" and obtain added functionalities beyond being accessible as autonomous products in a platform. Thus, for example, the camera functionality in a smart phone gains added value because the photo's exact location can be determined by the GPS facility and be uploaded through the phone and sent directly to friends or an archive. If successful, this strategy may threaten the existence of independent product markets. For example, Microsoft's integration of word processing with its Office package has to a large extent eliminated a distinctive word processor market. But perhaps more commonly, converged and stand-alone products may co-exist for extended periods. In IT security converged products did not eliminate autonomous products, but constrained their growth and relegated them to niche positions. Likewise, smartphones with embedded camera functionalities did not eliminate the camera industry but reduced the size of its mass consumer segments as reflected in the recent collapse of Kodak.

Converged products for centrally controlled platforms may eventually undergo divergence resulting in the formation of numerous more or less distinctive markets. The divergence of the IBM PC platform is the most well-known example. Another example relates to GPS-based route navigation. In the 1990s, navigation became embedded either in the telematics platforms of automobile manufacturers or in PDA platforms. However, they suffered from performance flaws from being integrated into overtly complex automotive systems or crammed into low-powered, small PDAs. In the early years of the new millennium, Garmin and TomTom pioneered radical divergence strategies implying the creation of stand-alone navigation products that did not suffer from the deficiencies in the previous contexts. They could be easily mounted on any car and more easily upgraded than was previously possible. This became a thriving product market for a few years until it was severely constrained due to a new wave of context-embedding convergence—this time linked to smart phones with navigation functionalities offered by, among others, Google and Apple.

5.7.2 Towards a Theory of Convergence and Divergence

Which are the contingencies or generative mechanisms that can explain the dynamics of convergence and divergence? In the following we shall first compare the economic mechanisms driving product markets into the shake-out and mature stages of the PLC/ILC with those driving product markets into convergence. Thereafter, we compare the economic drivers of convergence and divergence.

In the PLC/ILC model, the mechanism driving new product markets into the shakeout stage is *process innovation* that leads to substantial reduction in cost of production. Since the value of cost reductions is proportional to the volume of production, larger firms profit more from process innovation than smaller firms (Klepper, 1996, p. 565). As some firms grow and benefit from process innovation, industry prices are pushed down with the result that smaller firms are forced out of business ("shake-out") and entry barriers are increased (Klepper, 1997: 151). On this background two key factors can explain non-PLC/ILC dynamics. One is lack of or weak opportunities for scale economies through process innovations. Such a context is prevalent in software and many knowledge-intensive service sectors, including IT security. The other factor is that major product innovation may disrupt existing PLC/ILC and shape new patterns of development. This is a well-known phenomenon in contexts such as biotechnology, in which the nature and boundaries of products are less fixed at an early stage.

What then drives some product markets *away* from the PLC/ILC track towards convergence? We contend that a convergence trajectory is triggered when prospects for economies of scope[19] or synergy through product and technology integration *across* product markets or industries are perceived as higher than prospects for economies of scale or other specialization economies *within* the individual product markets. While economies of scope refer to cost advantages from joint development, production, sales etc., synergistic economies refer to added value from converged devices as compared to the value from stand-alone devices. While *process innovation* in the PLC/ILC track gives rise to scale economies, shakeout and lower entry, *systemic product innovation* in the convergence track—implying radical forms of open innovation—gives rise to disruption of the life cycle dynamics within the constituent product markets *and* to economies of scope and/or synergies. These economies are likely to be captured by those few first-movers that mobilize the boundary-crossing resources for systemic and open innovation and for integrative business development. This is especially likely to be the case if they not only have innovative assets, but also build strong commercial assets that can shape mass markets for their systemic innovations (Christensen, 1995; Teece, 1986). In manufacturing domains, commercial assets involve scale-intensive production systems, while in software and service domains they are rather linked to sales, distribution, and services, as well as marketing and branding. Even if this convergence process does not involve a shakeout in the same sense as implied by the PLC, the outcome, namely the dominance by few firms, remains the same.

Three empirical factors have contributed to create a tendency for new product markets to eventually converge rather than remain on the PLC/ILC track.

First, the central driver of the PLC/ILC, scale economies through process innovation, is especially prevalent in manufacturing sectors that account for a *decreasing* share of the national economies in the rich part of the world, while it is less prevalent in software and many service sectors that represent an *increasing* share of these economies (Möller et al., 2008). Second, the opportunities for interlinking functionalities of different products have significantly improved as a consequence of advancements in information and communication technologies and digitization of an increasing array of products. In particular, software-based interface systems have made the interlinking of such products technically feasible and increasingly cost-effective.

However, there is more to convergence than the interlinking of products by means of *interface technologies*. Convergence also involves the alignment of in-depth knowledge of *each* of the converging product markets, product functionalities and core technologies.[20] Since most firms that engage in convergence initially only have deep knowledge of one product market, namely their own, they will need access to core knowledge from "the other side."[21] The third factor that has enhanced the tendency for product market convergence is the improved effectiveness of markets for M&As, technology [22] and ideas.[23] M&As and technology licensing have become widely used tools for gaining access to complementary knowledge, hence for realizing convergence strategies. Both are management tools for open innovation, even if the post-acquisition and licensing innovation activities take place in-house. More broadly, we can also assume that convergence strategies have been accommodated by the growing experience that many firms have obtained in open innovation management beyond technology licensing and M&As.

We can hypothesize that convergence is more likely to be initiated via an M&A process than through internal development or technology licensing when two conditions apply. The first is that the competencies underlying the two (or more) converging product markets are highly dissimilar. The second condition is that the supply of small (affordable) firms in the complementary market is abundant, and this condition is more likely to exist in emerging product markets than in established industries. Technology licensing may especially drive or supplement convergence if the initiating firm needs access to a core technology that is patent-protected by another firm willing to out-license this technology. Technology licensing may also be relatively more attractive if no feasible opportunities for M&As exist. In the IT security case we have seen that technology licensing has played a significant role in enhancing entry of new firms into a diversity of product markets.

What then can explain the reverse dynamics of divergence? We contend that a divergence trajectory will be triggered when prospects for specialization economies, including scale economies and technological differentiation opportunities *within* component-based submarkets (e.g. microprocessors for PCs) or diverged markets (e.g. TomTom's navigation devices) become perceived as higher than economies of scope and synergy. Divergence may result from two partially interrelated processes, the weakening of scope or synergistic economies, and the enhancing of specialization economies. The former will take place when interfaces between components, reflecting relevant user functionalities, become subject to standardized modularity *and* accessible by other firms than the ones who created the interfaces. This may happen as a non-voluntary "natural" process of standardization, simplification and increasing opportunities for replication of what may originally have been proprietary standards (Chesbrough & Kusonoki, 2001). Or it may reflect a strategic decision to create open standards (Sanchez, 2008). Once interfaces have become standardized and open, specialization economies can more easily be accommodated especially when the "diverged" component/product possesses opportunities for high growth, scale economies, and radical technological progress. Large platform providers may open up access to such interface standards in order to stimulate innovation on their platforms. This creates scope for specialized innovators that may be less open in terms of their core technologies while being open in terms of cooperating with platform leaders.

5.7.3 A Convergence Life Cycle

Convergence and divergence can be depicted as alternative cyclical patterns to the PLC/ILC. We term such patterns the Convergence Life Cycle (CLC). In Figure 5.2 the conventional PLC/ILC and the new CLC are integrated into a unified framework. On the vertical axis it specifies three "life cycle" stages along a convergence/divergence track: Product market/industry formation, convergence, and divergence. The horizontal axis distinguishes two units of analysis or the key stages in the PLC/ILC: The emergent product market reflecting the embryonic stage, and the established industry reflecting the shake-out and mature stages. The CLC is illustrated as the movement from the top towards the bottom as a product market or an industry evolves, merges with other product markets or industries, and in turn, diverges into more specialized submarkets.

Product market formation (upper left box) involves the creation and commercialization of genuinely new products corresponding to the embryonic stage in the PLC/ILC. At this stage it is uncertain whether the product market will move into the subsequent stages of the PLC/ILC to form an industry (the arrow pointing into the upper right box), move into some convergence or divergence track—or eventually vanish.

Convergence is likely to be pursued under the economic and technological conditions discussed above. It takes place within an ecosystem that catalyzes opportunities for *systemic innovation* of converged products, hence the creation of new integrative product markets. Convergence may be driven by product market players, typically smaller firms (mid left box in Figure 5.2), or by larger industry incumbents (mid right box). As witnessed in the IT security case, convergence was initially driven by younger firms pursuing product bundling strategies. However, later on the initiative was taken over by incumbents in established industries seeking to embed products from distinctive product markets into their platforms. In other industrial domains we have witnessed product bundling taking place across already established industries, for instance the camera and the mobile phone industry. Divergence (the low row in Figure 5.2) implies the narrowing of the scope of product markets or industries and the resultant formation of new specialized product markets.

Units of analysis of the business environment

	Product market	Established industry
Key stages in product/innovation life cycle		
Product market and industry formation	Product market formation via product innovation (embryonic stage)	Industry formation via process innovation and incremental product innovation (shake-out stage)
Convergence	Convergence initiated by product market firms	Convergence initiated by industry incumbents
Divergence	De-bundling of product markets	Disintegration of industry

(Vertical axis: Stage of the convergence cycle)

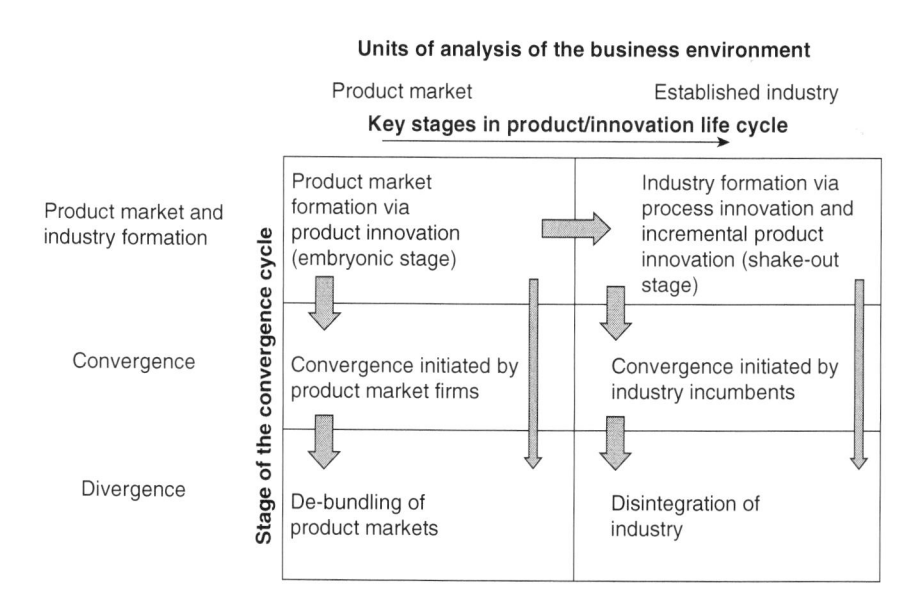

Figure 5.2 The convergence life cycle

Markets that are born out of divergence may again signal the beginning of a new PLC/ILC or CLC.

The CLC framework does not assume a fixed sequentiality of stages in industrial evolution. It is a contingency framework and a taxonomy for understanding various options for trajectories of product life or convergence/divergence cycles, rather than a model that assumes one universal pattern of evolution. As such, it can be operationalized for particular firms in particular markets as a framework for strategic management and innovation strategy.

CONCLUSION

For three decades strategic management has been strongly influenced by an industry-bounded paradigm associated with the Five Forces framework, the PLC, and the ILC. While this paradigm remains effective in explaining business dynamics in some fields, its scope of relevance has been substantially reduced due to the increasing prevalence of convergence and divergence and the underlying patterns of open innovation.

This chapter has presented a framework for understanding business dynamics that do not behave according to the industry-bounded paradigm. It contributes to scholarly and managerial work on strategic management and business dynamics in three ways. First, it offers a systematic way of analyzing boundary-crossing business dynamics. Second, it offers a taxonomy of industrial arenas reflecting the need to align arenas for competition (product markets and industries) and arenas for convergence (ecosystems). Third, it offers a contingency theory explaining "life cycles" that do not follow the classical prescriptions of the PLC/ILC but "derail" into convergence or divergence.

We argue that three factors have contributed to make convergence a more frequent and "patterned" phenomenon: 1) The decreasing importance of radical process innovation that is critical to the PLC/ILC, 2) the digitization of ever more products making interlinking of products increasingly feasible and cost-effective, and 3) the improved effectiveness of markets for M&A and technologies making external acquisition of even very dissimilar complementary knowledge easier and less costly. The increasing significance of convergence and divergence has implied increasing needs for firms to engage in these and other practices of open innovation. However, the competence-building in open innovation practices more generally over recent decades has most likely also contributed to further reinforce trends and opportunities for convergence and divergence.

Thus, the framework provides a contextual complement to the open innovation perspective. In a business world that is characterized by convergence and divergence, firms are faced with challenges of crossing existing boundaries, collaborating and competing with unfamiliar enterprises, institutions and knowledge domains, and these challenges require increasing capacities within and across firms to mobilize open and collaborative modes of managing innovation.

The proposed framework also throws light on the relationship between open innovation and dynamic capabilities (Teece, 2007) in the context of convergence. In the IT security case a few firms exercised dynamic capabilities, involving fairly dramatic features of open innovation, by changing their identity from being innovators within a single-product market to becoming convergence-oriented product-bundling innovators. This transformation required a commitment to move beyond product-specific core capabilities through a series of acquisitions of firms and technologies, a consistent engagement in systemic innovation, and the leveraging of commercial assets for brand-building and large-scale marketing and distribution. Through this diverse set of long-term investments, companies like Symantec and ISS managed to expand from specialized niche positions into broader positions as providers of product suites. Other kinds of dynamic capability were practiced by incumbents in established industries driving context-embedding systemic innovation. Their core competencies were grounded in other fields than IT security, but they were determined to build integrative competencies (Christensen, 2006), including capacities for systems integration, platform development, and complementary assets in distribution, marketing, and brand-building for mainstream markets. Context-embedding innovation aligned infrastructural knowledge situated in these firms with specialized knowledge of security products and technologies.

The firms driving the convergence trajectories in IT security demonstrated commitment to build *Dynamic Capabilities* by transcending existing boundaries and engaging in new challenges of systemic innovation and market creation. On the innovation side, this has involved an intricate balance and change between exercising different forms of open and more closed innovation. These firms' engagement in acquisition-based integration of specialized security knowledge initially reflects an *Open Innovation* practice in the form of searching, accessing, and acquiring specialized knowledge from external sources. But it also reflects the need for substantial in-house control over the subsequent innovation process—in other words the need to practice a more integral form of managing innovation.

Innovation research on other sectors has demonstrated that the need for in-house coordination of innovation is likely to decrease over time as interfaces across components/technologies become standardized, hence more ready for modularity and distributed forms of innovation (Chesbrough & Kusonoki, 2001, and Brusoni et al., 2001). This triggers a process of divergence creating new opportunities for specialist firms. During the last year of our investigation of IT security (2004), the leading drivers of context-embedding innovation had begun to encourage partnering strategies with IT security specialists indicating that the previous integral mode of managing systemic innovation was beginning to give way to a new form of open innovation that would rely more on partnership and outsourcing than on acquisitions and technology licensing.[24]

This chapter points towards a new paradigm of open business dynamics, not to replace but to supplement the hitherto dominant industry-bounded paradigm. The dynamics of open (and more closed) innovation seems to be closely linked to dynamics of open (and more confined) business dynamics. However, just as open innovation has become more prevalent than stable and introvert modes of innovation, open business dynamics have become more prevalent than stable industry-confined dynamics. And for both, the role of markets for firms (M&As), technologies, and more broadly ideas, have become crucial. Future research should address the relationship between open business dynamics and open innovation and in particular the trade-offs between using markets for technology and markets for firms (M&As) as a means of mobilizing outside-in knowledge flows in different stages of industrial dynamics, whether associated with the classical PLC/ILC, or with processes of convergence and divergence.

NOTES

1. The innovation life cycle was pioneered by Abernathy and Utterback in the 1970s and later further developed by Utterback and other scholars of innovation. See in particular W. J. Abernathy and J. M. Utterback (1978), P. Anderson, and M. L. Tushman (1990).
2. Mounting empirical research has over the last decade demonstrated that the classical life cycle models do not explain the evolution of a large and increasing number of industries. See S. Klepper (1997), F. Malerba, L. Orsenigo, A. Fosfuri and M. Giarratana (2007), and A. Bergek et al. (2008).
3. Business dynamics are also increasingly characterized by tendencies for convergence across technological paradigms (see D. J. Teece, 2008), an issue that will not be addressed in this chapter.
4. http://www.ifpi.org/site-content/statistics/worldsales.html
5. For a statistical account of the increasing product markets fragmentation of IT security, see Gambardella and Giarratana (2007). For empirical evidence of the innovative improvements of security products, see Christensen (2011).
6. One example is Certicom's invention of elliptic curve cryptography, see Arora and Nandkumar (2007).
7. Over time, complexity has not only become untenable to manage even for the most professional customers, it has also directly contributed to aggravate security vulnerabilities as it leads to flaws in technical systems coherence, and to configuration weaknesses in systems and applications.

8. Apart from the written sources referred to, these case vignettes are based on interviews with Kim Mikkelsen (Chief Security Advisor, Microsoft Denmark).
 9. While Cisco was a pioneer and leader in network-based security, it was not the only one. Juniper Networks, a leading Internet backbone provider, and 3Com, a leading network hardware supplier, likewise engaged, via large acquisitions of security specialists in building IT security functionalities into the network (Morgan Stanley, 2005).
10. In 2004, Microsoft released Service Pack 2 for Windows XP marking improvements in the security of the underlying software (Selzer, 2006).
11. In 2009, Microsoft announced the exit of the OneCare service implying an exit from paid consumer security. According to Hallawell et al. (2009), this step signified the difficulties Microsoft had in positioning this service against offerings from dedicated security vendors. OneCare was replaced by a free consumer anti-virus and anti-spyware offering which eventually contributed to spur the competition to increasingly offer free security products.
12. For an analysis of this perspective on the business environment (even if the term ecosystem is not used), see Pisano and Teece (2007). For a more explicit management perspective, see Adner (2006).
13. This notion of industry is close to the one Porter (1983) uses when exemplifying his Five Forces framework.
14. We do not pretend to monopolize the definition of ecosystem, only to specify the key aspects of relevance for understanding business dynamics that are *not* industry-confined. To be useful for management, a more rich and firm-specific operationalization is needed, see Iansiti and Levien (2004a).
15. User-based convergence may also take place in the absence of producer-based convergence. This may take two forms. First, coordination between producers of different products with respect to standardization of modular systems and interfaces may make user-based convergence easy and cheap without direct producer-based convergence (see R. Sanchez, 2008). Second, the users themselves may incur the development costs to make possible integrated use of two products provided by firms in different product markets or industries without producer-based efforts to make such integration feasible. In particular in the early stages of the IT security sector, producer-based convergence was often predated by user-based integration attempts.
16. For an overview see Iansiti and Levien (2004a).
17. For central contributions to platform analysis from a management perspective, see Eisenman (2008) and Gawer and Cusumano (2002).
18. Tom Tom, Company History (http://corporate.tomtom.com/history.cfm), and Tom Tom, Annual Report (2005).
19. Economies of scope implies a reduction in costs associated with integrated or coordinated production, development, marketing, and advertising of two or more products (spreading fixed costs across more product markets). This may give rise to new opportunities for revenue through selling complementary products as converged devices, bundled packages or platform-embedded systems to the same customers. Customers may experience benefits from "one-stop shopping," from

reduced interoperability problems and user training costs (see Cottrell and Nault, 2004; Tanriverdi and Lee, 2008).
20. "Core technology" is here used in contrast to "interface technology" to signify a product's or a component's central internal ("inside-the-box") technology.
21. We here assume that no modular and open (non-proprietary) standard interfaces exist across the product markets *prior* to the initiation of a convergence process. If open standard interfaces exist or are shaped *ex ante*, the alignment of product markets can take place as autonomous market processes by producers or customers "mixing and matching" diverse products/components without needing to have deep knowledge of each of the two products (Sanchez, 2008). This would thus make user-based convergence possible without further producer-based convergence, and at the same time allow for a range of divergence processes to evolve.
22. With respect to M&As, see Cantwell and Santangelo (2006) and Colombo and P. Garrone (2006). With respect to markets for technology, see Arora et al. (2001a); Davis (2008), and Lictenthaler (2007).
23. The market for ideas is increasingly provided by such platforms as Innocentive aligning "innovation seekers" and "innovation solvers."
24. For another example see Grunwald and Kieser (2007).

6

Exploring Open Innovation at the Level of R&D Projects

Wim Vanhaverbeke, Jingshu Du, Bart Leten, and Ferrie Aalders

6.1 INTRODUCTION

Firms are increasingly adopting open innovation strategies in their innovation activities (Huston & Sakkab, 2007; Kirschbaum, 2005; Van den Biesen, 2008). In the last two decades, several factors pushed companies to source technologies from external parties and monetize their unused technologies through licensing agreements or spin-offs. The increasing complexity of technologies (Brusoni, Prencipe & Pavitt, 2000), the (typical) over-utilization of own R&D personnel (Clark & Wheelright, 1990), the specialization of technology players such as universities and high-tech start-ups and the emergence of more effective technology markets with new types of intermediaries and technology services companies as main growth accelerators are important drivers of the popularity of open innovation among practitioners (Chesbrough, 2003a, 2006a).

Research on open innovation has been burgeoning in the wake of the increasing role of open innovation in companies. Yet, despite its popularity, the actual performance effects of open innovation are not well understood. R&D collaboration with external partners is an important element of outside-in open innovation activities. Over the past years, several studies have examined the performance effects of R&D collaborations. These studies have almost exclusively focused on the firm level. So far, no consensus has been reached in the literature (see Tsai et al., 2009 for an overview). Some studies show that R&D collaborations improve firm performance (e.g.: Shan et al., 1994; Dodgson et al., 2006; Sivadas & Dwyer, 2000; Tether, 2002; Becker & Dietz, 2004; Belderbos et al., 2004; Sofka & Grimpe, 2010), while other studies find no or negative effects of collaborations (e.g.: Campbell & Cooper, 1999; Knudsen &

Mortensen, 2011; Kessler et al., 2000; Bougrain & Haudeville, 2002; Schulze & Hoegl, 2008). Besides these conflicting findings, still some other studies argue that collaboration has a mixed effect on firm performance (e.g.: Laursen & Salter, 2006; Faems et al., 2010; Hopkins et al., 2011; Gassmann & Enkel, 2010), depending on a number of contingency factors such as the composition of the alliance portfolio and the performance dimension that is studied.

There are several possible explanations why researchers end up with different results, but one straightforward reason is that most studies aggregate different project level practices to general concepts at the firm level that are then linked to firm level performance indicators. As such, the firm has long been treated as a "black box", possibly leading to a number of seemingly contradictory findings on the effect of open innovation. In practice, a firm's performance can be influenced by many factors in the "black box" which are not related to its choice for open or closed innovation. A company may have developed poor business models for some of the technologies that it is working on: in that case an innovation will fail even though collaboration with innovation partners has been managed properly. In other cases, some R&D activities might not be in alignment with the needs of the business groups of the company (Chesbrough, 2003a). The breach between the R&D projects and the operational businesses will finally lead to suboptimal performance levels even when the company has developed best practices to team up with external partners in technology development. These are only a few examples of the many contingencies that might introduce noise in the relationship between open innovation and firm level performance. Therefore, it is not surprising that empirical research at the firm level may result in divergent outcomes.

One way to tackle these problems is by lowering the level of open innovation research from the firm to the R&D projects where open innovation activities take place. This allows controlling for the peculiarities of the R&D projects. Having not gained enough attention in the open innovation literature, R&D projects have been investigated in detail in the new product development (NPD) literature. The focus of this literature stream is however mainly, if not exclusively, on factors internal to the firm, such as project leadership, team composition, management support, process management, and cross-functional integration (Brown & Eisenhardt, 1995; Cooper et al., 2004; Kahn et al., 2006; Griffin, 1997; Carlile, 2002). With the exception of a few studies that look at customer/supplier integration in the R&D project (e.g.: Ragatz et al., 1997; Campbell & Cooper, 1999; Ma et al., 2012; Bahemia & Squire, 2010), the NPD literature has paid limited attention to external collaboration and partnership on project performance, particularly not to the involvement of technology-based partners, such as universities and knowledge institutes. Given the growing trend of companies to conduct R&D projects in collaboration with external partners, it is important to investigate the effect of open innovation collaboration at the project level (Du, Leten, and Vanhaverbeke, 2014). Yet, the NPD literature never systematically examined the effect of R&D collaboration with external partners on NPD performance, nor did it consider R&D collaboration as one of the critical success factors to project performance.

Furthermore, considering partnerships at the project level is also necessary given the limitations of the current NPD literature stream. Although operational improvements have been made over the past decades, the success rate of R&D projects has remained surprisingly stagnant (Page, 1993; Griffin, 1997; Cooper et al., 2004; Barczak et al., 2009). A possible reason for this might be that there are some factors which are critical to project success but have not yet been identified in the NPD literature. Opening up and collaborating with external partners on R&D activities might be one such factor.[1]

In sum, the NPD literature has been largely silent about the impact of external collaboration on project performance, while the open innovation literature has analyzed the impact of external collaboration mainly at the firm level. Little is therefore known whether and how collaboration with external partners affects performance at the R&D project level.

6.2 WHY ANALYZE OPEN INNOVATION AT THE R&D PROJECT LEVEL?

As discussed before, one way to develop a more thorough understanding of open innovation is to analyze it at sub-firm levels of analysis. Responding to the call of West, Vanhaverbeke and Chesbrough (2006: 287–301), the analysis of open innovation at the firm level needs to be complemented with analyses at other levels. R&D projects offer in this respect an interesting platform to study open innovation. We discuss some reasons below why this is the case.

First, firms typically organize innovation activities in R&D projects, and that increasingly more organizations switch to project-based forms (Hobday, 2000; Sydow et al., 2004). In practice, decisions on collaboration are taken on at the project level rather than at the firm level, based on innovation needs and missing competencies. To understand the benefits (or drawbacks) of open innovation, we have to understand how openness functions within R&D projects.

Second, R&D projects offer fine-grained information about the innovation activities in large firms. Apart from providing in-depth information on collaboration with external partners, R&D projects offer information on the peculiarities of projects. R&D projects may differ in many respects: the technology developed in projects may be different (radical vs. incremental, modular vs. architectural, etc.), projects vary in terms of budgeted resources, the leadership and team composition may be different and the innovation partner types might vary across projects. R&D projects show in detail how firms develop new technologies: When we study open innovation at the R&D project level, we no longer consider the firm as a black box where we argue – with the help of theoretical concepts (e.g. search breadth and depth, ambidexterity, absorptive capacity)—that openness will or will not lead to better firm level performance. At the R&D project level, we can control for the particularities of R&D projects and therefore we may monitor more directly the relation between open innovation and performance.

Third, aggregating information at the firm level always implies that there is a loss of valuable information. Relationships that are found may be driven by a few outlier projects. For instance, it may be possible that within a firm, the majority of projects are closed, but only a few projects are open. Assume that these few open projects generate very high financial returns or patent applications. In this case the overall collaboration intensity of the firm is low (because the majority of the projects are closed), but the performance can be high (because of the few projects that are open). An analysis at the firm level will mistakenly lead to the conclusion that a low level of open innovation is beneficial for a company. An analysis at the R&D project level will lead to opposite (but correct) conclusions.

Fourth, a detailed analysis of open innovation at the R&D project level extends the list of critical success factors which determine performance. Factors such as team composition, team leader profile, R&D project management techniques, and the way of interactions with external partners as well as with other departments/projects within the firm, may enrich our understanding of the boundary conditions under which open innovation can work. Therefore, analyzing open innovation at the R&D project level may facilitate project-level decision-making and improve management effectiveness. R&D projects face strict budget and time constraints, and R&D project managers should only decide to engage external partners when their input is really necessary. Therefore, if we understand how open innovation functions at the project level, we may help making better decisions in conceiving, setting up, and running R&D projects.

In the next section we give a detailed description how R&D projects are organized and managed in large companies. Researching open innovation at the R&D project level cannot be done without a salient insight in what R&D projects are and how external partners play a role during an R&D project.

6.3 ANALYZING OI R&D PROJECTS IN LARGE COMPANIES

We can only study how open innovation affects project performance if we have a good understanding of how large companies set up and organize R&D projects. In this section we give a detailed description of how R&D projects are structured and managed in large firms, exemplified by R&D projects developed in the central research units of Royal Philips. This description is instrumental to providing us a more nuanced picture about the differences of open innovation activities conducted at the R&D project level and at the firm level, and how and when open innovation improves innovative and financial performance (see also Du, Leten, and Vanhaverbeke, 2014).

R&D projects have a lifetime; they are typically terminated a few years after initiation. Proposals for new R&D projects can be specified by business groups in the company, top management, or by the central research unit itself. Only the most promising projects are selected, as annual budgets for R&D are restricted. Once selected, management agrees upon specific targets for each project, projects are budgeted, and a team of scientists and engineers are assigned to the project. A project leader manages the progress: most likely (s)he has been involved previously in other projects and preferably (s)he has been already a project leader in the past.

R&D projects are evaluated regularly (annually) and discontinued if they are not living up to expectations. In case of a negative evaluation, financial and human resources are released and reassigned to other more promising projects. A transfer takes place when the technological results of a R&D project are interesting enough for an internal recipient or "customer"—a business group, the central research unit, the IP department or one of the corporate incubators. A transfer of project results takes place when knowledge is purposefully disclosed to a customer of Philips Research under specific conditions:

- When the customer agrees to apply this knowledge in his/her business in (pre)-development projects, products, processes or services;
- When the customer recognizes this knowledge as adding value;
- When the customer takes action to absorb this knowledge in his/her operations to enable an application.

A transfer is only completed when the internal customer confirms these conditions. A project can transfer results to multiple business units using the same technology in different products, markets, or applications.

This short overview about how R&D projects are typically processed in large manufacturing firms raises automatically the question what success means in this context. We can rely on at least three indirect indicators that jointly provide an indication of the innovation success of research projects. The three indicators are *transfer volume, transfer speed*, and *the business value of transfers*.

Transfer volume measures the number of transfers from Corporate Research to the business groups. An R&D project could generate one or more transfers but many projects may not create any transfer to an internal business unit. Still, in that case, the company can generate extra income through licensing agreements. The total number of transfers is a first proxy for success when the analysis is done at the project level.

Transfer speed: This measure of open innovation success is defined as the elapsed time between the start of a project and the date of transfer of the technology to a business unit (development speed) or between the transfer date and the initial market sales (speed to market). The development speed can be divided in two parts: the first part is the elapsed time between the start of the project to its first transfer; the second part takes into account all transfers generated by an R&D project and is calculated as the average time between the start of the project and all the transfers a project generates.

Business success: Technology transfers are reviewed annually on their business success. In Philips, business success can have the following status:

- Business Success: delivering €25 million or more in turnover in a given year. Turnover is taken as a measure of success (and value) of a transfer;
- Potential Business Success: expected to become a business success in the foreseeable future (less than 5 years);
- Old Business Success: a previous business success, but no longer so;
- Inactive: the business opportunity is no longer pursued;
- Transfer: Transfer without a direct prospect of becoming a business success.

The status of a transferred technology is an interesting variable to get a better insight in which transfers (or projects) result in a (major) business success, how long it takes to reach business success, and whether (and why) some projects are more successful in generating new businesses based on the transferred technology.

When a company is recording its patenting activities meticulously, there is also a possibility to have an indication of the success of R&D projects in terms of *patent applications* or *patent grants*. Many large companies systematically patent technological inventions and most of them can be linked to one (or a few) particular R&D project(s). Therefore, *patent application or patent grants* can be used as an alternative (although different indicator) of the technology success of a project (besides transfers). Patent application has been a popular indicator of technical performance for decades. Despite its popularity, however, there has been great concern about the reliability of patenting as an output indicator (see, e.g. Basberg, 1987; Griliches, 1990). This concern stems from at least four aspects of consideration: the technological level and the economic value of patents are highly heterogeneous; the tendency to bundle patent claims together in one or more patents varies widely among countries; not all innovations are patented; not all patents become innovations. Patent applications only reflect the innovativeness of the invention, while giving little indication of the commercial value of potential applications. In large companies the majority of the patents do not make a contribution to a firm's performance.

Also, patenting can carry significant strategic considerations. Therefore, patents are rather a raw indicator of firms' innovative activities. In this sense, transfer related indicators might be a welcome alternative to estimate success in innovation studies.

Innovation performance is a multi-dimensional concept and should be measured in different ways. Moreover, the different indicators of success at the project level have nothing in common with the success metrics at the firm level: this already indicates that project and firm level investigations of the impact of open innovation on innovation success are complementary to each other. What these two levels of analysis consider as success are different concepts. So far we have not described how open innovation at the R&D project level has an effect on performance. Open innovation can be introduced in different ways: At Philips Research, partners are categorized as in Figure 6.1.

Two types of partners are distinguished: technology-based partners and market-based partners. In line with the flow of knowledge and technology, these are also referred to in the company as "upstream" or "downstream" partners, respectively.

R&D projects can be executed internally or in collaboration with external partners. At Philips, management makes a distinction between technology-based partners and market-based partners. Technology-based collaboration indicates that a project is executed in cooperation with academic institutes, government agencies, or organizations in other industries. Market-based collaboration indicates that a project is executed in cooperation

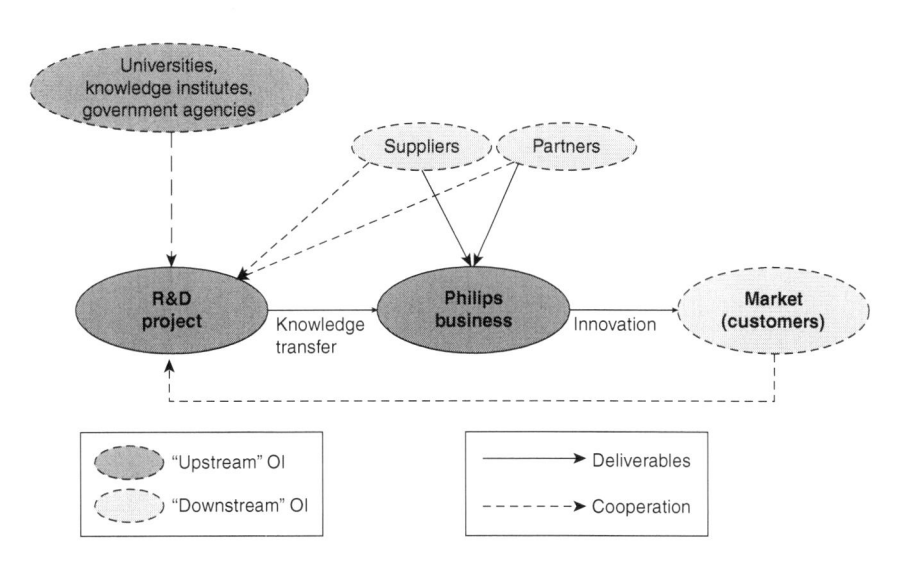

Figure 6.1 Graphical representation of Philips open innovation process

with customers, partners, or suppliers of Philips businesses. Depending on project needs, a choice is made between partners.

Open innovation can be operationalized in different ways depending on the availability of the information (archived data, reports, etc.). Some companies will only register whether or not they collaborate with partners. Other companies record the names of partners which may enrich the analysis substantially. When partners are identified, research can take into account the identity of partners, and examine the role of industry or technological distance between the focal firm and its partner(s) in a project, the role of the geographical and cultural distance between partners, the role of trust building when companies work together with partners with whom they have been working together before. It is well known that "familiarity breeds trust" between alliance partners (Gulati, 1995). Prior relationships with partners could increase the success rate of R&D projects. Good cooperation with partners in previous projects also explains why companies prefer to work again with the same partners rather than trying out new ones. It however may also entail a risk. New partners are more likely to come up with new technologies and business ideas, compared to existing partners. Therefore, it might be interesting to look for new partners when technology is changing rapidly. Furthermore, one could distinguish between different types of partners, such as collaboration with universities, or small start-ups compared to collaboration with large companies. Besides focusing on single types of partner, it would also be interesting to look at the different types of partners collaborating in larger innovation ecosystems.

When information about collaboration with external partners is gathered systematically over time, one can investigate the evolution of partnerships and examine the role of the duration of R&D collaborations, focusing on factors such as collaboration continuity. Time-varying variables introduce a whole wave of new research topics such as the optimal duration of collaboration and the role of simultaneous collaboration with different types of partners versus a sequential approach with different types of partners. In case a sequential approach is beneficial for the innovative performance of the company one has to look which sequence of collaboration with different actors leads to the best results.

There are several other ways to strengthen the inquiry of the role of open innovation in R&D projects. Two examples may illustrate this. First, governance modes play a crucial role in the innovative success of collaborative projects. Contractual arrangements, non-equity alliances, strategic supplier agreements, and joint development agreements are just a few examples how open innovation takes shape. Finding the right contractual arrangement with different types of partners is crucial in determining the success of R&D projects. This, however, requires the availability of data about contractual agreements. Second, the success of a project is not only determined by the collaboration with external partners, both closed and open innovation projects are tapping into internal capabilities of the firm. Projects can leverage intra-firm networks and benefit from internal collaborations with other departments. Some managers are more skillful than others to detect and mobilize internal resources for R&D projects. Especially the relationships with managers in different businesses of the company may be instrumental in generating one or more technology transfers. Data about internal networks in many cases are not recorded, but it is obvious that internal networks and support are as essential as the input from external partners. Proxies of internal networks could be generated via secondary data, such as information on co-ownership and citations in patents and publications. We hope that some scholars will take up this challenge in the next years.

6.4 HOW DOES OPEN INNOVATION AFFECT R&D PROJECT SUCCESS?

How does open innovation in large innovative companies have an impact on the success of R&D projects? The effect is likely to be influenced by a range of factors: the type of partners involved in collaboration, the phase of project development in collaboration, the organizational modes chosen for collaboration, as well as the technology fields involved in the collaboration, just to name a few. Project success can be measured in different ways (Swink et al., 2006): we use three proxies to represent project success: transfer volume, transfer speed and business success as explained in section 6.3. Below we describe how three types of factors—type of partners, phases in the project, and modes of collaboration—may influence the impact of open innovation on project performance. We cannot discuss all possible factors but we try to cluster them under several headings at the end of this section.

6.4.1 Type of Partners and R&D Project Success

An R&D project team may collaborate with different types of partners. Each type of partner has different capabilities and incentives to collaborate. For instance, market-based partners have expertise and knowledge on market needs (von Hippel, 2005; Prahalad & Ramaswamy, 2004a) and the latest technologies, parts and components that are available to satisfy these needs. They help a new product to establish a foothold in the marketplace (Appiah-Adu & Ranchhod, 1998) by eliminating the likelihood of product failures (Harrison & Waluszewski, 2008) and meeting customer satisfaction (Ragatz, Handfield & Peterson, 2002; Gruner & Homburg, 2000). Technology-based partners are experts in (basic) scientific research and provide project teams with knowledge on the latest scientific developments. Scientific knowledge may function as a "map" for scientific research and point R&D teams to the most profitable directions for applied research. Further, this knowledge may help teams to evaluate the outcomes of applied research (Rosenberg, 1990; Fleming & Sorenson, 2004; Cassiman et al., 2008). Because of their different roles, collaboration with these two types of partners will likely have a different impact on project performance.

Cooperation with technology partners may increase the chance of project success in the following aspects:

- Higher speed of knowledge transfer using scientific knowledge:

    Technology partners provide project teams with (basic) scientific knowledge, which is complementary to the applied knowledge of project teams. Partnerships with complementary partners allow for a partition of project tasks among partners and to benefit from a division of labor. Working in parallel on different tasks will likely result in a higher product development speed.

- Higher volume of knowledge transfer through more frequent science-related additions:

    Collaboration with technology partners may lead to the generation of new platform technologies that rely on the latest scientific insights. Platform technologies are cost-efficient as they allow for the generation of a family of derivative innovations or product line additions at low costs. Platform technologies will lead to a higher transfer volume per R&D project.

- Greater business value of knowledge transfer through higher speed and introduction of cutting-edge science:

    Collaboration with technology partners speeds up technology development, which gives an innovation firm the opportunity to be a first mover on the market, to outcompete competitors, and to have a larger market share in growth markets. Further, technology partners infuse innovations with the latest scientific insights. Innovations incorporating latest technologies are high-risk bets: they are risk laden but they may result in the most promising business opportunities. Exploring the technological frontier, and collaborating with technology partners, is one way for companies to create options for new business opportunities.

Collaboration with market partners may also increase the chance of project success through:

- Higher speed of knowledge transfer using market knowledge:

    Market partners have more and more specific information about customer needs, market trends, and foresight. Collaboration with these partners sheds light on the latest market knowledge. This increases the chance that developed technologies become a market success. The innovating firm can quickly transfer technologies if managers of business units perceive that the technological solutions are targeting real market needs.

- Higher volumes of knowledge transfers:

    Market-based partners inform the firm about market trends and customer needs, which makes the output of the research more valuable for the latter. Better market information and market preparation leads to a more correct estimation of business opportunities and fewer market failures. Therefore, we can expect that collaboration with market partners will on average result in higher business value of knowledge transfers.

- Downstream open innovation is not increasing the chance of greater business success through higher volume of knowledge transfer:

    Information about particular market opportunities from market-based partners will not lead to more frequent additions to technology. In contrast with collaboration with technology partners, collaboration with market-based partners will not increase the volume of transfers.

The joint use of technology-based and market-based collaboration may also be useful. For instance, business success may be better guaranteed when technology-based and market-based collaboration are combined, as the product is built on the combination of leading edge scientific insights and technologies and a thorough understanding of market trends and needs. However, R&D projects where both technology and market partners are involved may be more complex and harder to manage than closed innovation projects, or projects that involve one type of partner. Because of the distinct nature of these types of partners, their goals and working habits are likely to be different. Compared to closed innovation R&D projects or projects that only collaborate with one type of partner, communication and coordination of projects that collaborate with both types of partners can be more challenging. Therefore we expect that collaboration with the two types of partners will not be efficient for small projects. The larger the projects, the easier it is to deliver the extra investment related to collaboration with different partners.

6.4.2 Phase of Project Development in Collaboration and R&D Project Success

Not only the type of R&D partners but also the timing of collaboration may have an impact on the success of open innovation projects. Collaborations may take place at different time points of an R&D project. Projects dynamically evolve over time into further development stages, and in each phase, its goals, needs, and activities are different. Therefore, interactions with external partners may differ from phase to phase. However, the majority of studies have a static view. Success factors are considered to have the same impact on the success of R&D projects regardless their development phase (Pinto & Prescott, 1988).

The NPD literature has generally defined four phases of product development: initialization (also called "conceptualization" or "fuzzy front end"), planning, execution, and termination (e.g.: King & Cleland, 1983; Clark & Wheelwright, 1990). In the context of open innovation, R&D collaborations can take place in one or several of these phases. For each of these phases, project performance is likely to be influenced by external partnerships in different ways.

In the project initialization phase, project development is still in its fuzzy front end, the initial investment and commitment to the project is relatively small compared to the more expensive later project phases (Cooper, 1990; Van Oorschot et al., 2010). The project enjoys greater flexibility and multiple possibilities in making choices on its way to proceed (Pinto & Prescott, 1988). A large amount of information is needed to ensure the technology is feasible and market-opportunities are tangible. Moreover, costs for trials and experimentation are small in this phase (Van Oorschot et al., 2010). As external partnerships provide the R&D project diverse resources and insights from multiple sources, it is useful to team up with them during the initialization phase. A number of techniques have been proposed to source knowledge widely from the external environment, such as scouting (Rohrbeck, 2010), sourcing (West & Lakhani, 2008), as well as screening and signaling (Fontana et al., 2006). Future research may explore the detailed effect of collaboration in this phase, and the optimal combinations of different searching approaches as well as knowledge sources.

In the project-planning phase, the project successfully passes through initial selection and enters into a further development stage. The research direction and problem definition of the project become clear and committed resources are supposed to be in place. Since in this phase of development projects tend to rely more on internal decision-making and upper management support, overly relying on external partnerships may introduce noise into the decision-making and planning process. Moreover, numerous coordination and communication among partners may bring additional problems into the process. In sum, external partnerships may not be very beneficial for this stage of project development.

The third stage in the project life cycle is execution. In the project execution phase, the actual work of the project is performed. The involvement of external partners (both technology and market based) is supposed to be instrumental in solving project's problems in a timely and advanced manner.

The fourth phase of project development is termination. During this phase, the final outcome of the project is handed to its intended users (the business departments of the firm). Collaboration with internal business units in this phase may facilitate project transfer and smoothen the research result delivery. Most collaboration at this phase may relate to the identification of suitable business models and novel applications of the innovation that has been developed in the previous stages.

In short, choosing the optimal involvement of different types of partners at a right moment in R&D projects should improve their performance. These types of studies are promising but are not yet executed. We, therefore, encourage scholars to analyze the timing of R&D collaborations in the following years.

6.4.3 Organizational Mode Choices of Collaboration and R&D Project Success

Organizational mode choice of R&D collaborations will also influence R&D project success. Collaborations in R&D projects can be organized in different ways. Formal collaborations have received a lot of attention in the literature and they do play a considerable role in R&D project development. In formal collaboration, the targets and teams are clearly identified, and there is an agreement about the resources to be invested and the length of the collaboration period. Formal collaborations are in most cases contract-based agreements. Collaboration with partners is, however, not always formalized and "informal" relations play a crucial role in reality as well. Informal collaborations can take different shapes. Take for instance partnerships of firms in subsidized research and technology programs, scouting relationships, conference participations, partnerships in standard-setting organizations, long-term relationships with key technology partners, such as prominent universities and research labs, or key market players (key customers, first-tier suppliers, etc.), just to name a few. In other cases, collaboration is not formalized because of the nature of the collaboration such as crowdsourcing and online competitions. It is obvious that informal collaboration deserves more attention: First, it has been systematically underemphasized because of lack of reliable data, and second, firms are nowadays increasingly tapping into information from multiple informal knowledge sources (Tether & Tajar, 2008).

An R&D team can reach different objectives through different collaboration modes. Different mixes of formal and informal collaboration modes will best fit the needs of different R&D projects. Research on the choice of optimal collaboration modes in R&D projects has not been developed so far. There is a burning need to make progress on this research theme.

6.4.4 Other Determinants of Collaboration and R&D Project Success

There are many more factors that moderate or shape the impact of open innovation on R&D projects' outcome. One of them is the technology fields, and their state of development, in which R&D projects take place. Some technologies are emerging, others are established, some are relying critically on expensive R&D-infrastructure while still other R&D projects need the input of large communities of creative people. All these factors will determine how collaboration has to be organized to boost the technological and financial outcome of R&D projects.

Similarly, more propositions can be developed depending on the information embedded in the data about corporate R&D projects. We have mentioned before how information about the identity of the partners can substantially enrich our understanding on how open innovation works at the project level. The focal firm and its partners belong to particular industries, they have a specific position in the technology landscape and they are located in a specific country or city. With more detailed data about partners, we can seize new topics such as the role of geographical proximity or technological proximity in open innovation. Similarly, more detailed data about the R&D team leader and members or about the organization and management of R&D projects opens up new possibilities to investigate the conditions under which open innovation plays a positive role in improving the success of R&D projects. In Figure 6.2 we cluster several determinants under different questions. Each of these factors will influence how a company organizes for open innovation at the level of R&D projects, and how openness impacts on the innovative and financial performance of projects. We suggest that further research examines

- **Why**—the objectives or reasons why R&D teams collaborate with external partners. The alignment of the firm's objectives with those of its partners.
- **Who**—Types of partners; optimal mix of partners; identity of partners.
- **When**—Collaboration timing; R&D collaboration in different phases of project development.
- **What**—Technology fields involved into collaborations; emerging or established technologies; radical vs. incremental innovation; modular vs. architectural innovation.
- **Where**—Geographical issues of collaboration in R&D projects; choice and management of close and distant partners; influence of technological hotspots and location of existing R&D labs.
- **How**—Organizational modes of collaborations in R&D projects; management of collaborations in projects.

→ Organization of external collaboration per project

↓

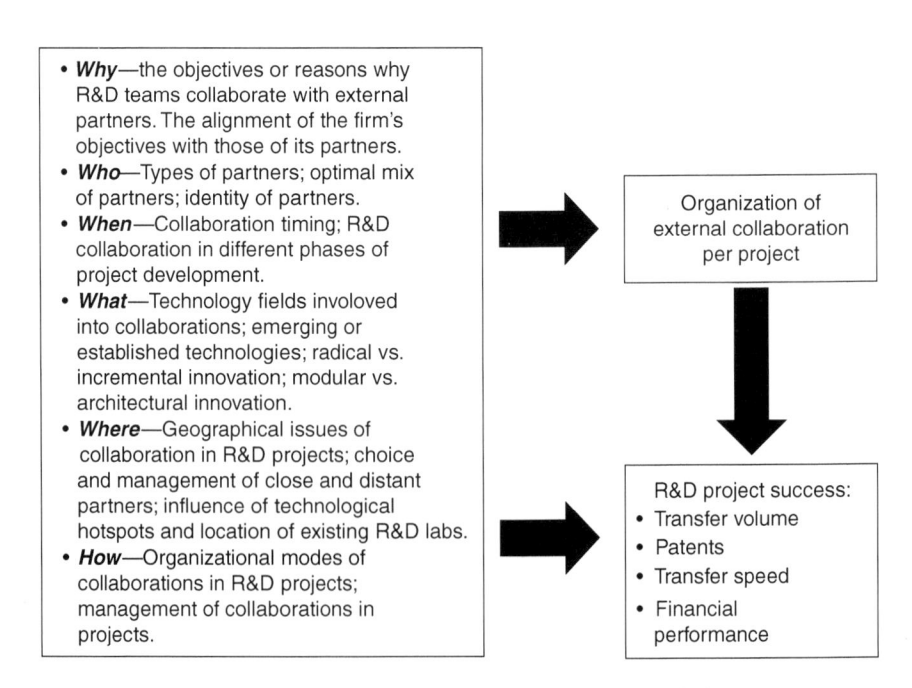

→ R&D project success:
- Transfer volume
- Patents
- Transfer speed
- Financial performance

Figure 6.2 Factors affecting the organization of R&D projects and their outcome

the relative importance of the suggested moderators of the open innovation project performance relationship.

6.5 PLACING R&D PROJECT RESEARCH INTO A BROADER PERSPECTIVE

In the previous sections we have shown that studying R&D projects is important to advance open innovation research. Analyzing open innovation activities and their effects at the project level, however, does not decline the importance of other levels of analysis. In fact, the role of open innovation in R&D projects can only be fully understood when the project level is linked to the firm and other observation levels. Examining open innovation activities in R&D projects can lead to great insights about the mechanisms of how collaboration with different partners enhances the technological and commercial success of projects, but we should also study how decisions about open innovation at the firm affect open innovation at the project level and vice versa. Think for instance about a firm's corporate growth strategy, where management may decide to explore growth options in a particular new technical domain. Linking up with external partners in R&D projects to explore new opportunities in new technological fields may have to be organized in a different way than open projects that serve ongoing innovation for the mainstream businesses. Open innovation at the R&D project level should thus be related to corporate strategy and the ambidexterity literature to understand why managers open up R&D projects and which partners they select to obtain specific strategic objectives.[2]

Likewise, we should not look at individual projects in isolation from each other but take the portfolio of R&D projects into account. R&D projects are embedded in the organizational context of the firm and, consequently, their value has to be derived from their position within the network of R&D projects in the firm. Firms not only set up a range of R&D projects, they also coordinate and integrate internally developed and externally sourced knowledge across projects. Each individual project develops a piece of technological knowledge but a firm should also develop mechanisms to disseminate the knowledge and integrate it in the overall technology and business developments of the firm. Hence there is an urgent need to connect the project and the firm level to each other for two major reasons: First, we can only fully understand why firms engage in open innovation projects if we can position them within a firm's portfolio of projects and connect them to the overall innovation strategy of the firm. Second, one can only understand (firm level) concepts such as technology depth, breath, orientation, or absorptive capacity if they are related to open innovation activities in R&D projects. An optimal level of breadth of technology search at the firm level for instance is after all the outcome of a mix of open and closed R&D projects. The question is: how do companies decide on the mix of these projects? What are the reasons behind the choice for open or closed innovation in each project, and how is this choice affected by a company's prior experience with open innovation and the open innovation culture that it had developed previously? The most interesting research in open innovation could be developed at the intersection of these different levels of analysis. We badly need a multilevel analysis of open innovation to advance research in this field.

The interaction with other levels of analysis deserves more attention too. Success of open innovation in R&D projects is most likely dependent on the quality and experience of individuals both in the R&D team of the focal company as well as the individuals with whom they interact in the partnering organizations. Studying the role of individuals in open innovation is still uncharted territory, and interaction with the openness in R&D project level investigations is not touched upon yet. A notable exception is the (yet) unpublished PhD thesis of Meijer (2013) that is investigating the role of individuals and team composition in alliance teams: this research pays attention to the teams in both companies that are establishing a technology alliance. The success of open R&D projects is here further analyzed at the level of their components. Moreover, R&D projects are no longer investigated as projects of an innovating company but as a concept that has to be explored as a joint management initiative of the innovating firm and its partner.[3]

6.6 CONCLUSION

Research at the project level can improve our understanding of how open innovation is implemented in large companies and how open innovation affects the technological and financial performance of firms. Open innovation activities mostly take place in R&D projects, which differ in many respects, such as the type of collaboration partners (technology vs. market partners), collaboration modes, timing of R&D collaboration, technologies under development, strategies, the size and composition of project teams and the way projects are managed. Lowering the level of open innovation research from the firm to R&D projects opens up possibilities to examine a broader set of factors that may determine how open innovation translates into a superior performance at the project and firm level. This will increase our understanding of the boundary conditions under which open innovation can work.

Several large companies record detailed information on the organization, management and open innovation practices of R&D projects. Philips Research serves as a nice example. Starting from 2003, they recorded annually the open innovation practices of all R&D projects executed by the central Philips Research departments, resulting in a panel dataset on the open innovation activities of several thousands of R&D projects. Collaborating with companies to analyze available databases is suggested as an interesting route for academic researchers to move open innovation research further. A detailed analysis of these data also provides managers new insights on how R&D projects have to be organized to generate more transfers, speed up product launches, and seize bigger market opportunities. Both academics and practitioners can win by opening up large-scale databases about R&D project management.

Lowering the level of analysis to projects does however not imply that analyses at other levels are unimportant. There are clear links between decisions that are taken at the project level and other levels of analysis, such as individuals, R&D units, firms, R&D networks, sectoral, national and regional innovation ecosystems. Multi-level analyses that take into account the relationships of decisions that are taken at multiple levels could increase our current understanding of open innovation strategies.

NOTES

1. Cooper and Edgett (2012) assert that the stage gate process can easily be modified to incorporate open innovation. However, they do not operationalize what criteria to add or change in the stage gate model to do this.
2. This theme is for instance related to managerial considerations in opening R&D projects in core and non-core technologies. Interesting implications are formulated by Chesbrough and Schwartz (2007). We refer to Vanhaverbeke et al. (2012) for an empirical analysis of the role of alliances in core and non-core technology.
3. It is also possible to link open innovation practices in R&D projects to innovation ecosystems or R&D networks (Nambisan & Sawhney, 2011; Adner, 2012; Leten et al., 2013).

Part III

New Application Fields for Open Innovation

7

Exploring Open Innovation in Small and Medium-Sized Enterprises

Sabine Brunswicker and Vareska van de Vrande

7.1 INTRODUCTION

Over the past years, research in open innovation has flourished and the analysis of open innovation has been extended to various topics. However, a large proportion of open innovation research of the last decade concentrates on large and multinational firms while innovation has become a more level playing field. Therefore, one would assume open innovation to be beneficial for large firms as well as for small and medium-sized enterprises (SMEs) (Chesbrough, 2006c). Prior work on innovation and SMEs emphasizes the role of inter-organizational relationships and external sources of innovation, and indicates that innovation in SMEs "by nature" has an external focus (Baum et al., 2000; Edwards et al., 2005). Nevertheless, SMEs have been excluded from the mainstream discussion in open innovation research (Lee et al., 2010; Wynarczyk et al., 2013).

Only recently, researchers have started to investigate the relevance and the specific nature of open innovation in SMEs (Lee et al., 2010; van de Vrande et al., 2009b; Vanhaverbeke, 2012; van der Meer, 2007; Spithoven et al., 2013). These studies confirm that small firms do engage in different types of open innovation practices and that openness drives innovation performance in SMEs (van de Vrande et al., 2009b; Parida et al., 2012; Brunswicker, 2011). Moreover, recent work provides a clear indication of the specific nature of how SMEs can benefit from opening their business models and using inflows and outflows of knowledge (Lee et al., 2010), suggesting that existing findings on open innovation in large firms cannot be directly transferred towards the SME sector. These first lessons learned provide the argument for studying open innovation in the SME sector in more detail. In addition, the economic relevance of SMEs further emphasizes the importance of research on open innovation strategies in SMEs. In Europe, for example, more than 60%

of private sector jobs are in the SME sector and more than 90% of all businesses are SMEs (European Commission, 2005; OECD, 2009; Acs & Audretsch, 1987). As discussed next, it is the specific nature of open innovation in SMEs that opens up new research avenues.

7.2 THE SPECIFICS OF OPEN INNOVATION IN THE SME SECTOR

7.2.1 The SME Sector and the Nature of Innovation in SMEs

As the term suggests, small and medium-sized enterprises (SMEs) are organizations that are characterized by their "smallness," which is usually measured with an upper ceiling for number of full-time employees, yearly turnover, and/or annual balance sheet total.[1] In practice, the term SME is regularly associated with high-tech start-ups, new small firms, and entrepreneurial firms. However, there are different "subpopulations" of SMEs (de Jong & Marsili, 2006; Leiponen & Byma, 2009), and SMEs subsume more than just young technology entrepreneurs and science-based ventures from high-tech sectors (Gans & Stern, 2003), and also include established SMEs that are at a later organizational lifecycle stage (Koberg et al., 1996). However, studies on SMEs and entrepreneurship literature in particular have a bias towards young and small firms (de Jong & Marsili, 2006; Macpherson & Holt, 2007).

It is widely recognized that SMEs make a significant contribution to our economies and that SMEs, compared to large firms, also have the capacity for innovation (Acs & Audretsch, 1988). It can even be observed that SMEs have been increasing their expenditures for R&D (National Science Foundation, 2006). However, it is not just high-tech start-ups that innovate. Low-tech SMEs, as well as established SMEs that have successfully passed the critical lifecycle stage of 8 to 15 years, also play an important role in today's innovation landscape (Koberg et al., 1996; Santamaría et al., 2009). While some may continue to engage in radical innovation, and eventually become market leaders, others remain "small" and compete in market niches (de Jong & Marsili, 2006). However, despite the considerable amount of literature on SMEs at the aggregated level, research on the firm-level innovation activities in SMEs is limited. Prior studies suggest that innovation processes and models in SMEs are quite different compared to large firms (Edwards et al., 2005): They are usually flexible, fast decision makers, and quicker in reacting to changing market demands (Vossen, 1988). At the same time, they face limitations in terms of material, human, and resource factors (Acs & Audretsch, 1987; Vossen, 1988; Harryson, 2008). Further, an owner-manager, a group of partners, or the members of a family dynasty dominate most small businesses (Roper, 1999). Moreover, they generally have less formalized R&D procedures.

Due to the liability of smallness, SMEs cannot cover all innovation activities required to successfully realize an innovation. Thus, innovation in SMEs regularly has an external and boundary-spanning component. Indeed, there has been a long tradition of research on the role of external relationships and networks in SMEs (Birley, 1985; Edwards et al., 2005; Macpherson & Holt, 2007). Prior work on young biotechnology SMEs indicates that strategic alliances enable them to innovate. Dyadic partnerships and multi-actor alliances help them to get access to critical resources, to extend their technological competencies, and also to build legitimacy and reputation. SMEs that are involved in multiple ties are also more innovative than those that use only one type of tie (Baum et al., 2000). Further, existing literature on SMEs and social capital emphasizes the preference of entrepreneurs and their systems towards informal and social contacts that may provide opportunities and at the same time shape the development of a firm (Macpherson & Holt, 2007). In fact, SMEs that belong to formal and informal networks are more innovative than others. One factor driving this positive association is the presence of a large variety of relationships among network members; *personal networks* support the diffusion of innovation within networks of SMEs (Ceci & Iubatti, 2012). Social and personal relationships are often strongly embedded in the economic actions of SMEs and are therefore not "used" in a purposive manner. Thus, SMEs may regularly lack the capability to proactively articulate their needs for external knowledge (Bessant, 1999). Even though they could build upon strong external relationships and interpersonal networks, SMEs often don't have the internal capabilities required to do so (Bougrain & Haudeville, 2002). Further, organizational and social relationships can act as a barrier to innovation as such ties may close opportunities (Macpherson & Holt, 2007). SMEs even run the risk of becoming too dependent upon their relationships.

Overall, literature indicates that inter-organizational linkages and networks are important drivers of innovation in SMEs. However, existing studies reveal a "paradox". Even though SMEs regularly have strong inter-organizational ties, they struggle with making the best use of these ties. Studying open innovation in SMEs should provide insights in "how" SMEs can use network relationships and social capital by purposively using inflows and outflows of knowledge. If SMEs become proficient in applying and managing open innovation, they can use their relationships in a positive manner rather than becoming dependent upon them. As the locus of innovation regularly resides at the network level, open innovation in SMEs naturally is quite specific and different from large firms; it postulates researchers to explore the unique challenges in leveraging and managing open innovation in SMEs.

7.2.2 Mapping the Field of Research on Open Innovation in SMEs

Prior studies have addressed open innovation in SMEs from different angles[2] (an overview of selected studies published since 2003 can be found in Table 7.1). Some studies empirically investigate the role and the performance impact of open innovation in SMEs, providing a high-level insight into open innovation in SMEs (e.g. Laursen & Salter, 2006; van de Vrande et al., 2009b; Drechsler & Natter, 2012; Parida et al., 2012), while others take a more in-depth perspective and provide insights into the specifics of open innovation in SMEs (Vanhaverbeke & Cloodt, 2006; Lee et al., 2010; Vanhaverbeke, 2012). In the following chapters, we provide a brief overview of the existing literature.

Table 7.1 Overview of empirical studies on open innovation in SMEs

Author(s)	Year	Objective and focus of the study	Type of study
Barge-Gil	2010	Empirical examination of the role of the degree of openness in small and large firms	Quantitative study based on PITEC database (2004–2006) of Spanish firms
Bianchi et al.	2010	Investigation outbound open innovation (out-licensing) in SMEs with a focus on a managerial instrument for the identification of out-licensing opportunities and alternative application areas of a technology	Qualitative study (design science); single case study on an Italian SME
Brunswicker & Vanhaverbeke	2010	Exploration of different inbound sourcing strategies and internal managerial facilitators of open innovation in SMEs	Quantitative study based on firm-level data on European SMEs
Christensen et al.	2005	Analysis of the strategic mode of open innovation from an industrial dynamics perspective based on an in-depth study of the transformation of sound amplification within the consumer electronics system of innovation	Qualitative study based on primary and secondary on firms and R&D pioneers in the class D amplification around the Technical University of Denmark

Author(s)	Year	Objective and focus of the study	Type of study
Classen et al.	2012	Investigation of the differences in the diversity of cooperation partners used for innovation-related activities (i.e., search breadth) between family and nonfamily SMEs	Quantitative study based on SMEs from Belgium and Netherlands (Belgium Belfast database and the Dutch Chamber of Commerce database)
Cosh et al.	2011	Investigation of the nature of open innovation in British large and small firms from different sectors	Quantitative study based on survey among more than 12000 UK firms from 15 sectors performed in 2010
Drechsler & Natter	2012	Investigation of the underlying drivers of openness in small and large firms	Quantitative study based on the German Community Innovation Survey 2005 (CIS IV)
Gardet & Fraiha	2012	Exploration of the coordination modes used by an SME project bearer in a network	Longitudinal single case study on a SME in a network in France
Gruber & Henkel	2006	Investigation of how three key challenges of venture management—the liabilities of newness and smallness of start-ups and market entry barriers— affect new ventures in OSS	Empirical study on large scale survey of 268 embedded Linux developers and 30 personal interviews with industry experts
Huang & Rice	2006	Empirical investigation of the interaction between open innovation strategies and absorptive capacity	Quantitative study based on the Australian Bureau of Statistics' Business Longitudinal Survey; 292 manufacturing Australian SMEs
Laursen & Salter	2006	Empirical explanation of the impact of search strategy an innovative performance in small and large firms	Quantitative study based on the UK innovation survey 2001 with 2707 manufacturing small and large firms
Lee et al.	2010	Conceptualization of the role of an intermediary in open innovation and descriptive analysis of the success of Korean SMEs working with an intermediary	Qualitative study on Korean SMEs based on the report of Survey Science and Technology Policy Institute (STEPI) published in 2005

(continued)

Author(s)	Year	Objective and focus of the study	Type of study
Parida et al.	2012	Investigation of the effects of four inbound open innovation practices (both monetary and non-monetary) on innovation performance of high-tech SMEs	Quantitative study based on survey data of 252 high-tech SMEs; survey performed in Sweden in 2009
Spithoven et al.	2013	Investigation of the importance of four open innovation practices (focus inbound) and estimation of their performance effect with a focus on the differences between large and small firms	Quantitative study based on the Community Innovation Survey 2006 in Belgium
Theyel & Cosh	2012	Investigation of the differences in the importance and performance effect of different inbound and outbound open innovation activities in young versus established firms	Quantitative study based on survey data from 1202 firms from the UK; survey was performed in 2010
van de Vrande et al.	2009b	Exploration of the incidence of and apparent trend towards open innovation among Dutch SMEs addressing eight different practices for inbound and outbound open innovation	Quantitative study based on survey data of 605 SMEs in the Netherlands performed in 2005
van der Meer	2007	Descriptive analysis of the adoption of open innovation in Dutch firms (including small and large firms)	Empirical analysis of survey-data of 814 SMEs in Dutch firms based on the Dutch National Innovation Survey 2003 and 28 interviews with highly innovative Dutch firms
Vanhaverbeke & Cloodt	2006	Exploration of open innovation from a value network perspective focusing on the role of interorganizational networks in the commercialization of new product offerings based on technology breakthroughs in the agriculture biotech	Qualitative case study example of the role of value networks for open innovation in SMEs in the agriculture biotech sector

Author(s)	Year	Objective and focus of the study	Type of study
Vanhaverbeke	2012	Inductive investigation of the strategic role of open innovation in SMEs considering the interdependences between the firm strategy, business model, and open innovation	Qualitative study based on ten inductive case studies on Belgium SMEs

7.2.2.1 The Adoption and Prevalence Open Innovation in SMEs

As argued before, the adoption of open innovation in large firms and SMEs differs significantly. Survey-based studies indeed suggest that firm size positively influences a firm's openness (Drechsler & Natter, 2012), while others find an inverted U-shaped relationship between firm size and search breadth (e.g. Barge-Gil, 2010) or indicate that although large firms appear to be more open, SMEs have in fact a higher open innovation intensity (Spithoven et al., 2013). Moreover, SMEs have increased their open innovation activities over the last years (e.g. van de Vrande et al., 2009b), with inbound open innovation practices being far more diffused than outbound open innovation. Regarding inbound open innovation activities research indicates that SMEs have a preference for non-monetary activities such as networking, over complex transaction-based ones, such as acquisitions and in-licensing (van de Vrande et al., 2009). There are different open innovation types among SMEs. While some SMEs engage in a large variety of different practices, others tend to use a subset of potential practices only. Moreover, SMEs also differ in how they combine different types of sources of external knowledge. Some open up only along the value chain while others heavily draw upon universities and research organizations (Cosh & Zhang, 2011; van de Vrande et al., 2009b; Brunswicker & Vanhaverbeke, 2010).

Outbound open innovation, on the other hand, is hardly adopted in the SME sector (van de Vrande et al., 2009b) even though prior literature on the "market for ideas" indicates that technology-driven and venture-capital backed entrepreneurial firms do consider out-licensing of know-how and technologies as an alternative to developing a product and selling it on the market (Gans & Stern, 2003). Apparently, the identification of potential opportunities for out-licensing is challenging for SMEs. The role of non-monetary outbound innovation activities also receives little attention in existing literature on open innovation in SMEs. A notable exception is a study by Gruber and

Henkel (2006) who indicate that free revealing—a non-monetary form of outbound innovation—may enable SMEs to overcome the liability of smallness.

Finally, prior studies have pointed to a number of factors that influence SMEs' open innovation decisions and activities, such as the firm's need for funding, and the characteristics of the innovation system. (Christensen et al., 2005) Indeed, market-related motives are a primary reason for SMEs to engage in open innovation (van de Vrande et al., 2009b). However, a firm's knowledge gap (market and technological knowledge) and ineffective IP protection mechanisms prevent firms from being open (Drechsler & Natter, 2012). A closer look into the SME-specific organizational and leadership characteristics reveals that the ownership structure also shapes the adoption of inbound open innovation in SMEs. Family-owned SMEs have a lower search breadth than their non-family counterparts. Furthermore, the level of education of the CEO and nature of the top management team affect the openness of SMEs (Classen et al., 2012).

7.2.2.2 *The Performance Impact of Open Innovation in SMEs*

First evidence on the performance impact of open innovation in SMEs suggests that sourcing positively affects innovation performance, both in developed and developing countries (Laursen & Salter, 2006; Chen et al., 2011). In addition, Spithoven et al. (2013) found that besides search, innovation collaboration and external R&D contracts also have a positive effect on SME's innovation performance. There are also differences between young and established firms with regards to openness and the performance impact of open innovation. Firm age and a firm's stage in the organizational life cycle affect its openness and the performance impact of open innovation— open innovation has greater benefits for young firms than for established firms (Theyel & Cosh, 2012). Moreover, the impact of openness on innovation performance also depends on the SME's sourcing strategy and how they combine different external knowledge sources. Some combinations have a negative effect on innovation performance and thus represent a potential risk (Brunswicker & Vanhaverbeke, 2010; Brunswicker, 2011). Interestingly, different open innovation practices have a differential impact on incremental and radical innovation. While technology scanning is more important for incremental innovation, sourcing (accessing technology via licensing) is of higher relevance for radical innovation. Somewhat counterintuitive, vertical collaborations along the supply chain with customers and end-users positively affect radical innovation in high-tech industries (Parida et al., 2012).

7.2.2.3 *The Network Dimension of Open Innovation in SMEs*

Prior research has provided extensive evidence that networks play an important role in explaining the performance of start-ups and SMEs (e.g. Baum et al.,

2000; Bruderl & Preisendorfer, 1998; Rogers, 2004). In addition, SMEs also frequently participate in more informal knowledge networks (Buchel & Raub, 2002). These knowledge networks are often regional initiatives aimed at knowledge sharing and networking among a broad range of parties, such as start-ups, incubators, venture capitalists, experts, etc. (Collinson & Gregson, 2003). Indeed, network relationships shape performance of SMEs and influence their strategic actions. The strategic role of networks in SMEs has also implications on the nature of open innovation in SMEs. Open innovation in SMEs is directly linked to the business strategy and the firm's overall strategic objectives (Vanhaverbeke, 2012). While large firms can implement open innovation without a strategic change, the shift towards open innovation in SMEs goes hand in hand with a strategic change (Vanhaverbeke, 2012).

As open innovation in SMEs is embedded in and directly linked to strategy, it also directly links to the SME's position in the value chain and its value creation relationships with partners. When SMEs engage in innovation, they regularly lack complementary assets and resources to commercialize a new product or service. For example, SMEs working on new technologies face the challenge to commercialize it and involve new value creation partners to access complementary assets (Vanhaverbeke & Cloodt, 2006; Lee et al., 2010). Thus, open innovation and boundary-spanning innovation activities are not just important in the "front-end"—that is in R&D—but also in the commercialization phase. For SMEs, interorganizational ties relate to the value network (Nalebuff & Brandenburger, 1996) as well as to the "front-end" innovation network (Vanhaverbeke, 2012).

Due to the importance of networks in open innovation, existing work also touches upon the intermediation and management of such networks. Lee et al. (2010) study possible models for SMEs to profit from networks and discuss the role of intermediation to support the commercialization phase and to establish different types of network relationships. In addition, the network dimension of open innovation also has implications for the managerial capabilities of SMEs. Coordination capabilities are required to benefit from innovation and value creation networks and to access the required resources and capabilities, especially if they play the role of a "hub" firm (Gardet & Fraiha, 2012). SMEs need to diligently implement different coordination tools through elaborating on their implementation mode, and consider their interaction to act as a "hub." For example, they need to decide on whether they rely on informal, semi-formal, or formal communication, whether they can rely on trust, how they divide the benefits, and how conflicts are resolved. Further, the project evolution (phase) and the dependency on the partners seem to be important antecedents of the coordination mode (Gardet & Fraiha, 2012; Harryson, 2008).

At the network level, there are also new managerial capabilities required, especially if the network is large and there is no "hub" firm managing it. Novel, less centralized organizational arrangements for open innovation, such as SME innovation networks, employ boards to effectively manage joint research and development activities. The network board of such SME innovation networks holds a central role as it is responsible for motivating the participants, fostering collaboration, and ensuring that R&D activities are implemented in the best interest of the network's members. Gronum et al. (2012) find a U-shaped relationship between the network board continuity, describing the rate of renewal of network board members, and the innovation performance within large networks (but not in small networks). Under certain circumstances the renewal of network board officers may be an important determinant of the innovation performance of members in the network; even though this may be at odds with a SME's preference for trusted and long-term relationships (Gronum et al., 2012). In other words, the composition and the managerial capabilities of network boards of SME network arrangements for open innovation requires careful consideration; it is worthwhile studying the managerial dimension of open innovation at the network level.

7.3 PROPOSITIONS FOR FUTURE RESEARCH

Overall, existing literature on open innovation is emerging and shows that open innovation in SMEs is a fascinating research topic. SMEs do engage in open innovation and also seem to benefit from doing so. However, existing studies often do not capture the full picture as they address only a subset of potential open innovation practices and do not provide sufficient insights into why and how SMEs make use of inflow and outflows of knowledge. Further, they only marginally address the specific nature of open innovation in SMEs.

As a consequence, there is a range of topics that either haven't been explored sufficiently or that require more detailed investigation. Out of those, we consider four topics of high priority for future research: First, the existing literature stresses the importance of inbound open innovation in SMEs to overcome their liability of smallness. However, new open innovation practices such as IT-enabled crowdsourcing for involving a large number of unknown "outsiders" have not been explored in the context of the SME sector. Second, the importance of networks—both R&D and value networks—is highlighted in existing discussions on open innovation in SMEs. Prior literature on SMEs also points to the importance of social capital and dependency of SMEs on network relationships. So far, it is not fully understood how SMEs can make use and manage different kinds of relationships and networks, also personal ones, when engaging in open innovation (Birley, 1985; Macpherson & Holt, 2007; Ceci & Iubatti, 2012). Third, the interplay of IP management and open innovation is hardly addressed in existing work. Prior studies show that the lack of appropriate IP protection mechanisms prevents SMEs from engaging in open innovation (Drechsler & Natter, 2012). While IP management is important in open innovation activities both in large firms and in SMEs, IP management in SMEs is subject to its own set of issues. Finally, existing literature clearly indicates that open innovation requires SMEs to build new internal capabilities when engaging in open innovation, particularly since open innovation is often directly linked to a strategic change in a firm's overall business model. Thus, the internal dimensions and the change process when engaging in open innovation provide an interesting avenue for future research. In the following sections, we will elaborate on these four themes in more detail and discuss the particular nature of open innovation in SMEs with regards to these four themes.

7.3.1 Theme 1: Beyond "Traditional" Search: Crowdsourcing in SMEs

Over the last years, open innovation scholars and practitioners have become increasingly interested in the potential of internet-enabled inbound open innovation practices such as online innovation contests, online idea competitions, and other practices which fall in the category of crowdsourcing or broadcast search (Lampel et al., 2012). Crowdsourcing and broadcasted search are similar concepts and describe the act of outsourcing as allocating a task in the problem solving process to a "crowd," rather than to a designated "agent" (an organization, informal or formal team, or individual), such as a contractor, in the form of an open call (Howe, 2008; Afuah & Tucci, 2012; Jeppesen & Lakhani, 2010). The diffusion of the Internet and social networking technologies has opened many opportunities for implementing these sourcing strategies which go beyond the "traditional" sourcing strategies of a firm and tap into the power of many unknown agents. Prominent case studies such as the Netflix competition, in which the company crowdsourced the task to develop an algorithm to further improve its recommender system via an open call to the world, depict how crowdsourcing enables firms to overcome the problem of "local search" and to identify solutions that are much superior to the internal ones (Afuah & Tucci, 2012; Lakhani et al., 2006). In some cases, Internet-enabled crowdsourcing may be a better search strategy than more common alternatives (such as a club of well-known partners or internal resources); in some cases it won't. Recent theoretical contributions argue that the success of crowdsourcing is influenced by various factors such as the characteristics of the problem, the knowledge required for the solution, the characteristics of the crowd, and characteristics of the solutions to be evaluated (Afuah & Tucci, 2012). Indeed, observations of existing practices indicate that not all crowdsourcing activities provide satisfying results. For example, some crowdsourcing activities lack participation. Further, practitioners proclaim that they receive only incremental solutions for the "problems" or "challenges" distributed via an open call to the crowd which raises new questions about the design and management of crowdsourcing activities from a socio-technical perspective, particularly for SMEs.

While firms may organize crowdsourcing activities on their own, they can also organize them through new Internet-enabled intermediary services for open innovation in order to overcome the difficulties in accessing the market for ideas (Hossain, 2012; Chesbrough, 2006a). Open innovation intermediaries such as Innocentive.com, yet2.com, Ninesigma, YourEncore, and IdeaConnection have received high interest in open innovation discussions (Lakhani et al., 2006; Hossain, 2012). These providers offer open innovation services to their clients—mostly large firms—and help them design and implement sourcing strategies. For example, they design the challenge, help to communicate it widely among the potential solvers, and play a brokering role in the technology transfer process. SMEs may rather rarely engage with those global service intermediaries which provide a sophisticated IT infrastructure and hold a large service portfolio due to the relatively high service fee. However, more "traditional" intermediary organizations that facilitate innovation at the regional or network level, such as Technology Transfer organizations (TTOs), can also act as broker and connector between multiple parties (Howells, 2006; Lee et al., 2010; Spithoven et al., 2010). Some of those organizations have already expanded their service portfolios and integrated Internet-enabled crowdsourcing activities in their traditional brokering activities.

Despite the interest in crowdsourcing and IT-enabled open innovation practices, existing literature on SMEs has not explored whether and when Internet-enabled crowdsourcing practices make sense for SMEs and help them to overcome the liability of smallness. As highlighted in the case of the small photonics firm Ocean Optics (see Case Example 1), they provide instant access to a greater "innovation workforce" without having to put them on the payroll. However, they also have their own unique managerial challenges, especially for SMEs, as they regularly lack the reputation and brand value of a large firm needed to attract external actors, have difficulties in articulating and defining the problem, and may lack the capabilities and resources to manage the process and evaluate ideas appropriately. In particular, it raises the question whether and when crowdsourcing is superior to alternative sourcing strategies in SMEs. How can SMEs attract external solvers and achieve high quality submissions? How should they manage the inflow of ideas and how should they deal with IP issues in crowdsourcing? How should they design and use information systems to benefit from crowdsourcing? Do IT-enabled open innovation intermediaries really work for SMEs, and if so which ones and under which conditions?

CASE EXAMPLE 1

Ocean Optics is a 25-year-old US-based photonics technology SME, with about 200 employees and more than 50 million dollars in sales.[3] They invented the first miniature spectrometer, a disruptive technology that enables field analysis of samples. Over time, Ocean Optics' R&D focus has switched from breakthrough to incremental innovations. Its intensive and long-term R&D partnerships with universities were not creating really new growth opportunities. To strengthen its R&D breakthrough capacity, CTO Jason Eichenholz implemented a significant shift in the firm's open innovation strategy. They designed a "Blue Ocean Grant" program to create breakthrough innovation opportunities. The Internet-enabled crowdsourcing program was implemented in two-stages: Phase 1 represented an open call to the "world" for submissions to provide support funding of $10,000 for about 10 teams to work on a 6-months proof-of-concept project. In phase 2, Ocean Optics planned to fund one R&D project with $100,000. The program exceeded their objectives in terms of participation and innovativeness of the results. Overall, it increased their R&D capacity by at least four times within the first year, made a positive impact on the firm's brand value, and drove the firm's strategic change. The unique IP management approach, the strong engagement with the teams throughout the program combined with a strong engagement internally, and social and personal relationships of the open innovation team externally within the photonics research community and internally with other functions were undeniable key enablers of the program. Motivated by the success, Jason Eichenholz initiated a larger open innovation membership-based initiative, called Open Photonics (www.open-photonics.com).

Source: Brunswicker (2013)

7.3.2 Theme 2: Network Relationships and Coordination

As indicated earlier, SMEs are by nature embedded in a variety of formal and informal interoganizational networks which can be a driver of innovation performance (Ceci & Iubatti, 2012; Edwards et al., 2005). Network relationships play an important role in open innovation in SMEs: SMEs can use interorganizational networks to create value in the earlier stages of the new business development process to extend their technological competences (Edwards et al., 2005; Vanhaverbeke, 2012), while SMEs can also team up with downstream partners to get access to marketing and sales channels (Lee et al., 2010). As such, network relationships serve to accommodate the two sides of the open innovation coin: SMEs collaborate with other firms to enhance their innovation processes and to create value, while they may also need external partners to capture part of that value as they often lack complementary assets to commercialize a new product or service (Gans & Stern, 2003). Open commercialization, where SMEs collaborate with other companies for the commercialization of their technologies, ideas, products, or services, is therefore an important topic. It implies value constellations (Norman & Namirez, 1993) linking firms with different assets and competencies together in response to or in anticipation of new market opportunities (Vanhaverbeke & Cloodt, 2006: 259). As such, open commercialization is central to high-tech and venture capital backed SMEs that form ties with partners to commercialize their technologies (Gans & Stern, 2003). However, open innovation researchers remain relatively quiet about the role of open commercialization in other sectors, other "subpopulations of SMEs," and more collaborative rather than purely market-driven relationships between SMEs and their partners.

Despite the abundant literature on the importance of networks for SMEs, little is known about the process that underlies the formation and coordination of these networks (Gardet & Fraiha, 2012). In particular, the management of different types of innovation networks is a crucial factor in extracting value from these networks (Nambisan & Sawhney, 2011). So, how can SMEs act as "hub" firm and "orchestrate" their innovation network relationships? Given their limited resources and the dominance of "informal and personal relationships," what role can SMEs play in the management of innovation leverage, innovation coherence, and innovation appropriability? Finally, many authors have stressed the interplay between small SMEs and large MNEs as a driver behind achieving superior innovation performance (e.g. Rothwell & Dodgson,

CASE EXAMPLE 2

Isobionics is a Dutch biotechnology company, active in the flavor and fragrance industry. The basis for their products is a technology developed by DSM, a large Dutch company active in the Life Sciences and Performance Materials industry. Using this technology, it is possible to produce flavors and fragrances synthesized by micro-organisms, which is considerably cheaper than using traditional methods. DSM, however, had decided not to pursue this technology, but instead was open to external partners to commercialize the technology. In 2008, Isobionics was established and in 2010 the first product, BioValencene, was introduced in the market. In developing this product, Isobionics worked closely together with DSM and other innovation partners. The collaboration between Isobionics and DSM is a good example of a development and commercializing partnership, in which a start-up works together with a large, established firm. Through this collaboration, Isobionics is able to build a successful company through the licensing of this product and tapping into the resources of DSM, while DSM benefits not only from the licensing revenues, but also has first-hand access to potential new discoveries and applications of the technology.

Source: Vanhaverbeke (2012)

1991). Also in the commercialization phase, large firms play a prominent role. However, we know less about the role of specific characteristics of these large MNEs. For example, how important are market power, prominence, reputation, etc. in the innovation and commercialization phase of these collaborative efforts? And what is the key to successful collaboration between the two? Evidently, small and large firms are different in the way they operate and in their motivations for the collaboration (e.g. Blomqvist et al., 2005). Therefore, as highlighted in our case on the Dutch SME Isobionics (see Case Example 2) these asymmetric partnerships are challenging for both parties and require both contracts and trust to foster successful collaboration. This raises an additional question: what role can SMEs play in the successful attraction of and collaboration with, larger multinational partners?

7.3.3 Theme 3: Open Innovation and Intellectual Property (IP) Management in SMEs

Intellectual property (IP) protection is of major concern for firms engaging in open innovation. Common methods for IP protection include formal mechanisms such as patents and trademarks, as well as more informal mechanisms such as trade secrets and lead-time. However, while large firms may have many possibilities for both formal and informal means of IP protection, SMEs appear to be more limited in this respect.

Prior studies have indicated that most SMEs find patents to be less efficient than more informal IP protection mechanisms (e.g. Kitching & Blackburn, 1998). Obtaining and maintaining a patent is usually a costly and complex process, with many regulations and procedures. Moreover, given the limited power and resources of SMEs, the costs associated with patent enforcement make patents less attractive for small firms (Lanjouw & Schankerman, 2004; Penin, 2005). Hence, it is no surprise that small firms tend to prefer more informal IP protection mechanisms, such as speed to market or secrecy (Arundel, 2001; Leiponen & Byma, 2009).

Nevertheless, formal means of IP protection play an important role in open innovation and may actually facilitate knowledge flows in open innovation. For example, if knowledge is protected by means of a patent, the transfer of the underlying knowledge becomes much easier as patents help to define the intellectual property rights explicitly (Alexy et al., 2009; Leiponen & Byma, 2009). As such, patents enable technology and IP trading and help to modularize knowledge. Firms engaging in cooperative R&D are therefore also more likely to favour patents over secrecy (Arundel, 2001). In addition, formal IP protection may also serve as a signalling device, demonstrating their technological capability. Particularly for small, start-up firms, having a patent is almost a prerequisite to receive any kind of VC funding or for larger firms to be willing to cooperate (Gans & Stern, 2003). In this case, formal IP protection not only facilitates knowledge sharing, it may actually be a precondition to engage in open innovation and can form the basis of negotiations with VCs and potential alliance partners (Alexy et al., 2009). Particularly in horizontal or vertical cooperation efforts, informal means such as secrecy appear to be not very effective, mostly because secrecy is more difficult to maintain in joint projects (Leiponen & Byma, 2009).

However, despite the importance of formal IP protection in open innovation, some authors argue that free revealing enables small firms to overcome the liability of smallness (Gruber & Henkel, 2006). Free revealing relies on the idea that firms may benefit from selectively revealing some part of their intellectual assets for free usage by others (Harhoff et al., 2003). This is a common practice in open source software development, but has also been adopted by large organizations in other sectors. For example, pharmaceutical companies such as Novartis and GlaxoSmithKline freely reveal some of their patents in an effort to embrace the larger research community to further understand particular diseases, such as diabetes and tropical diseases. It would be interesting to study whether selective free revealing offers specific benefits to SMEs, as it may reduce entry barriers and sunk costs for SMEs.

Consequently, there is no one-size-fits-all approach towards IP management in open innovation. Rather, open innovation requires effective IP management where different strategies may prevail in different situations. However, the conditions that favor particular types of knowledge protection mechanisms differ substantially among industries and types of firms. In particular, firms active in the service industry are much more likely to rely on speed to market, while SMEs from R&D intensive sectors more regularly engage in patenting. Thus, it is important to recognize that the role of IP management in open innovation is dependent on contingency factors such as the technological environment and the knowledge distribution therein (Alexy et al., 2009). According to these authors, patents are particularly useful in calm environments where knowledge resides with a few players, whereas turbulent environments with distributed knowledge may benefit more from free revealing.

To conclude, IP management plays an important role for SMEs engaged in open innovation, not in the first place to avoid unintended knowledge spillovers, but much more as an accelerator and facilitator of knowledge exchange and partnership formation. However, prior research has been inconclusive on many aspects of the interface between legal IP management and open innovation. Given the preference of informal IP management in SMEs, what is the role of informal means of IP management in open innovation in SMEs? Moreover, do formal and informal means of IP protection perhaps play different roles for different open innovation practices and in relation to different sources of external knowledge? And if so, when is legal IP protection hindering or enabling value creation and capture? In addition, legal IP protection, secrecy, and free revealing are not mutually exclusive (Arundel, 2001). The case of Collabra Software Inc. depicts that in some cases it may be beneficial to selectively reveal IP (see Case Example 3). This raises the question as to what are appropriate IP management practices throughout the innovation process? And how do these affect the open innovation practices? In other words, do SMEs indeed rely on secrecy in the early stages of the R&D process and how does that affect their willingness to engage in open innovation in that stage? Or are there alternative informal mechanisms of appropriation, such as trust and social norms, which allow IP protection and utilization at the interorganizational level even without patenting? Finally, although most of the IP protection debate has focused on manufacturing firms, many SMEs are in fact service firms. Future studies may investigate the specifics of IP management in services SMEs. IP management in open innovation in SMEs is a great area for research with manifold and exciting research questions that haven't been explored yet.

CASE EXAMPLE 3

Collabra Software Inc. was a US based software company, founded in 1993 and acquired by Netscape in 1995. Collabra's main product, CollabraShare, allowed users to organize and share information in electronic forums, and thereby to jointly create and edit documents. During the first year after the company was formally founded, Collabra primarily used NDAs as a way to protect its intellectual property. The advantage of doing so is that it was able to operate low profile until the release of its first product. Moreover, while Collabra was particularly open to its customers and third-party developers, it was much more reluctant to share information with analysts and competitors. Having customer to buy the product and third party developers that created products that were complementary to Collabra's offerings is of course essential to enhance to value of the company. On the other hand, sharing information with Lotus Notes (their main competitor) and Microsoft was less straightforward as they didn't want to reveal their product just yet. However, in order to compete successfully with Lotus Notes, Collabra needed an ally and therefore decided to approach Microsoft, again using an NDA to protect its IP. Collaborating with Microsoft allowed Collabra to gain marketing presence and publicity, and three months later they introduced their first product, CollabraShare to the market. In the case of Collabra, they decided to selectively reveal their knowledge to attract partners that were needed to successfully compete against Lotus Notes.

Source: Chesbrough (2006a)

7.3.4 Theme 4: The Internal Dimension: Managing Open Innovation in SMEs

Open innovation poses new managerial challenges—not just for large firms but also for SMEs. Indeed, firms that shift from closed towards open innovation experience several difficulties in actively managing the open innovation processes (Lichtenthaler, 2011). Today many firms, including SMEs, still rely more on a trial and error process rather than established organizational practices to manage open innovation internally (Gassmann et al., 2010). Both scholars and practitioners agree that open innovation requires new internal capabilities (Chiaroni et al., 2011; Laursen & Salter, 2006; Spithoven et al., 2010). Existing discussions highlight two facets of such internal organizational capabilities. On the one hand, it is crucial to understand the internal organizational practices, systems and routines for managing open innovation and related knowledge flows in SMEs. On the other hand, the transition from closed towards open innovation implies some kind of organizational change which usually spans different phases (Chiaroni et al., 2011; Teece et al., 1997). It is also important to understand how SMEs can manage the transition from closed towards open innovation, which we assume is quite different from that of large firms. As we discussed above, open innovation in SMEs is regularly directly linked to the business model and implies a strategy change.

The first perspective links back to the seminal work of Cohen and Levinthal (1990) on absorptive capacity. Firms require the ability to absorb external knowledge in order to benefit from it (Cohen & Levinthal, 1990). Absorptive capacity is a pre-requisite for inbound open innovation and is built through formal R&D. In line with this argument a range of studies on inbound open innovation, and especially on sourcing of external knowledge, indicate that inbound open innovation activities do not substitute internal R&D; rather they are complementary (Dahlander & Gann, 2010; Laursen & Salter, 2006); In SMEs, R&D is usually not a formal process; however, formal R&D seems to impact the SME's ability to profit from open innovation (Huang & Rice, 2009; Spithoven et al., 2013). Given their limited resources, SMEs may also call upon third parties to support them in building absorptive capacity, such as collective research centres (Spithoven et al., 2010).

Even though absorptive capacity is important for open innovation, it concentrates on using external knowledge internally only and neglects other important organizational capabilities which are required in open innovation; neither does it address all dimensions of managing knowledge flows in open innovation, nor does it acknowledge the distributed character of knowledge in open innovation. For example, absorptive capacity does not capture the specifics of outbound open innovation. It also does not address the question of how to apply innovative knowledge and means to turn it into successful outcomes (Bianchi et al., 2010; Robertson et al., 2012). Recent theoretical contributions propose additional capacities (groups of capabilities) for managing different knowledge processes in open innovation, which complement the construct of absorptive capacity (Lichtenthaler & Lichtenthaler, 2009; Robertson et al., 2012). While there are new knowledge capacities required for managing the acquisition and retention of knowledge at the intrafirm and interfirm level, open innovation also implies new capacities for applying knowledge, and turning external and internal knowledge into successful outcomes. Examples of such knowledge capacities for managing open innovation are accessive, adaptive, and integrative capacities (Robertson et al., 2012). Further, these knowledge capacities do not function "automatically" and therefore firms need some sort of a higher order capacity to guide these capacities; thus, *knowledge management* and *innovation management* capacities represent relevant "facilitators" for open innovation in SMEs; however, they are regularly lacking in SMEs (Brunswicker & Vanhaverbeke, 2010; Robertson et al., 2012; van der Meer, 2007).

To establish such knowledge and managerial capacities, SMEs require new systems, processes, and routines (Huizingh, 2011; Pavitt, 2002). For example, SMEs may establish new processes for inbound open innovation to manage knowledge integration as proposed by Wallin and von Krogh (2010) in their five-step model, namely (1) define the innovation process, (2) identify innovation-relevant knowledge, (3) select an appropriate integration mechanism, (4) create effective governance mechanisms, and (5) balance incentives and controls (Wallin & von Krogh, 2010).

However, innovation is organizationally pervasive and therefore the required innovation management capacity relates to different managerial levels such as the strategic, the operational, the cultural and the network level (Adams et al., 2006; Müller-Stewens & Lechner, 2005). Particularly the network level is an important one in open innovation in SMEs (Gardet & Fraiha, 2012; Vanhaverbeke, 2012).

In addition, capacities are usually perceived as organizational and collective-level constructs. To shed light on those, it is inevitable to study their microfoundations and the underlying intentional actions, experiences and preferences of *individuals* (Felin & Foss, 2009). Open innovation may require new individual skills and may be supported by preferences and interests of different individuals and groups within and outside the organization.

Given the importance of internal knowledge and managerial capacities in open innovation, a range of questions emerge related to the SME sector. Besides absorptive capacity, what organizational capacities—including knowledge and managerial capacities—are required in open innovation? What kinds of capacities are required for open innovation in low-tech and process innovation in SMEs? What managerial systems, processes and routines support these capacities? Given the resource constraints of SMEs, how can SMEs embed organizational routines and practices for open innovation within their existing organization routines? What are the microfoundations of such organizational capabilities and what individual skills and interactions build required capacities?

The second perspective of managing open innovation in SMEs is about the transition from closed towards open innovation over time. As highlighted in prominent case studies on large firms, such as the case study on Procter & Gamble, this transition implies significant organizational change and transformation (Huston & Sakkab, 2006; Dodgson et al., 2006). Regularly, a first open innovation project triggers a more fundamental and strategic change (Gassmann et al., 2010). Chiaroni et al. (2011) describe the change process from closed towards open innovation, highlighting the important role of the top management in enabling the change and the need for a champion promoting the change along different managerial levels. Further, they show that in large firms the starting point of the transition is a change at the organizational structure level. The establishment of a new independent open innovation unit (or role) represents an important trigger for change and sends signals to other organisational units (Chiaroni et al., 2011). In SMEs there might be different triggers. For example, in the small software firm CAS, a strategic engagement

---

CASE EXAMPLE 4

CAS is the market-leader in the field of customer relations management (CRM) software for SMEs in Germany. The company was founded in 1986 and employs approximately 430 people today. In recent years it has won numerous innovation awards and has shown a constant and double-digit growth in sales over the last years. Recently, the company has successfully transformed itself from a closed towards an open innovator. Along this journey it went through five different phases of change in open innovation: In the first phase they established R&D partnerships with universities and research organizations. In the second phase, management started to strengthen collaboration with other partners including firms offering complementary products and services, sales partners, and also competitors. The third phase was about formalization of routines and managerial systems supporting open innovation. They established structures for managing open innovation and related knowledge flows including new roles and functions such as "network promoters," and implemented an information system structure for open innovation. Today, CAS can be described as a "platform player" which builds upon a modular organizational design to best align its business model with the business model of their innovation and value chain partners. The case highlights the role of innovation management capacity in integrating different knowledge flows in open innovation and turning them into a successful outcome. It also underlines that the transition is directly linked to the changes in the business model.

Source: Brunswicker and Ehrenmann (2013)

in R&D partnerships with universities represented a starting point of the transformation which was linked to a business model change (see Case Example 4). This provokes research questions for studying SMEs in particular: What triggers the change towards open innovation in SMEs? What managerial levels are affected by that change and in which sequence? What internal factors enable the various change phases and what is the role of entrepreneurship in enabling change? And how are the change and transformation activities different from those in large firms? How does the change process interplay with the changes of the SME's business model?

7.4 CONCLUDING REMARKS

Open innovation in SMEs is an exciting and promising area of research. The sector is of high economic relevance and SMEs increasingly adopt different types of open innovation practices. However, utilizing and managing open innovation in SMEs is quite specific in nature. For example, in open innovation in SMEs different kinds of network relationships play an important role. Thus, SMEs require capabilities for managing these network relationships, which they regularly don't have and which are different from those of large firms. However, this is just one facet of the specific nature of open innovation in SMEs.

Today, we have only marginal insights into the specific nature of open innovation in SMEs and there are manifold research questions to be explored. We proposed four research topics with manifold research questions to stimulate future open innovation research. At this stage it is too early to draw any conclusions related to potential outcomes of research in these four research areas. Further, we don't proclaim that they address all relevant research topics and questions on open innovation in SMEs. For example, we did only marginally touch upon specifics of open innovation in subpopulations of the SME sector such as SMEs in the services sector, venture capital backed growth-oriented start-ups, SMEs in the emerging markets, or export-oriented SMEs (Wynarczyk et al., 2013). There are great opportunities for exploring the specifics of open innovation in these subpopulations. Thus, we invite future research to build upon and extend our proposed research questions within our four research topics, or even propose an additional one.

We would like to see both theoretical and empirical research to address open innovation in SMEs. Empirical research, in particular, will help us to gain a deeper understanding of open innovation in SMEs. However, implementing empirical research comes with significant challenges, especially quantitative ones. Firm-level data on SMEs are difficult to access, and survey-based research methods are not easily implemented. Further, cross-sectional data will not allow researchers to answer questions related to the change and transition processes within open innovation. For some questions, only longitudinal analyses of SME case studies may allow a "deep dive" into the specific nature of open innovation in SMEs. Overall, we are convinced that research on open innovation in SME will benefit open innovation scholars as well as researchers from adjacent domains such as entrepreneurship, small business, and innovation policy research.

NOTES

1. Referring to the official definition of SMEs laid down in the European Commission Recommendations 2003/361/EC, they employ fewer than 250 employees. In addition to the headcount ceiling, an enterprise "officially" qualifies as SME if it meets either the turnover ceiling of less than € 50 million or the annual balance sheet ceiling € 43 million but not necessarily both (European Commission, 2003).
2. We performed a literature review by searching scientific database on publications relevant to this chapter. Further, we also added working papers, book chapters, and reports which considered as relevant
3. http://www.tampabay.com/news/business/workinglife/article1163254.ece

8

Open Innovation in Multinational Corporations

New Insights from the Global R&D Research Stream*

Kazuhiro Asakawa, Jaeyong Song, and Sang-Ji Kim

8.1 INTRODUCTION

Rapid technological innovation in today's business world has made it nearly impossible for any firm to sustain its technological supremacy without utilizing external knowledge and technologies. Examples abound. Procter & Gamble (P&G) has adopted a policy of "Connect & Develop" to emphasize the importance of sourcing external knowledge and ideas to achieve innovation (Dodgson, Gann, & Salter, 2006). The "Osaka Connection" is a famous example of P&G's successful open innovation strategy. P&G's Japanese technology broker (open network) discovered a sponge with the property of removing any dirt only with water, Basotect, in Osaka and connected it to P&G. In two years, P&G succeeded in the co-development and introduction of "Mr. Clean Magic Eraser" with BASF (German chemical firm).

As another example of a notable success of open innovation in an international context, Samsung Electronics of Korea developed its blockbuster Galaxy Note, a smartphone/tablet computer hybrid, in collaboration with Wacom, Japan. The device was innovative enough to be called "phablet" by Forbes and has been sold more than 10 million times in less than a year. One of the most distinctive features of the Galaxy Note was its stylus, which Samsung calls "S Pen." The stylus can be used in a variety of applications, including functions such as writing sticky notes with drawing/handwriting, text input, and pictures. Wacom had a world-class stylus pen technology and digitizer system that results in accurate, pressure-sensitive input. Samsung and Wacom collaborated to modify Wacom's existing technology to be more appropriate for smartphones/tablet computers.

IBM adopted its open innovation approach by managing its overseas R&D centers as "collaboratories," i.e., laboratories run in the form of collaboration through which the firm can efficiently source key external knowledge from external research organizations, such as universities, research institutions, and venture firms (Hamm, 2009).

Air Products and Chemicals (APD), a U.S. gas and speciality chemical firm, emphasizes "Identify & Accelerate" and has actively made alliances with R&D ventures, universities, and laboratories in emerging economies such as Russia, China, and India since the early 1990s to enjoy the low labor costs of R&D human resources in those emerging countries.

These are just a few examples of open collaborations around the globe. An overall trend is that firms put more emphasis on the external sourcing of technologies (Rigby & Zook, 2002). Open innovation enables firms to access new external knowledge efficiently (Chesbrough, 2003a), thus allowing them to maintain operational efficiency and flexibility (Pisano, 1990).

Despite the abundant examples of open innovation by multinationals (MNCs) on a global scale, open innovation literature only pays scant attention to the international dimension, revealing a remarkable research gap. In contrast, the recent development of global R&D research manifests a growing interest in and attention to external knowledge sourcing and leveraging (Kuemmerle, 1997; Doz, Santos & Williamson, 2001; Frost, 2001; Asakawa, 2001; Asakawa & Lehrer, 2003; Schlegelmilch, Ambos & Chini, 2003; Song & Shin, 2008; Frost & Zhou, 2005; Song, Asakawa & Chu, 2011).

Obviously it is important to note that the geographic dimension of open innovation is not totally neglected in the open innovation research (Simard & West, 2006). Chesbrough (2003) specified the "erosion factors" that undermined the logic of closed innovation in the late twentieth century. The factors that account for the emergence of open innovation include the mobility of human resources, the quality of university research, the presence or absence of venture capital, and the strength of IP protection. Apparently, they substantially vary by geography. Dealing with the geographic dimension of the erosion factors, we could deepen our understanding of open innovation which is now globally emerging.

For example, a geographic factor such as the presence of industry or technology clusters can influence the mobility of qualified and talented human resources. There is frequent inter-organizational mobility of workforces within industry or technology clusters (Casper, 2007). When the mobility rate is high, knowledge within a company can be easily spilled out to other companies (Song, Almeida, & Wu, 2003). A higher mobility rate of labor market promotes open innovation.

In the case of labor mobility across borders, there is high mobility between countries that have developed similar or related industries. Talented human resources also flow into countries where companies offer competitive salaries, excellent benefits, and better work environments. Although labor mobility causes knowledge transfer or spill-over, the personal networks of the workers in a former organization can serve as collaboration linkages for open innovation. By observing the direction of labor mobility, we can explain locational choice of global open innovation or open innovation of certain emerging countries.

Open innovation research will be rich with international geographic dimensions to which the global R&D research stream has paid attention. Meanwhile, global R&D research stream would also benefit by dealing with geographically varying erosion factors. The focus of global R&D shifted from in-house to open. It is clear that erosion factors played roles in the shift. Delving into erosion factors and their geographic differences will help to study the new pattern of global R&D.

The purpose of this chapter is therefore to elucidate how global R&D literature and open innovation literature can complement each other. The chapter is organized as follows. After reviewing the characteristics of open innovation research and global R&D research, potential contributions by global R&D research to open innovation research are discussed, followed by the potential contribution by open innovation research to global R&D research. We suggest some promising areas for future research that may facilitate the cross-fertilization of these two research areas.

8.2 COMMON CHARACTERISTICS OF OPEN INNOVATION RESEARCH

Open innovation has some distinctive characteristics that differentiate it from the overall innovation activities featured in the innovation literature (Chesbrough, Vanhaverbeke, & West, 2006). Several characteristics of open innovation are summarized in Table 8.1 and discussed below.

First, open innovation theory captures multi-directional flows of knowledge: outside-in, inside-out (Gassmann & Enkel, 2004), and coupled process (Enkel et al., 2009) albeit with insufficient attention to the geographic dimension of knowledge flows. Firms with superior scientific discoveries can benefit from monetizing their scientific knowledge for commercial purposes, i.e., inside-out process to host countries in which the scientific knowledge is appreciated. In contrast, firms with limited advanced technologies can benefit from sourcing key technological knowledge from the outside, i.e., outside-in process.

Second, the open innovation literature widely covers partnering with various external parties. These partners include university and research institutes (Fabrizio, 2006; Enkel & Gassmann, 2008; Asakawa, Nakamura, & Sawada, 2010), venture firms, suppliers (Un, Cuervo-Cazzura & Asakawa,

2010; Takeishi, 2001; Dyer & Nobeoka, 2000), customers or clients (Enkel & Gassmann, 2008), lead users (von Hippel, 1988; Prahalad & Ramaswamy, 2004b), and competitors (Hamel et al., 1989; Brandenburger & Nalebuff, 1996), which constitute a common set of partners of open innovation.

Third, various forms of open connectedness are examined in the open innovation literature. These forms range from equity-based alliances, M&A, and contract research to more informal joint research. Other forms of open collaboration include cross-licensing, corporate venture capital investments, and collaboration with lead users (Vanhaverbeke, Du, and von Zedtwitz, 2013).

Fourth, open innovation literature puts an emphasis on the importance of absorptive capacity within a firm for engaging in open innovation. The importance of absorptive capacity for engaging in open innovation can be validated by abundant examples of successful firms engaging in open innovation that have already accumulated sufficient firm-specific capabilities necessary for searching, acquiring, integrating, and leveraging such new knowledge (Lichtenthaler & Lichtenthaler, 2009; Laursen & Salter, 2004), as represented by IBM and P&G.

Fifth, open innovation literature clearly underlines business model innovation which is significant for sustaining open innovation (Chesbrough & Schwartz, 2007). Business model enables a firm to create value from its internal and external knowledge and to capture a portion of the value by definition (Chesbrough, 2007a). There is no economic value in a technology unless it is commercialized and makes profits via a business model. Moreover, different business models will yield different returns with the same technology. Business model innovation is a process for a firm to find more appropriate business models which enable it to make more profits or create new value from a technology. To improve or renew the existing business model, a firm needs to go through extensive business model experimentations (Chesbrough, 2010; West & Gallagher, 2006). Furthermore, successful business model innovation calls for organizational leadership necessary for overcoming organizational barriers to changes which a firm faces during the process of business model experimentation (Amit & Zott, 2001; Chesbrough, 2007a; Chesbrough, 2010).

In spite of such diverse issues that have been covered in the open innovation literature as surveyed above, little attention has been paid to the international geographic dimension of open innovation. Open innovation research should benefit from considering the geographic aspects of global innovation, that affect the effective accessing, integrating, and leveraging of external knowledge. For example, partnering with nearby suppliers differs from that with distant suppliers in terms of the challenges and opportunities related to open innovation. Sourcing local knowledge requires absorptive capacity both at the headquarters and the subsidiaries, depending on where the knowledge sourcing takes place. We argue that the global R&D literature can complement open innovation theory by filling such gaps in the literature.

Table 8.1 Common characteristics of open innovation research

Wider Scope of Innovation	Success Factors of Open Innovation
• Multi-directional flows of knowledge • Various external partners for open innovation • Various forms of open connectedness	• Absorptive capacity • Business model innovation
Lack of attention to international geographic dimension	

8.3 WHAT GLOBAL R&D RESEARCH CAN CONTRIBUTE TO THE OPEN INNOVATION RESEARCH

Global R&D literature has evolved in the past several decades to accommodate changes in the directions of open innovation and emerging country innovation. Global R&D theory is particularly useful for the open innovation literature in that the former can provide a much more fine-grained approach to the geographic aspects of open innovation. Here we propose multiple areas in which global R&D research could advance our knowledge on open innovation.

8.3.1 Locational Decisions Matter for Open Innovation

Global R&D theory can offer detailed explanations regarding where open innovation takes place. Global R&D activities are located in countries with a strong research base in relevant technology fields, reflecting technology-sourcing motives (Kuemmerle, 1999; Belderbos et al., 2006). Particular attention has been paid to the merit of agglomeration, which influences locational decisions regarding R&D sites. Agglomeration economies attract firms in similar industries to form an R&D cluster to enjoy the benefits of externalities (Krugman, 1991; Belderbos & Carree, 2002).

Thus, locational decisions regarding R&D investments have been a central issue for MNCs intending to relocate their R&D abroad. The motivation behind internationalizing R&D can be classified into a market-seeking vs. a technology-seeking logic, which corresponds to the two types of overseas R&D laboratories: home-base-exploiting (HBE) and home-base-augmenting (HBA) as proposed by Kuemmerle (1997) or competence-leveraging and competence-creating as proposed by Cantwell and Mudambi (2005).

Doz et al. (2001) argue that knowledge is increasingly dispersed across the globe in a somewhat unexpected way: potentially valuable knowledge lies in peripheral locations that were often ignored by decision-makers within a firm.

Thus, global R&D literature can bring the following two perspectives to the open innovation theory by articulating the importance of the geographical location: whether knowledge is sourced from inside the innovation cluster or not, and the geographical/cultural distance (Ghemawat, 2001) that determines the effectiveness of external knowledge sourcing.

Technological distance across locations plays an important role in determining the locus of innovation. Song and Shin (2008) find that regarding home-base-augmenting R&D, an MNC tends to source knowledge from a host country where technological capabilities measured by USPTO patent filings are higher relative to those of its home country. Such a finding suggests that an MNC is sourcing overseas R&D in a host country where the technological capability is high compared to the MNC's home country, which motivates a firm to learn from the host country environment. Iwasa and Odagiri (2004) also show the importance of the technological capability of a host country for enhancing the R&D performance, although they have not explicitly captured technological distance in their study.

At the same time, a persistent trend of non-globalization of R&D can be observed (Patel & Pavitt, 1991), especially regarding core knowledge and technologies that remain susceptible to the erosion of intellectual property rights, as examined in the context of wireless telecom (Di Minin & Bianchi, 2011). Thus, locational decisions need to be comprehensive, such that both globalization and non-globalization (thus centralization) of R&D remain valid options.

8.3.2 Extending the Levels of Analysis for Global Open Innovation

Global R&D literature can also be helpful for open innovation researchers to extend the level of analysis. In fact, global R&D research has been conducted at multiple levels of analysis, ranging from a national to an individual human level. At the national level, the sourcing pattern of overseas R&D knowledge depends on the relative competitiveness of the knowledge source and recipient countries (Song & Shin, 2008). Open innovation research can incorporate such a macro-level influence of national competitiveness into its research domain. At a sub-unit level (i.e., R&D subsidiary level), global R&D research has investigated the effect of subsidiary initiatives, subsidiary capabilities, the subsidiary's host country environment as well as the HQ influence on the role and mission of a subsidiary (Nobel & Birkinshaw, 1998; Cantwell & Mudambi, 2005). Global R&D literature explicitly investigates open innovation at the overseas subsidiary level where each subsidiary is assigned a different role by the headquarters and engages in locally-specific innovative activities (Kuemmerle, 1997; Nobel & Birkinshaw, 1998; Cantwell & Mudambi, 2005). Global R&D research is also conducted at the team level (Ambos & Schlegelmich, 2004).

Research at the individual level has shed light on communication patterns among researchers across geographic distance, most typically between the HQ and foreign R&D subsidiaries, or among foreign subsidiaries (De Meyer, 1991). The research also captured the human resource dimension of global R&D (De Meyer & Mizushima, 1989; Cheng & Bolon, 1993) and includes issues such as recruiting and retaining talented foreign researchers abroad, training, transfers to the HQ, and incentives such as compensation.

While shedding light on the role of overseas R&D subsidiaries for open innovation, global R&D literature tends to be sensitive to other levels of analysis. For example, extant literature examined the associations between subsidiary-level R&D, national competitiveness and public policy (Lehrer, Asakawa, & Behnam, 2011). Some studies (Asakawa, 2004; Lehrer, Asakawa, & Behnam, 2011) featured home-base-compensating R&D as a situation in which MNCs compensate for comparative weaknesses in their home-country R&D by locating core R&D activities in foreign countries with a stronger R&D base.

8.3.3 Attention to Capability for Overseas Knowledge Sourcing

Global R&D scholars indicate positive effects of technological capabilities on the performance of overseas R&D and the knowledge sourcing from foreign locations. Iwasa and Odagiri (2004) indicate the importance of R&D investments for the performance of overseas R&D by showing the contribution of overseas locations' technological capabilities to the performance of R&D laboratories. Penner-Hahn and Shaver (2005) indicate the importance of the technological capability of MNCs in increasing the performance of international R&D, even though the locus of technological capability within MNCs (i.e., whether at the local laboratory or at the firm level) was beyond the focus of their research. Song and Shin (2008) went a step further by showing the relative impact of the parent company's and the local/home country's technological capabilities on overseas knowledge sourcing, among other factors. The authors propose a capability-motivation paradox in that MNCs with strong home-base technological capabilities could source overseas knowledge; however, such MNCs often have limited motivation to do so (Song & Shin, 2008). Their study indicates the importance of absorptive capacity but also considers enhanced motivation on the part of the parent (i.e., knowledge-sourcing unit) to source local knowledge as a sufficient condition. To extend this work, Song, Asakawa and Chu (2011) investigated what factors determine knowledge sourcing from host locations to overseas R&D laboratories and identified the existence of an optimal level of absorptive capacity of local laboratories for facilitating local knowledge sourcing. Knowledge sourcing from the host country environment increases as the capability of a local R&D laboratory improves up to a certain level, beyond which it starts to diminish, largely because the local laboratory starts to shift its role from a local to a global innovator. A global-innovator laboratory can be considered as sourcing knowledge from all over the world; thus, the relative importance of the local host country environment would plummet (Song, Asakawa & Chu, 2011).

8.3.4 Managing Internal and External Networks

Global R&D research includes extensive literature on the management of global R&D networks, both external and internal to the firm. Regarding the management of *internal* networks, MNCs comprise multiple subunits such as headquarters and foreign subsidiaries. Engaging in open innovation on a global scale requires managerial skills to coordinate these subunits, in addition to the various external partners for R&D collaborations (Lehrer & Asakawa, 2003). Global R&D literature captures management challenges pertinent to such complex network management. As Cantwell and Mudambi (2005) indicated, managing complex intra-firm networks provides MNCs with cross-border learning opportunities. It requires managerial efforts to overcome various managerial difficulties such as the tension between headquarters and R&D subsidiaries (Asakawa, 2001; Birkinshaw & Hood, 1998; Florida & Kenney, 1990). Global R&D literature has long examined the optimal level of autonomy and control (Behrman & Fisher, 1980) as well as external and internal connectivity. While such managerial tension is related to open innovation in general, overseas R&D management involves much more tension and uncertainty (Asakawa, 2001).

Regarding the management of *external* networks, the extent of sharing and disclosing the firm's core proprietary knowledge with external parties becomes an issue. This issue is delicate, because cross-border collaborations, especially with global competitors, entail a much higher level of uncertainty than with their domestic counterparts. IBM disclosed its core technology such as the program source code and a circuit diagram, its core technologies of a computer design. IBM ended up exiting the PC market due to the open strategy, surrendering the leadership to Microsoft and Intel. In contrast, Apple only opened peripheral technology and outsourced manufacturing tasks to Foxconn. Apple also opened the right to develop applications for Apple products; however, the distribution of the applications is restricted to Apple's App store. Global R&D management literature captures a persistent trend of non-globalization of R&D (Patel & Pavitt, 1991). DiMinin and Bianchi (2011) by studying the case of wireless telecom show how MNCs attempt to avoid the erosion of intellectual property rights in the area of their core knowledge and technology.

Sharing core knowledge with local external partners requires nurturing trust to minimize the risk of opportunistic behavior on both sides. The assumption is that embeddedness in a local environment should generate social capital, which should enhance the mutual trust necessary for partners to exchange proprietary knowledge (Inkpen & Tsang, 2005; Granovetter, 1985; Uzzi, 1996). Local embeddedness is found to foster subsidiary innovation (Andersson, Forsgren, & Holm, 2002) and to source valuable locally specific knowledge (Almeida & Phene, 2004; Lehrer & Asakawa, 2002; Hakanson & Nobel, 2001; Song, Asakawa & Chu, 2011).

Open innovation research can benefit from insights into the global R&D context, which stress the importance of high-quality network management, whether internal or external. Because growing numbers of open innovations cross national borders, global R&D literature provides valid suggestions.

8.3.5 Emerging Countries Innovation

Recent global R&D research indicates an explicit interest in emerging countries, especially China and India, as local sites (Asakawa & Som, 2008). Global R&D scholars focused their research on identifying the extent to which location matters as a source of innovation. To what extent do firms achieve innovation in emerging countries compared to their innovations in the developed countries? Until recently, global R&D literature treated emerging countries primarily as a market for existing products developed and launched in developed countries. Recent improvements in technological competencies in developing countries such as India and China have changed this picture, and the main sources of technologies and knowledge lie now within emerging countries.

While the fundamental principle of open innovation may remain the same, where open innovation takes place could change the whole picture. In terms of inbound open innovation, MNCs typically source external resources that are unique to the particular location. MNCs typically source state-of-the-art knowledge from the most-advanced innovation clusters in the developed countries while sourcing resources from developing countries at lower cost. Adding the geographical dimension is crucial for a better understanding of open innovation.

Global R&D literature has also begun to highlight MNCs' innovation patterns that originated in emerging countries, which imply an alternative model of global innovation to the existing model of firms in developed countries. Most of the MNCs from developing countries neither have a sufficient level of home-country advantages nor the ownership advantage of the headquarters. For example, Mathews (2002) presented his model of dragon multinationals to illustrate how they globalize without a home-country advantage. Firms in emerging countries manifest a "scaling out" pattern of global innovation (*The*

Economist, 2010) in that they resort to open sourcing of external resources due to a lack of internal resources, whereas MNCs in developed countries can tap into their home-country resources.

Recent attention to the reverse innovation phenomenon as represented by the case of GE's low-price ultrasound scanner (Immelt, Govindarajan, & Trimble, 2009) can be explained by the global R&D theory. From the standpoint of global R&D, reverse innovation can be considered as a case of transition from local-for-local innovation to local-for-global innovation (Bartlett & Ghoshal, 1990). Reverse innovation can also be considered as a derivative of a home-base-augmenting (HBA) type of laboratory (Kuemmerle, 1997) in that low-cost, disruptive products that initially had been anticipated to fit only the emerging country context turned out to be appreciated even in the home-country markets.

Open innovation patterns should vary depending on the nature of the home country. Many MNCs from developed countries can actively engage in "inbound" open innovation based on their already existing absorptive capacity within the firm. These MNCs are also ready to engage in "outbound" open innovation if they have high levels of technologies within their firm, if they are prepared for this engagement, and if they find opportunities to monetize their own technologies. In contrast, many MNCs from developing countries are not ready for either inbound or outbound open innovation concerning advanced technology due to a lack of absorptive capacity within their firm; thus, sourcing high-standard knowledge becomes more difficult. Instead, these MNCs are more likely to engage in open innovation to lower innovation costs by recruiting low-cost engineers locally or procuring low-cost raw materials. MNCs from developing countries in general have more difficulty to engage in outbound open innovation due to the insufficient level of competencies within their firm.

While open innovation literature does not sufficiently examine the geographic dimension of knowledge sourcing, global R&D literature brings in fine-grained aspects of locational effects on knowledge sourcing.

8.4 WHAT OPEN INNOVATION RESEARCH CAN CONTRIBUTE TO THE GLOBAL R&D RESEARCH

Obviously, global R&D research has limitations. Global R&D research covers open innovation in a global context but mostly related to inbound open innovation with little attention to the outbound type of innovation (Vanhaverbeke, Du, & von Zedtwitz, 2013). In that respect, open innovation literature can complement global R&D literature by providing theoretical frameworks related to outbound open innovation—see also Table 8.2. The reality shows an increasing degree of outbound open innovation taking place across national borders; therefore, it makes sense for global R&D researchers to adopt an outbound open innovation framework.

Global R&D literature also captures the effect of R&D on product innovation, implying that the literature covers multiple stages of the value chain beyond the narrowly defined R&D function. However, other functions such as marketing and sales are obviously beyond its primary focus. In this respect, open innovation research can add value to the global R&D literature. Of course, focusing on the R&D function allows a much clearer understanding of open innovation at the more-upstream phase of innovation; however, such a focus obviously puts lower priority on downstream innovation such as user-driven innovation. The significant role of lead users for innovation (von Hippel, 1988) cannot be ignored. The knowledge input from competitors should also be useful for innovation, particularly for benchmarking, because competitors typically share very similar contextual knowledge with the focal firm (Asakawa & Un, 2012). At the same time, focusing on R&D reveals some of the most distinctive traits of open innovation, i.e., the challenge of overcoming the "NIH syndrome" (Katz & Allen, 1982) prevalent in open sourcing of external R&D knowledge. Because R&D is often considered as the core intellectual property of the firm, sourcing R&D knowledge from the outside creates psychological and socio-political tensions on the part of the scientists and engineers within the firm.

Another advantage of the open innovation literature lies in its wider coverage of interested parties as key actors in the innovation activities, including intellectual property department, corporate venture department, incubators, strategic alliance department, among others (Vanhaverbeke, Du, & von Zedtwitz, 2013). Global R&D literature does not consider collaboration with all these parties in depth, if occasionally touching upon them.

Moreover, open innovation literature considers external partners such as lead users, user communities, innovation intermediaries (Vanhaverbeke, Du, & von Zedtwitz, 2013), beyond the brick-and-mortar type of overseas R&D centers, which are taken into extensive consideration.

Furthermore, open innovation literature adds value to global R&D literature by elaborating on the development of R&D capabilities much more fully. For example, Chesbrough and Schwartz (2007) present core, critical, and contextual R&D capabilities which feature different aspects of capabilities. Global R&D manifests increasing degrees of external collaboration to aim at building capabilities. But much more fine-grained classification of capabilities is desired. Open innovation literature complements that by filling the gaps in the literature. Here we draw on insights from open innovation literature, in that different capabilities require collaboration with different types of external partners, with appropriate type of governance mode (Vanhaverbeke, Du, & von Zedtwitz, 2013).

Table 8.2 Cross fertilization of open innovation research and global R&D research

What Open Innovation Research Can Contribute to the Global R&D Research	What Global R&D Research Can Contribute to the Open Innovation Research
• Adopting outbound open innovation framework • Covering multiple stages of innovation activities • Covering wider scope of external parties involved in innovation activities • Introducing fine-grained classification of R&D capabilities	• Considering locational decisions • Extending the levels of analysis for global open innovation • Drawing attention to capability for overseas knowledge sourcing • Managing internal and external networks • Considering emerging countries innovation

8.5 CONCLUSION

Considering that open innovation is becoming increasingly global and that global R&D is becoming increasingly open, such cross-fertilization is the most natural direction. Nevertheless, the research heritages of these two fields are different. It is our aim to suggest primarily to the open innovation research community how global R&D research can add value to the further advancement of open innovation research, in the context of globalization. We argue that the global R&D literature can contribute to the open innovation literature by offering a much more fine-grained approach to the geographic aspects of open innovation. More specifically, we proposed various research areas in which global R&D research could advance our knowledge on open innovation, such as locational decisions, distance and geographic scope, role of overseas subsidiaries, extending the level of analysis to global open innovation, organizational capability of managing internal and external networks, and managing open innovation in emerging countries.

Global R&D research has limitations, some of which open innovation literature can complement. For example, primary attention to the inbound type of open innovation by the global R&D literature can be complemented by the open innovation literature which captures the outbound type of innovation sufficiently. Global R&D literature's exclusive focus on R&D function can be complemented by much wider scope of attention by the open innovation literature to various functional areas that include sales and marketing.

Although the open innovation phenomenon has expanded rapidly in business and increasing numbers of management scholars have paid attention to this phenomenon over the past decade, rather limited attention has been paid to the international dimension of open innovation. Future research can shed more light on both the benefit and limitations of global R&D literature for investigating the nature of international dimension of open innovation much more in depth.

9

Open Social Innovation[*]

Henry Chesbrough and Alberto Di Minin

9.1 INTRODUCTION

Innovations that get to the market create change in society. However, as students of management, we focus primarily upon the private benefits of innovation: to consumers, to producers and to investors; and treat the overall social benefit as an endnote for our papers. The phenomenon of open innovation has also followed this path, as prior research has tended to overlook its impact outside of the private sector.

In this chapter, we hope to start the process of closing that gap in prior open innovation research. We will examine the role of open innovation in organizations that seek to achieve positive social change as their primary mission, rather than private sector organizations who regard positive social change as a byproduct (Drucker, 2001; Porter & Kramer, 2011).[1]

These organizations operate in the public sector and in the non-profit sector (NPO). According to the definition of the World Economic Forum "Social innovation refers to the application of innovative, practical, sustainable, market-based approaches that achieve transformative social and/or environmental change, with an emphasis on under-served populations."[2]

Social innovations can be the result of the work of individuals, as well as groups and organizations. While innovators in the business world measure their success and effectiveness in a rather straightforward way (i.e. profits and ROI), social innovators by definition need to account social change as the ultimate goal of their strategy. According to a recent report published by NESTA and the Young Foundation (Murray et al., 2010), the innovation process can be characterized by six very distinct phases. In each of these phases, social innovators respond to different actors and stakeholders, and have different targets to achieve, but ultimately social innovation is about achieving systemic change.

What can open innovation contribute to such organizations going through the transitions shown in Figure 9.1? What conditions would allow open innovation strategies to be usefully applied for organizations whose mission is to achieve social change? In examining these questions, we came to a new construct we term Open Social Innovation (OSI).

We define Open Social Innovation (OSI) to be the application of either inbound or outbound open innovation strategies, along with innovations in the associated business model of the organization, to social challenges. To the best of our knowledge this is the first time that the open innovation framework is being applied to the social sector. We therefore take here an exploratory approach to develop the relevance of the Open Social Innovation paradigm. We find that the Open Social Innovation framework is particularly useful to accessing prototypes (stage 3), sustaining innovative efforts (step 4), and scale-up activities (step 5) within either the current business model or a potentially novel business model to meet the needs of under-served target populations that pure-market mechanisms are not able to address.

Given the nature of this study we here present three case studies from the social and public sector. We have sampled for variance, searching for innovative actors within the public and social sector. Thus, we do not claim that our sample is representative of the larger public or social sector. Instead, we seek to uncover some of the processes within organizations we consider to be quite innovative within the social context. Our cases involve two NGOs that share a similar drive for social change, have analogous size, age, budget, global reach, and are both founded and currently led by two charismatic leaders. Also we focus on the City of Birmingham, England and its efforts to reform its government model.

In spite of many similarities, Emergency from Italy, Ashoka from the U.S. and the City of Birmingham, England, operate in very different sectors, address very different societal challenges, and adopt very different strategies.

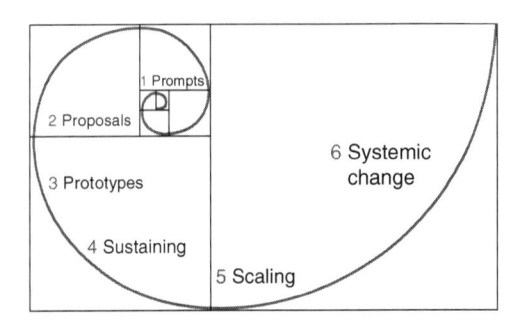

Figure 9.1 Social innovation as seen by NESTA and the Young Foundation

They also share very little in common in terms of their organizational model and control systems. Yet each organization has adopted certain practices that illustrate the Open Social Innovation framework we develop inductively at the end of this chapter.

Emergency's mission requires it to deliver top-quality emergency medical care in war-torn environments. At the center of Emergency's delivery system lies a hospital, or a state-of-the-art clinic, operating according to the best European practices, and providing first class medical treatment. Around this core, consistent with an outside-in OI perspective, various other activities complement what is done in the surgical room, and end up engaging resources available in the local communities.[3] In line with an inside-out OI perspective, the final exit strategy for an Emergency intervention is to become redundant, by transferring know-how and best practices to local institutions and by spurring local imitation. In order to achieve such result, Emergency needs to be extremely embedded in the local community it serves.

Ashoka's mission is to catalyze social change in developing countries through the cultivation of indigenous social entrepreneurs. Ashoka actively recruits, trains, funds, and supports these social entrepreneurs (Ashoka fellows), who operate in a wide range of problem areas. These entrepreneurs are in turn responsible for implementing Ashoka's mission. Ashoka's fellows receive direct financial support from Ashoka, and at the same time they become part of a tightly connected network, which can constitute a formidable asset to implement their ideas.

In describing Ashoka's model from an OSI perspective, we argue that its strategy combines various projects, people, resources and ideas, with the result that Ashoka does not only financially support, but also ends up co-creating the initiatives of its fellows.

When one thinks of innovation, large urban cities usually do not spring to mind. The typical public bureaucracy is not generally considered to be a wellspring of social innovation. The City of Birmingham is quite typical of many large cities. Its citizens confront a variety of social challenges, and its children in particular often find themselves in abusive situations at home, or themselves demonstrate antisocial behavior in school or in the city. Yet the City of Birmingham has managed to innovate some novel initiatives to address the social problems of child neglect and child abuse. In so doing, the City transformed its business model, employed some state-of-the-art data analytics, and realized superior outcomes for its citizens. It even offers this transformation as a service for other public agencies, a classic inside-out open innovation approach.

We turn now to a more detailed discussion of each organization's innovation challenges, and how they were met. As we shall show below, each employed aspects of an Open Social Innovation framework.

9.2 EMERGENCY

It all started from an idea, originating from an informal network of Italian professionals, who realized that it was necessary to bring surgical help where both standard HMOs and international organizations were not able to intervene. This was particularly relevant in areas of prolonged civic and military conflicts, and the hundreds of informal wars that are plaguing the world. Emergency was founded as an Italian NGO under the leadership of Gino Strada in 1994, as it entered the campaign against antipersonnel landmines.

The original mission of Emergency was to treat victims of war, creating the facilities and the organization that were necessary to deliver world-class quality surgical treatment, even in areas that were in the center of wars and conflicts. The functioning model of the organization was based on a large voluntary basis, donations and the involvement of trained physical doctors and nurses. Today Emergency, whose 2010 budget was EUR 30 million, has subsidiaries in the U.S. (established in 2005), UK (since 2007), Switzerland and Japan (as of 2011). Emergency has worked in 16 countries and has provided medical and surgical treatment to slightly less than 5 million individuals. At present the NGO is running its own hospitals in six countries: Afghanistan, Cambodia, Central-African Republic, Iraq, Sierra Leone, and Sudan. In the past, Emergency had also run operations in Chad, Congo, Djibouti, Egypt, Eritrea, Ethiopia, Iran, Rwanda, Somalia, and Uganda.

It is part of Emergency's style to understand and adapt to the rules, traditions, and culture of the region where it is located. While remaining absolutely neutral with respect to the conflict, and tenaciously firm when it comes to preserve conditions of safety for its employees, Emergency enters in a negotiation with local civilian, military or religious authorities, to establish clear rules of conduct, and boundaries that both parties agree to respect. Quite challenging had been in particular the case of Afghanistan, Emergency's second largest program (16% of the total budget for 2010), where we can clearly see the advantages of Emergency's strategy. Emergency is present in Afghanistan since 1999, it runs now three surgical centers, a maternity clinic, a network of 30 first-aid posts, health centers and a medical point in the largest prison in the country. In the medical center of Anabah, Emergency employs 214 local staff, and performed so far over 17,000 surgical operations. Opening and running a free maternity service under the Taliban regime had also been challenging, but right now approximately 300 babies are born every month, and parents can take classes and free consultations. Cecilia Strada, daughter of the founder, Gino, told us that these results were achieved exactly thanks to Emergency's ability to interpret the situation on the ground, and to blend in with a type of intervention that was both respectful of norms and customs, as well as safe for medical staff and patients.

With operations scattered around the globe, Emergency remains a centralized organization, which tightly controls activities from the Milan and Rome headquarters. The rules of engagement in very complex conflict zones are to be understood by everyone in the organization. Both volunteers and professionals need to be conscious of local traditions and respectful of religious and cultural norms. Emergency currently employs 260 people in Italy and it is supported by a network of approximately 4,000 volunteers worldwide. Doctors and nurses deployed internationally, are trained, supervised, and paid directly by Emergency. Emergency is extremely strict when enforcing procedures and rules of conduct: not only when negotiations are concluded with local authorities, but also with its own staff. What is at stake is not only the reputation of Emergency but also the safety of local employees. Hence room for maneuver is extremely limited.

Within this framework, which seems not to allow much room for innovation, Emergency had been on the contrary able to adopt various forms of open social innovation. One core principle of open innovation is embodied in Bill Joy's expression, "not all the smart people work for you." In order to provide free, high-quality medical treatment to everyone who is in need, Emergency does not only deploy nurses and doctors but it has to identify and rely on local suppliers and staff to support its activities. Hiring and sourcing to external contractors in certain areas of the world, and within extremely unstable environments is a complex endeavor, which requires advanced operation management. Still, relying on external resources is a quintessential component of Emergency business. Local competencies are acquired through trusted intermediaries with an in depth knowledge about the local environment. The deep and extensive involvement of local external resources allows Emergency to become deeply rooted in the complex fabric of communities in war-torn conditions, and that often are plagued by high levels of corruption.

A second aspect of Emergency's organization, which can be linked back to the OSI framework, can be defined as business model flexibility. Open innovation requires careful consideration of one's business model, and how a successful model can sometimes limit the ability of an organization to innovate (as Chesbrough, 2003a showed in the case of Xerox). Emergency must constantly examine its business model in light of the conditions it encounters in every new conflict. Political consent and legal regulations are the boundaries within Emergency needs to operate. Trust between Emergency and external partners is quintessential, especially when society and politics in the country are in heavy turmoil. Emergency has to respect fragile cultural rules and habits with care, without compromising aspects that are necessary for its operations (i.e. hygiene, safety, etc.). In the course of her interview, Cecilia Strada emphasized that flexibility and steadiness are extremely difficult to balance. Under certain conditions it turns out to be impossible, and Emergency had to give up its ideas and objectives for certain regions of the world. Or it was forced to take strong stands, such as closing down the operations of its women's clinics in Afghanistan, on issues related to the employability of its female staff.

A third aspect of Emergency operation can be linked to the inside-out OSI framework. Let's consider the case of Iraq. Emergency had been present in Iraq since 1996, going through wars, a regime change and a never-ending conflict. Emergency had to negotiate with local authorities, identify potential local partners, and run its operation in a very difficult and unstable environment. Still, early results and commitment positioned Emergency in a privileged position to expand its mission in the country, and address the pain of more than 10,000 unemployed victims of war. Treating patients, perhaps with prosthesis, artificial legs or arms, might not be the solution if the environment around these amputees does not allow them to go back to an active life. Even though medical rehabilitation gives a patient physical autonomy, undesirable consequences of their mutilations or disabilities prevent these people from contributing to their communities or simply going back to their job. The network of relationships established by Emergency in the country allowed the organization to identify almost 300 cooperatives and associations that could provide assistance and help find new jobs to its patients, starting with six months training courses, and continuing with other forms of help. In Iraq, Emergency is really going beyond the surgical room, realizing that some of its relationships and assets allow the organization to address pain in a more complete way.

Let us focus on a fourth aspect of Emergency's model: its exit strategy. The ultimate goal for Emergency's intervention is to become redundant, and to contribute to the improvement of a local health system that can function once Emergency no longer operates in the area. Transferring knowledge, training doctors, nurses, and technical staff, is therefore part of Emergency's mission, and critical to its ability to exit eventually.[4] In this regard, Emergency is able to report significant success stories. For example, while the organization is still active in Iraq, various local health units are now completely autonomous and run independently by Emergency. Local employees maintained their job, and in most cases the quality of the service provided did not suffer, once Emergency made its exit.

Emergency started from an idea: that even in the most desperate, war-torn areas, access to medical care remains a critical human right. The organization went through incredible growth that was unprecedented among Italian NGOs. Emergency is now recognized worldwide as a leading provider of medical treatment and it is well regarded in some of the most troublesome areas of the world. Commitment to the ideals and the guiding principles of the organization has allowed Emergency to engage resources well beyond the surgical room. Through its embrace of external resources, its careful revision of its business model, and its ability to transition itself out of conflict environments, it demonstrates the value of open social innovation in a rather extreme environment.

9.3 ASHOKA

When in 1980 Bill Drayton set off to start Ashoka his goals were very ambitious. He wanted to make the difference into society through the activities of a network of social entrepreneurs.[5]

Social entrepreneurs, rather than international organizations or the public sector have the motivation and the knowledge, understand the local context to tackle crucial problems plaguing communities, but often lack the scale and access to complementary assets needed. How to engage a potentially global and dispersed network of people of goodwill?

Ashoka was founded in 1981 with a clear mission: "to shape a global, entrepreneurial, competitive citizen sector, one that allows social entrepreneurs to thrive and enables the world's citizens to think and act as changemakers," as we can still read on Ashoka's website. Still, what makes the Ashoka case such an interesting example of OSI is that its success arrived later when the organization truly opened up its original functioning model. Let us consider the details of such transition from a *fellow-centered* to a *collaborative-entrepreneurship based* model.

Since the very beginning, at the core of Ashoka's operations we find a fundamentally open innovation idea. Indeed, Ashoka accomplishes its mission providing direct financial support to its fellows, rather than building a permanent staff of employees. This fits well with Bill Joy's admonition above that not all the smart people work for you. In Ashoka's case, they look for people already active in social entrepreneurship, and then seek to support and enhance their work.

Approximately 30% (i.e. $10.8 million) of Ashoka's budget is still today allocated to provide financial support to its 3000 fellows operating in 70 different countries around the world. The initial focus was on Bangladesh, Brazil, India, Indonesia, Mexico, Nepal, Thailand, and African countries to support the start-up phase of social enterprises. In the 1990s, Ashoka introduced new initiatives, and in a couple of years tripled its size. Recently its programs expanded also in Western Europe, East Asia, and the Middle East (Meehan, Koehane & Levenson, 2012).

The original tool to assist fellows had been a two years stipend provided directly by Ashoka. Such support allows fellows to devote themselves completely to the realization of their social ventures. Also, during the 1990s, Ashoka focused on providing individual services to each one of its fellows, and on guiding them through to make sure their activities were successful. Ashoka was providing direct support with special training, fundraising practices, managerial consulting. Although this model brought sound results, and Ashoka's reputation grew significantly, still the system was too expensive, not scalable, time-consuming, and its impact on Askoha's fellows work was intermittent rather than continuous. In the language of Figure 9.1 above, Ashoka's initiatives too often were stuck at stages 3 and 4.

As time passed, the Washington-based organization realized it was possible to organize its activities in a different way and Ashoka turned into a truly open social innovator. While the financial support continued to be the main form of assistance, Ashoka realized that it was able to provide other services, tapping into what was becoming the organization's main asset: its constantly growing crowd of social entrepreneurs.

Obviously, comradeship and collaboration among fellows had always been a very important component of Ashoka's model, but the organization felt it needed to go beyond the sporadic support and exchange of suggestions. In the course of the years, Ashoka perfected the concept of "collaborative entrepreneurship" (Drayton, 2011), whose implementation was based on access to resources beyond the network of donors and fellows, and revolved around a close interaction with companies and public institutions (the very same organizations which Drayton claimed were not successful at addressing social problems). The key was to realize that social entrepreneurs and companies or public institutions could complement each other and had clear incentives to work together. Indeed, on one hand that most of the fellows were embedded in their own regions, deeply passionate and open to learn from other examples, share information, but still extremely busy to carry on their operations in their chosen regions, and with little to no time nor resources to expand their operations, and hence scale-up to what Figure 9.1 suggests is the final goal of a social innovation. At the same time, private companies or institutions might be interested in seizing an opportunity or solving a social problem, but they were not able to do so, lacking the very valuable and context specific information in the hands of social entrepreneurs on the ground. Ashoka shifted its focus from individual support to system integration, discovering new opportunities for fellows, achieving linkages and collaboration beyond the network of 3000 fellows, with the society at large, and engaging fellows in collaboration with all the various components of the local environment where they were operating.

Collaborative entrepreneurship's goal was identified as the definition of solid partnerships between the business sector, NPO, and public sector, in order to align business models and serve a common goal. Under this new perspective, Ashoka's global network not only comprised its fellows and donors, but also corporations, entrepreneurs, policy makers, academics, and journalists, sharing interest in social problems, but with very different goals in mind.

In other words, according to the fellows-based model, Ashoka was impacting society through its network of fellows. Under this new collaborative entrepreneurship-based model, Ashoka was addressing social needs through a larger coalition. At the center we still find the direct effort of a social entrepreneur (and Ashoka's fellowship system), that is however empowered through more cohesive forms of interactions. The ultimate goal becomes to change the social/political/business infrastructures through collaboration within the citizen sector.

In Figure 9.2 inbound and outbound OSI blend together in Ashoka's concept of the Hybrid Value Chain.

Through the Hybrid Value Chain (Figure 9.2), Ashoka claims that its role is to help social entrepreneurs to close the gap between corporations and civil society, address a social problem and eventually scale up the operations on a local community to reach a much broader constituency. Let's see how, starting first with the type of collaboration between social entrepreneurs and the business sector.

As we have already suggested, Ashoka fellows tend to have a deep knowledge necessary to tackle a local social problem. These correspond to levels 1 and 2 in Figure 9.1 above. From a business perspective they have the potential to significantly lower transaction costs and barriers to enter a new market. They have the time and the energy to explore the needs of a new market segment, which at this point sits outside the reach of any single business organization. Working with them, entrepreneurs and corporations get access to new knowledge and talent, necessary to turn a market failure into a real market opportunity and hence into profit (for a recent analysis, see Drayton & Budinich,

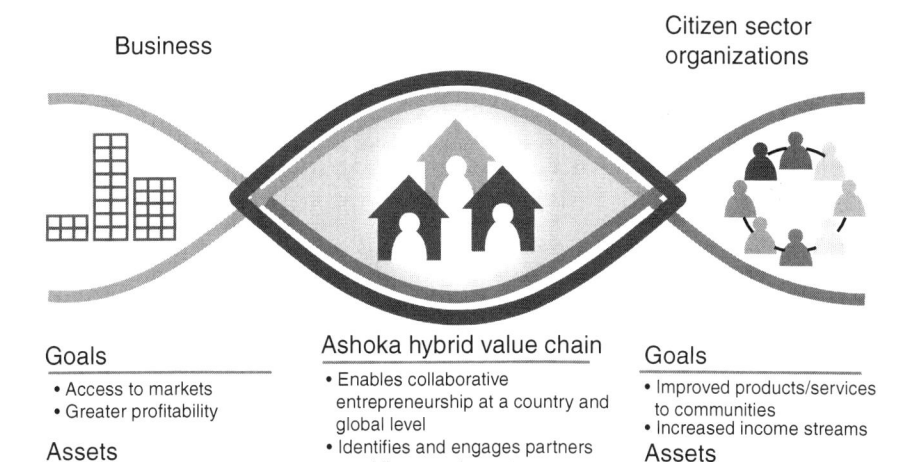

Business

Goals
- Access to markets
- Greater profitability

Assets
- Vital goods and services
- Ability to operate at scale
- Investment capacity
- Operational capacity
- Infrastructure and logistics

Ashoka hybrid value chain
- Enables collaborative entrepreneurship at a country and global level
- Identifies and engages partners providing complementary competencies
- Demonstrates how to deliver integrated solutions at scale
- Spread knowledge, learning and specific know-how emerging from HVC applications

Citizen sector organizations

Goals
- Improved products/services to communities
- Increased income streams

Assets
- Deep knowledge of communities and consumers
- Ability to aggregate demand
- Social networks
- Behavior change capacity

Figure 9.2 Ashoka's Hybrid Value Chain.

Source: Ashoka's website (http://fec.ashoka.org/

2010). This corresponds to levels 3 and 4 in Figure 9.1. At the same time, tapping into the resources and infrastructures of large corporations, social entrepreneurs have the opportunity to get access to those complementary assets to scale their operations, and eventually to deliver value much beyond their individual reach. They can empower their idea, maintaining loyalty to their values, but still operating through the mechanisms of the market economy. This corresponds to levels 5 and 6 of Figure 9.1. At the same time, companies are in a better position to replicate a successful model in other regions, where they perceive similar needs and opportunities.

These alliances require close monitoring, screening of potential partners and fine-tuning in the implementation phase. Multinationals and social entrepreneurs speak very different languages, and some form of intermediation is needed. At the same time, it is important to stress-test the level of commitment to a social cause of business organizations, and to identify potentially hidden agendas, which might be in sharp contrast with the fellow's ideals. Individual fellows are not in the bargaining position, they don't have the scale, or the skills to screen adequately partners and projects. Here is where Ashoka comes into the picture. Ashoka becomes the trusted intermediary for both partners, the shared platform where social entrepreneurs and the private sector can attempt to align models. Ashoka provides fellows a rich selection of pre-screened alternatives, and engages even the largest multinational as a peer, sharing a common language. Not only is Ashoka able to convince the industry to consider a new business opportunity based on its fellows' knowledge, scouting and connections on the ground, it is also in the position to negotiate the rules of engagement between a business partner and potentially interested fellows.

From an OSI perspective, we can see Ashoka acting as an intermediary, between different social entrepreneurs and between social entrepreneurs, companies, and institutions. Ashoka's brand and reputation does not only become useful for interactions with the business sector, but also when an entrepreneur enters in negotiations with the public sector. Fellows are addressing true social needs, and they are working under extremely difficult conditions in very troublesome areas. They are addressing situations characterized not only by a market failure but also by a policy failure. In these scenarios, an alliance among corporations and social entrepreneurs might be the solution only if the public sector intervenes and provides some basic infrastructures and allows operations to run smoothly. Once again, fellows might be in a very good position to grasp the fundamental issues at stake, but they might not be in the right position to negotiate with the public sector, or worse, they can fall victims to incompetence and corruption. Ashoka's role here is to smooth the process, identify the key policy makers, and to facilitate the flow of information for the hybrid value chain to work properly.

The Housing for All project can clearly exemplify the process. Housing for All worked successfully in various countries such as Brazil and India, and proved that this OSI strategy could clearly be applied under very different local contexts. Many were the stakeholders involved in a process, which ended up providing clean water, decent housing and new jobs in some of the poorest slums of these two countries. Ashoka fellows were indeed at the center of the network. They identified the opportunities and acted as the main social entrepreneurs, devoting full time to the completion of the project. Their role was functional to point everybody's attention on the little details that could hamper the completion of the project. The collaboration with local planning authorities was also very important to clear land and review zoning, and political authorities shared the success for improved quality of life of their constituencies. A new market opened up for construction companies, real estate developers, financial institutions, and local craftsmen. With transparent rules of engagement, and a clear commitment, business thrived under the guidelines of Ashoka's hybrid value chain. The program was first successfully introduced in Brazil and then expanded in India, Colombia and Egypt.[6]

The OSI approach and the hybrid value chain in the Housing for All project led to scalable and replicable solutions. The role of the partners involved was to understand the local environment and to address the needs through a strong network. While solutions were tailor-made, the processes applied in the various countries were similar and supervised by Ashoka through a global team of entrepreneurs and support staff.

Fundraising for Ashoka is a complex operation to run. While corporations and large foundations are responsible for 35% of Ashoka's budget, more than 65% of the budget comes from individuals and small business contributions. As a consequence, Ashoka's financing model is extremely distributed. Operations of a $39 million organization are managed from the global HQ in Arlington (Virginia), and Ashoka is present in 40 different countries with 25 regional offices. Ashoka is currently employing around 450 staff members, 125 located at the HQ. Fundraising, support to fellows and other operations are highly decentralized, but still the level of integration and control of operations is necessarily extremely tight.

In a recent piece on the *Stanford Social Innovation Review* (Clay & Paul, 2012), Roshan Paul, who joined Ashoka in 2003 and helped running some of the most successful initiatives of the organization, identifies Open Innovation as an enabler for community participation and more in general for scaling the operations of NGOs. Together with Alexa Clay, Paul writes about turning beneficiaries into co-creators, moving from an enterprise-centered view of a project to an ecosystem view, identifying free licensing schemes to replicate innovation in the social sector, sparking entrepreneurial spirit within and beyond a single organization, and finally allowing for the emergence and evolution of business models. The potential is huge, as founder Drayton suggests in a recent article in *Harvard Business Review* (Drayton & Budinich, 2010: 58): "collaborations between corporations and social entrepreneurs can create and expand markets on a scale not seen since the Industrial Revolution." Ashoka is now becoming an example for other NGOs to follow and, as a true Open Social Innovator, is not shy to share its views and models with other partners around the world. Quoting once again Bill Drayton: "the more eyes we have on society's problems—and opportunities—the better our chances of coming up with valuable solutions."

9.4 CITY OF BIRMINGHAM

The City of Birmingham (COB hereafter) has a proud municipal history of social innovation, dating back to the nineteenth century, when Birmingham became known as "the best governed City in the World." However, by 2003 Birmingham was a long way from being the best local government. Under a UK inspection scheme, which rated all local authorities on a scale of one to four "stars," Birmingham was deemed to be just one star, with uncertain prospects for improvement. It had fallen from the top to near the bottom in the century that had elapsed. The spark of social innovation lit in the late nineteenth century appeared to have gone out.

What changed to ignite a new cycle of social innovation? The leadership of the COB decided to embark on a bold project to improve and eventually transform its provision of public services. This led to a fundamental re-thinking of the organization's business model, one of the core constructs of open innovation. An offsite meeting intended to discuss the mission of social service provision in just an hour or so turned into an all day debate instead, which led to a fundamental distinction: instead of delivering services, the new COB mission was to improve the health and well-being of its citizens. That shift subtly changed the focus from inputs (delivering services) to outcomes (health and well-being). Focusing the mission on outcomes, in turn, prompted some radical thinking.

In the area of childcare services, the COB realized that it was expensive and inefficient to wait to provide services once child abuse had arisen. Instead, the COB worked closely with an external vendor of database software, and utilized the wealth of data it had collected on its clients over many years. Working with independent outside experts, they then developed algorithms that predicted which households were at high risk for abuse. These data allowed the City to create new service offerings for these high-risk households *before* any abuse had been reported. In this way, the chances of parents abusing children was reduced by addressing underlying causes, and providing alternatives (such as anger management programs for the parents) that reduced the chance of incidences of abuse in those households.

Most importantly, these programs allowed the children to remain in their homes (since no abuse had yet occurred), rather than enter into foster care or other institutional care arrangements. The data analysis had shown that children who were taken out of their homes into these more institutional arrangements were also at higher risk of later antisocial behavior. These were important changes in the way COB did business, really a change in its business model. They were precipitated by the decision to intensify collaboration with external software and analytics vendors, delivering results the City would not have obtained if it relied solely on its internal resources.

In Open Innovation terms, the COB reconceptualized its business model it employed to redefine its service offerings, and alter the way and the timing to deliver its services to its "customers," the residents of the city. This required a new definition of its value proposition to its clients, and it required a governance process to accomplish the changes.[7]

The governance of this business model transformation was also challenging. The COB decided to take the governance of business transformation outside its existing departmental structures. A special set of Program Boards was established to oversee the individual transformations (the shift in child care services was only one of a number of initiatives; each had its own Program Board). These Program boards were accountable directly to the Cabinet (the most senior decision body in the COB), reporting annually on progress and plans. Only the Cabinet could establish a new program or close down an existing one.

Further, it was recognized that a program of this degree of change would require a significant level of coordination across the different initiatives. A new role was therefore created on the corporate management team to oversee the whole business model transformation, to manage inter-program interactions, to be responsible for the transformation process "roadmap," and to monitor the delivery of benefits from each initiative.

As a result of these efforts, the COB has developed a process capability for managing these transformations. In the best inside-out open innovation tradition, the COB decided to enable other municipalities and public sector organizations to learn from its experiences. It now offers its methodology under the name of CHAMPS2 (CHAnge Management for the Public Sector). This methodology is now mandated by the COB to be used for all business transformation programs in the city. It is illustrated in Figure 9.3. In summary, it is an eight-phase approach to transformational change. Each phase consists of a number of stages, and each stage of a number of activities. In total, the methodology comprises over 500 activities, each with supporting information including examples, templates and "how to" guides.

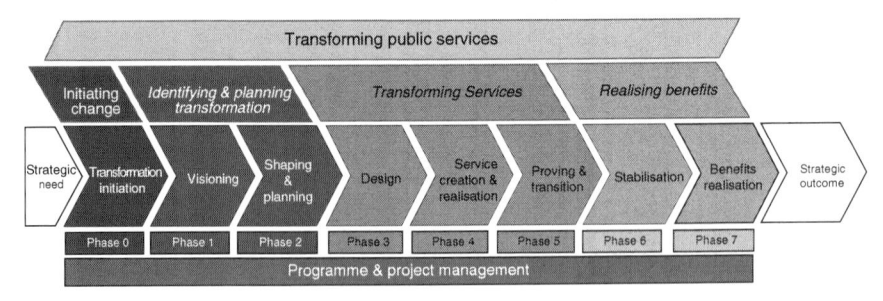

Figure 9.3 CHAMPS2: Transforming public services

One of the distinguishing characteristics of CHAMPS2 is that it has been developed for the public sector, where it is essential to balance increased demands from citizens for improved services with the need for public sector efficiency savings.

The whole CHAMPS2 methodology has been placed in the public domain and is available at www.champs2.info. It is shown in Figure 9.3, and maps closely to the six stages of social innovation detailed in Figure 9.1. However, this is more of a process map for how a public agency can realize a strategic innovation.

9.5 DISCUSSION

In contrast to Emergency and Ashoka, which are non-profit organizations, the city of Birmingham is a public entity. There are many aspects of its mission that differ from these other two organizations as a result. Emergency and Ashoka are focused organizations, dedicated to a single issue, or a narrow scope of issues. The City of Birmingham must respond to all of the demands of its inhabitants, as expressed through its city council, its elections, and its governing institutions.

There is a second aspect to this difference as well. Both Emergency and Ashoka position themselves at the front end of social change, corresponding to levels 3 and 4 in Figure 9.1. Emergency is often the first organization on the ground in a new conflict. Ashoka seeks out and supports entrepreneurs who are testing new approaches to create social change. Public entities like the COB, by contrast, follow behind such organizations in the progression of social change. These public entities are the ones who administer programs for millions of people on a daily basis, often for decades. In many ways, they are the final repository of systemic change sought by the non-profits as shown in the sixth stage of Figure 9.1. If and when a social change becomes part of an established program by a public entity like the COB, the cycle of social change has fully run its course.

A third difference lies in the sustainability and scalability of these organizations. Non-profits like Emergency and Ashoka depend on donors to cover their expenses. Their "business models" imply that the recipients of their services cannot and do not pay for the full cost of their provision. In many cases, the recipients pay no money towards the costs of providing services. Donors must make up the rest of the funds required. As such, these organizations need to cultivate long-term donor relationships, and must protect against "donor fatigue," as we are seeing now in Haiti. There, after years of charitable support to rebuild Haiti after the devastation it received from a hurricane in 2010, donors are now shifting attention to other crises in other places, such as the devastation from the more recent Hurricane Sandy.[8] These factors limit the ability of non-profits to sustain their missions, and likewise limit their ability to scale the provision of the services offered within their mission.

Public agencies, by virtue of their power to tax and the monopoly they hold on service provision, do not have these worries. They can sustain their programs more readily, and can marshal the resources to scale up their program offerings. But they have a requirement that the non-profits do not. No one expects Emergency to provide health care services throughout Iraq. Ashoka entrepreneurs initiate new services where they can, with no expectation of universal coverage. But public agencies do face this expectation; to provide their services to everyone at scale. In many ways, it can be worse for a public agency to have services available for some, but not all, of the residents of a city than not to provide those services at all. The partial availability means that the agency must have a process to determine who does, and does not, receive the services. Residents who hear about the availability of services, but find they are not able to receive them, are disappointed or even angry. And news media are only too happy to publish such stories, to the embarrassment of the public authorities. Voters can and sometimes do punish public representatives for these errors. So public agencies face unique political pressures to scale up their offerings for universal access.

9.6 HOW OPEN SOCIAL INNOVATION CONTRIBUTED TO SOCIAL INNOVATION

Having discussed about the differences between these three organizations, let us now consider how the Open Social Innovation (OSI) framework can be used to interpret the strategies implemented. The table offers an overview of these strategies, which we are going to discuss below.

Table 9.1 Open social innovation in the three non-profit organizations

	Emergency	Ashoka	City of Birmingham
Outside-in	• Identify and rely on local suppliers and staff to support its activities. • Hire and source to external contractors. • Acquire local competencies through trusted intermediaries with an in depth knowledge about the local environment. • Obtain permissions from local power brokers required to operate in the conflict region.	• Shift its focus from individual support to system integration, discovering new opportunities for fellows. • Extend linkages and collaboration beyond the individual fellows to a larger network of past, and present peers and other resources. • Engage fellows in a collaboration with all the various components of the local environment where they were operating.	• Use external technology to develop a comprehensive database of social services provided to citizens. • Use externally sourced analytics vendor and methods to obtain insights about factors that were associated with higher risks of child abuse.
Inside-out	• Establish a network of relationships that allows the identification of local cooperatives and associations. • Exit Strategy: transfer knowledge to local partners, training doctors, nurses and technical staff. • Exit Strategy: ensure that local employees will maintain their job, and the quality of the service provided will not suffer upon departure.	• Build a global network that not only comprises its fellows and donors, but also corporations, entrepreneurs, policy makers, academics and journalists, sharing interest for social problems. • Become the trusted intermediary for both partners, the shared platform where social entrepreneurs and the private sector can attempt to align models. • Become able to negotiate the rules of engagement between a business partner and potentially interested fellows.	• Offer its methodology under the name of CHAMPS2 (Change Management for the Public Sector). • Enable other municipalities and public sector organizations to learn from its experience. • Authorize this methodology to be used for all business transformation programs in the city.

Emergency's Open Social Innovation strategy combines inbound and outbound Open Innovation to create tightly controlled but deeply embedded platforms to achieve the twin objectives of first-class treatment and locally supported and sustained services. Its Inbound Open Innovation happens as most of the skills and resources outside the surgical room have to be sourced locally and they constitute a fundamental part of Emergency's job. The core staff of Emergency comes from the West, and they act as the coordinators, integrators, and quality control for the range of services delivered. But the actual delivery of services, outside of the intensive care surgeries, is provided by local partners. This increases the employment Emergency provides to the local community, and likely increases its support within the community.

Emergency's Outbound Open Innovation happens at various levels. First of all, Emergency seeks to transfer back to society an active citizen, self sufficient and ready to contribute to the renewal of a devastated community, rather than an amputee. Also, Emergency provides extensive on-the-job training to its local partners in the course of its mission. As a result, some of these partners leave to engage their newly advanced skills outside Emergency's facilities. In this way, Emergency becomes a role model for other local healthcare institutions, which is entirely consistent with Emergency's mission. (Indeed, it helps the community prepare for the time when Emergency withdraws from its facilities.) While these spillovers are difficult to control, when for example trained nurses or technical staff, leave to join local profit-oriented private clinics, they help generate consensus among different factions in the local community around the entire operation. Indirectly, they help protect Emergency from the caprices of local warlords.

In the case of Ashoka, OSI strategy combines inbound and outbound elements of Open Innovation, and this is made possible by an organization running a centrally administered but globally dispersed network. Ashoka's presence throughout the world, and in the community it reaches, is mediated by Ashoka's fellows. Inbound Open Innovation at Ashoka is clearly witnessed by the diversity of projects supported by the organization. For Emergency, and also for Ashoka, the absorptive capacity to incorporate external ideas and technologies into their offerings is extremely important. Constantly seeking to expand the network of fellows, Ashoka is looking for ways to attract the most talented social entrepreneurs out there. Instead of bringing in young volunteers from the West (think of the Peace Corps, for example), Ashoka has determined that it is best to focus on local, indigenous people to become the agents of change. Once these fellows are chosen, they embark on training by Ashoka, much of it provided by previous fellows who have direct experience in social change themselves. In addition, the fellows are introduced to a large network of current and former fellows, who become valuable resources for these entrepreneurs as they seek to enact the proposals for change that Ashoka selected.

Outbound Open Innovation happens in Ashoka as the organization acts like an intermediary. Ashoka transfers its value not only by selecting and financially supporting fellows, but also through combining ideas present in the network, creating linkages across organizations and between fellows, professional staff and institutions, around individual projects. Armed with the results, both good and bad, from the many entrepreneurial initiatives of their fellows, Ashoka can accumulate and then disseminate effective social entrepreneurship practices in developing country environments.

The City of Birmingham did not play the role of an intermediary. Instead, it was the provider of the services. Nonetheless, it too followed elements of the OSI framework. With regard to outside-in open innovation, the City needed external technology to develop a comprehensive data base and associated analytics necessary to obtain insights about factors that were associated with higher risks of child abuse. This helped the City decide where and when to intervene, and in turn stimulated the provision of new, preventive services that were more likely to keep families intact and reduce the incidence of child abuse.

Another aspect of the OSI framework is the need for a business model to sustain the provision of services. In the case of Birmingham, they refocused their business model from that of a supplier of inputs (social services) to a provider of outcomes (enhancing the well-being of Birmingham citizens). This shift in focus highlighted the need for a deeper understanding of the causes of child abuse, and what the City might do to address those causes (in other words, addressing the causes of abuse, rather than managing the resulting symptoms of that abuse). As private sector firms shift from product-based businesses to service-based businesses, they often shift their focus from inputs (products) to outputs (services) as well.

A third part of the OSI framework is the inside-out branch of open innovation, where stranded or unused internal ideas and technologies are allowed to go to the outside. This is a potentially powerful part of the OSI framework that could dramatically boost the impact of social change. Since public agencies like the City of Birmingham often don't face the same kind of competition that private sector (and in some cases, social sector) organizations do, they can afford to be quite open about sharing successful methods and practices that have proven to be effective. We see this in the COB from its methodology for managing its change process to arrive at these new outcomes. This CHAMPS methodology is now available for other public agencies to use in their own initiatives. As other agencies embrace these methods and perhaps further improve upon them, a community of learning could emerge that could drive the social impact of these changes to new heights.

9.7 CONCLUSION

The Social Innovation Council at the 2013 World Economic Forum in Davos posed a very distinguished panel a challenging question: how can social innovation generate impact? The answer was: social entrepreneurs, companies, institutions, NGOs, and the public sector need to scale-up their operations. Partnerships are tremendously important to achieve positive social change.

In this chapter we have argued that the ideas of in-bound and out-bound open innovation, and the integrating role of the business model, are relevant well beyond the business world. At the heart of the Open Innovation framework lies the idea that for an alliance to work, partners need to align their business models (Chesbrough, 2006a). In so doing, incentives and goals are declared, and the definition of shared-resources codified. As for the business world, a comprehensive view of open innovation strategies can be very relevant for social entrepreneurs, for at least three reasons. First of all, as the cases shown in this chapter suggest, tapping into the resources of partners beyond a single organization's borders is quintessential for implementing the mission of the social enterprise. Second, social entrepreneurs seek to achieve a goal, which is not possible to measure merely through financial accounts, and therefore aligning different objectives is fundamental and difficult for the organizations involved. Finally, in order to achieve systemic change, models and practices need to be sustainable economically as well as socially, even when serving the needs of segments of the population that the market is not able to address.

NOTES

* Authors are grateful to Ceclia Strada, Paolo Busoni, Simonetta Gola with Emergency, Alan Landis, Darlene Damm, Konstanze Frischen, Paul O'Hara and Beverly Schwartz with Ashoka, and Glyn Evans with the City of Birmingham, for the time they spent with us during the interviews. We would like to acknowledge the work of Barbara Gilicze and Francesca Lazzeri, as they assisted us in the preparation of this and earlier drafts. We received important feedbacks and ideas also from Paolo Busoni, Devin McIntire, and Thomas Osburg, as well as from the editors and various authors of other chapters in this book. While we recognize their valuable comments and contributions to these projects the two authors of this chapter remain responsible of the opinions expressed in these pages.

1. We also note that more and more frequently, for-profit organizations are interested in the social impact of their operations and identify strategies to maximize the positive social impact of their activities. In a special report by *The Economist* in 2001, Peter Drucker claimed that today more than yesterday and even more in the future, the key challenge for corporations worldwide will be to prove their social legitimacy. Ten years later, Harvard professors Porter and Kramer (2011) also called for creating shared values to rethink capitalism, and focused exactly on the social impact of an organization's activities.

2. See here the work of the Global Agenda Council on Social Innovation 2012 (http://www.weforum.org/content/global-agenda-council-social-innovation-2012). Authors are grateful to Prof. Peter Russo and Dr. Susan Muller at the Center for

Social Innovation and Social Entrepreneurship at EBS for guiding our first steps in understanding social innovation.
3. In Iraq, for example, Emergency is retraining amputees, introducing them to new professions; in Afghanistan, Emergency provides a complete range of services and support to mothers and their children. This is only possible by engaging a local community into a dialogue, and into a negotiation respectful of local habits, rules and religious mandates.
4. Emergency also must guard against too much labor mobility. This can happen when Emergency-trained local staff depart Emergency and set up or join competing private practices in the area. While this is to be desired in the long run, it can seriously disrupt Emergency operations if not carefully monitored.
5. While today the concept of social entrepreneurship has become mainstream, Dayton's intuition, more than 30 years ago, during his studies in Harvard University, was that there was a highly unexploited potential in the civil sector. New models had to be identified to correct disparities created by the ups and downs of a turbulent economy, and enter into areas where real needs existed, but no clear market solution was addressing them effectively. Social entrepreneurs, if properly guided could provide such solutions.
6. The numbers of the Housing for All project are impressive. The first pilot ran in Brazil in 2007, and the organization today realizes this was a very important test to achieve the right balance between grassroots commitment and technical support. By 2011, $1.6 million were invested to provide new housing, and the plan is to serve by 2014 between 40,000 to 60,000 families. The project has expanded in India in 2010, and more than 10,000 new housing units were available by mid 2011, with an estimated market of $100 million. Sales of $11.7 million are estimated in Colombia with 28,000 new homes; and finally more than 18,000 units are now being improved or constructed in Egypt.
7. This business model change process was itself controversial with the COB. The subject of public administration differs in many important respects from running for-profit businesses, and even differs from running non-profit businesses as well. Many city workers see themselves as working for a public interest, not for a business. Their self-image and sense of identity means that one must tread carefully when advising a public agency to "be more like a business." Moreover, there are significant objective differences between a business and a public agency. The COB has the power to tax, and wields a monopoly over the provision of its services. There is no direct competition to speak of, and service recipients have little recourse if they are unsatisfied with the services they receive—except to the ballot box and the next election.
8. Source: http://www.nytimes.com/2012/12/24/world/americas/in-aiding-quake-battered-haiti-lofty-hopes-and-hard-truths.html?pagewanted=all.

Part IV

Managing and Organizing Open Innovation

10

Open Innovation and Intellectual Property

A Two-Sided Market Perspective

Henry Chesbrough and Roya Ghafele

10.1 INTRODUCTION

A recent review of research on intellectual property revealed that out of a body of roughly 9000 articles on intellectual property, fewer than 20% were not legal in character.[1] An even smaller fraction of articles addresses how open innovation relates to intellectual property.[2] Traditionally research in intellectual property has been driven by the quest to understand the scope of substantive intellectual property law with a continuous search to grasp whether IP regimes should be "weak" or "strong" in character (Gould & Gruben, 1996; Lanjouw & Lerner, 2000). The limited economic research in this area did not break with this paradigm and sought to grasp what type of regime ("weak" or "strong") best promotes growth (Maskus & Reichman, 2004; Helfer, 2004; Hassan & Tucci, 2010).

A much more recent viewpoint is the role of intellectual property as an enabling mechanism for innovation, as a means to promote the open exchange of innovation inputs. This has only been addressed by a very few scholars like Arora, Mazzoleni or Merges (Arora, 1995; Feldman & Florida, 1994; Mazzoleni & Nelson, 1998; Merges, 1999). These authors have started to ask how licensing arrangements can promote the growth and efficiency of markets for technology and how IP needs to be subsequently managed to achieve these goals. Because this type of approach essentially incites a paradigm shift in how we think about IP, the literature building upon this work is to a large extent still concerned with the very simple question "how do you actually do that?" (Gollin, 2008; Holmes, 2009; Hurmelinna et al., 2007).

In this chapter, we build upon this second perspective, and connect it directly to open innovation, a particular type of innovation that has become more salient in recent years (Chesbrough, 2003a). We will discuss how IP can inhibit open innovation, or, if properly managed, can enhance its effectiveness. We sketch out some illustrative examples to demonstrate our arguments, and make them more concrete for the reader.

10.2 WHERE ARE THE MANAGERS?

The audience for mainstream legal research on IP is usually other legal scholars. In this writing there is little of direct relevance for most practicing managers who must make choices about whether, when, and how to protect intellectual property that arises from invention discoveries. Indeed, perhaps the single biggest criticism of the existing approach to the role of patents in innovation is that it omits entirely any role for managers of industrial firms in the innovation process. This neglect of any role for management in overseeing the innovation process is a glaring deficiency that deserves to be redressed.

Here are a few of the omissions in the body of research to date on IP management that trouble us. In the accounts of IP management in these studies, we do not know how innovation researchers were hired, nor the allocation of resources and incentives to inventors. We don't know how they were paid. We don't know what incentives they were offered for inventions reported, or for patents issued, vs. for scientific publications they authored. Most important of

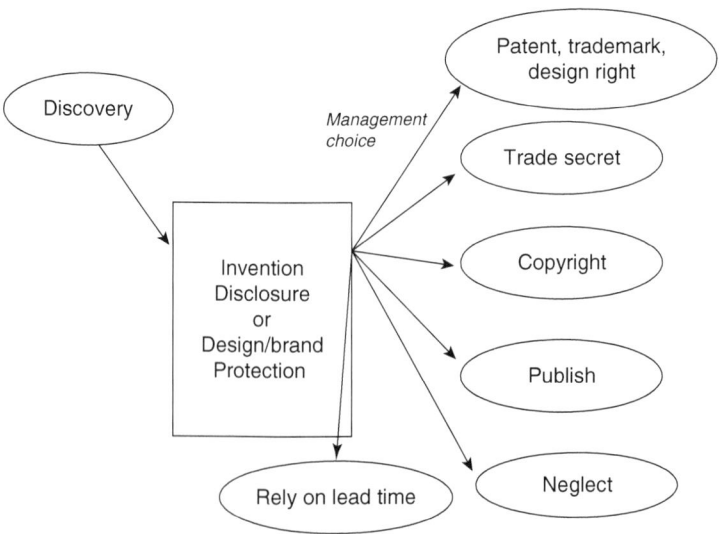

Figure 10.1 The many possible paths for IP Creation

all, we don't know what the strategy of the firm was towards the technologies being patented. All of these important influences are determined by managerial decisions. None get mentioned in the vast bulk of the academic literature on IP. This cannot stand. Managing innovation, after all, is complex, contingent, and connected to the strategy of the firm in question.

Figure 10.1 shows that even within the domain of protectable knowledge, managers might strategically choose not to protect the knowledge with a patent, but instead to pursue an alternate course. These alternate vehicles for protecting knowledge include: trade secrecy, copyright, licensing,[3] neglect,[4] reliance on lead time or even publication. This last mode is used to ensure that others cannot assert claims over useful knowledge that the firm seeks to use, but might prefer not to patent. As the first-to-invent criterion in the US patent system shifts to first-to-file, in harmony with practice in the EU, Japan, and most other countries, publication may become an even more attractive alternative to patenting.

10.3 THE EVOLUTION OF THE PRACTICE OF IP MANAGEMENT

Prior to the 1990s, the management of intellectual property was a small niche that was managed by either the in-house attorney of the firm, or the external patent counsel if the firm was sufficiently small. There seemed to be little in this area to interest top management.

Then, a number of business events occurred that caused skeptical managers to sit up and take notice. One such event was the issuance of the so-called Kilby patent to Texas Instruments (TI) in 1986. This patent gave TI the right to exclude others from many aspects of semiconductor design (Kilby was an early inventor of the original semiconductor who worked at TI and assigned all rights to his invention to TI). Over the next several years, TI generated a substantial portion of its entire corporate net income from royalties it received for this patent (Grindley & Teece, 1997). Another such event was Polaroid's successful suit against Kodak in 1989 in which Polaroid got the largest settlement ever awarded by a US court to that point in time (over $900 million) for Kodak's infringement of its patents. A third, more gradual event, was IBM's enormous success in creating a stream of patent royalties from its IP, starting at some few millions of dollars in the early 1990s, and growing to $1.9 billion in 2002 and continuing to generate hundreds of millions of dollars in royalties annually.

Astute business observers took note of these events, and struggled to imitate them. A group of managers formed The Licensing Executives Society, to exchange ideas and best practices in licensing out patents and other IP. Commentators like Petrash (1997), Sullivan (2000), and especially Rivette and Klein (2000) called attention to the profit opportunity latent in licensing out or selling intellectual property. This last effort, titled *Rembrandts in the Attic*, promised great riches to those who, as the title implied, would dust off their moldy IP, bring it down from the corporate attic, and offer it for sale to others.

These efforts provide useful facts to managers and executives charged with leading this activity. Yet the larger context, and especially the connection between a company's IP and its overall business strategy was lacking. For example, nowhere in these observers' accounts of these new and apparently profitable practices was there a rationale for why companies would *buy* these assets. Sure, companies might wish to sell their IP, but why would anyone ever want to buy someone else's IP? For these assets were not Rembrandts, at least not to most people. There needed to be a rationale for companies to want to buy someone else's IP if there was to be a market for these assets.

In *Open Innovation*, Chesbrough (2003a), provided just such a rationale. Useful knowledge, the book argued, was now widely diffused, so that no company had a monopoly on knowledge in their field. At the same time, the quality of work at small companies, universities, and non-profit institutions was increasingly high. So, instead of inventing it all yourself, you could innovate effectively by accessing excellent work from an outsider. But what to sell, and what to buy, and what to publish instead? Says Dreyfuss (2011): "On the one hand, more and more segments of the knowledge domain are becoming the subject of IP rights. At the same time, open innovation is flourishing. The puzzle is this: How can these trends be going on simultaneously?"

Those choices depend on the company's business model, the way in which a company creates value and captures some portion of that value for itself (Chesbrough & Rosenbloom, 2002). Opening up to external sources of knowledge may accelerate time to market, fill technical gaps in internal R&D, or reduce the total cost of innovation for a firm. IP rights, in turn, enable markets for IP to function, and foster revenues for those pursuing an inside-out open innovation approach. We turn to a more detailed consideration of motivations to both sell IP and to buy it.

10.4 THE TWO SIDED MARKET FOR IP: BUY AND SELL SIDE MOTIVATORS

10.4.1 The Rise of Intermediate Markets

One key force that is affecting the market for intellectual property is the growth of what Ashish Arora and his colleagues called "intermediate markets," or markets for innovations (Arora et al., 2001a). In the closed innovation model, companies had to take their new discoveries to market themselves, both because they would obtain more money that way, but also because there weren't many other companies who knew enough to utilize the technology successfully. Innovation markets in the closed innovation system were sparse, and IP was managed defensively, to preserve the freedom to operate. In an open innovation world, where useful knowledge is widespread, there are many companies with many potential ways of using a new technology, and many potential technologies that might be utilized in a company's business model. No company can hope to exploit all of the many ways a new technology might be used, and no technical lead lasts indefinitely, so temporary technical advantages should be exploited both internally and externally. In addition, open innovation companies typically license technologies liberally to other companies.

An economy full of technologies being licensed for others to use is one in which one can say there are highly developed intermediate markets for those technologies. These markets are termed "intermediate," because one firm initiates a technology and develops it to a certain extent, and then a different firm might carry that technology from that point through to the market. The presence of these intermediate markets expands the number of ways a new technology can be used, and promotes specialization among the different participants in the market. So some companies specialize in creating new technologies, others specialize in developing new products, and still others focus on special niches, services, or applications along the way.

As Arora and his colleagues found, a pronounced division of innovation labor has emerged in the chemicals industry. When new chemical plants are built, the company building the plant typically hires a specialized engineering firm to design the new facility. These specialized firms work on virtually all of the new chemical plants being constructed around the globe, so they are up-to-date on the latest ideas and techniques for making the plants as efficient as possible. Since these plants are extremely expensive, amounting to billions of dollars each, no one chemical company builds them very often. So the specialized firms are able to accumulate knowledge and learning much faster than even the biggest of the chemical companies.

Another example of this specialization of innovative labor can be seen in the semiconductor industry. Back in the 1960s, the major semiconductor firms were captive subsidiaries of product firms, such as IBM or AT&T. There were markets for the final product systems, but no markets for the components of these systems. By the late 1970s, independent firms like Intel and Texas Instruments specialized in making chips, and selling them to product companies, who used these chips to create new computer systems, or cell phones, or videogame players. Markets had emerged for chips, which were purchased and integrated to make systems products. By the 1980s, the manufacturing function in developing chips became separated from the design function, as semiconductor fabrication companies (known as "fabs") like TSMC built chips that were designed by so-called "fabless" companies, who effectively outsourced their manufacturing. Now there were markets for semiconductor manufacturing capability, and associated markets for assembly, packaging, and testing capabilities. In the 1990s, companies like Qualcomm and ARM Holdings began selling intellectual property such as tools and designs to the companies that were designing and building chips. So now a company could buy a design from ARM, model it using tools from Cadence or Synopsys, have the design built by TSMC, and then offer it to the market, creating a market for semiconductor designs themselves.

Surrounding this vertical separation of functions in the semiconductor value chain are still more companies offering design tools (e.g., Cadence or Synopsys above), test equipment, and other services to the industry. This specialization has migrated around the world. In China alone now, there are more than 600 specialized semiconductor design houses, and a number of new manufacturing facilities are being built as fabs for other companies around the world to use to build their chip designs.

Yet another example of this innovation specialization comes from the life sciences. Thirty years ago, drugs were discovered, developed, tested, and marketed by large pharmaceutical manufacturers. By the 1980s, however, specialized biotechnology firms began to discover and patent new compounds. They would then partner with a pharmaceutical company who would take the compound through the clinical trials required by the Food and Drug Administration, and then sell the drug to prescribing physicians. More recently, there have emerged a group of contract research organizations that partner with the biotech and pharmaceutical companies to conduct the clinical trials for them. In the 1990s, Millennium Pharmaceuticals began doing contract research for pharmaceutical clients, but reserving residual fields of use for a compound for itself, and began developing new drug applications for these compounds in the year 2000. Still other firms offer specialized equipment, tools, tests, and other services that assist in the drug development process.

This specialization of innovation also is emerging in the consumer products sector. Procter & Gamble has had a long tradition of internal science-driven innovation, which it has used to create differentiated products to offer to its customers. More recently, though, P&G has realized that its core strengths are not in science, but in its ability to create strong brands. In some of its new brands, such as the Spinbrush and the Swiffer, P&G has created new and large businesses with technologies that it acquired from outside the company. Through its new innovation processes, which it calls Connect and Develop, P&G seeks to exploit the market for external technologies, as it seeks opportunities to create new brands for its customers.

This innovative specialization needn't be based upon products per se. There are intermediate markets that have developed for services too. Based on a longitudinal study of the US mortgage banking industry, Jacobides (2003) found that intermediate markets became a powerful force in the mortgage securities market. As with the chemicals industry, he found that markets arose through firm efforts to exploit gains from specialization of different intra-organizational functions and trade with different firms. This in turn led to the standardization of information and the simplification of coordination between firms. Unlike chemicals, the government also played a role in disintermediation of the mortgage market, through creating an information standard (in the case of US mortgage banking, this standard was the Federal Housing Administration regulations for conforming loans).

10.4.2 Managing IP in Intermediate Markets

While the intense specialization from intermediate markets has unleashed a lot of innovation, the intellectual property aspects resulting from the specialization can be quite complicated to manage. When a company brings in an external technology to incorporate into its business, it must carefully assess whether it has the legal ability to use that technology without infringing the legal rights of another company. The protection of a particular technology is unlikely to cover every aspect of its usefulness. If the entity licensing or selling the technology has patented the technology, for example, the scope of that patent may or may not cover the uses that the acquiring firm wishes to provide. In turn, the protection of a technology may involve claims that inadvertently infringe on some aspect of another company's technology.

So intermediate markets for technology in a world of open innovation profoundly change the management of IP. On the one hand, a firm cannot acquire and utilize an external technology unless they are confident that they have the legal right to practice the technology(ies) that they wish to use. To be sure, this ability to practice a technology also was a concern with technologies in a closed innovation world. But in that world, the company knew the entire history of the internal technology. In this more specialized world, where technologies flow across the boundary of the firm (perhaps multiple times), obtaining the ability to practice one's technology without incurring an infringement action by another firm is more challenging, because the full history of the technology's development is not as well known. There is always a concern that there may be blocking IP out there that precludes deployment of a technology being bought or sold.

On the other hand, secondary markets provide the opportunity to greatly increase the utilization of internally-owned technologies, by offering them to other firms for use in their business. Not only does this increase the utilization of a given technology, but it also increases the number of areas in which a technology might be applied.

But secondary markets for innovations present other challenges. Before a company identifies a promising technology, it must interact with many companies, and explore a variety of possible technologies, in order to have any hope of finding a useful technology. As the old fable goes, "you have to kiss many frogs, in order to find a prince."

This raises an old, but very important, problem first noted by economist Ken Arrow: I as a customer need to know what your technology can do, before I am willing to license it. But once you as a seller have told me what the technology is, and what it can do, you have effectively transferred the technology to me without receiving any compensation! And that is not all. If the customer discusses possible technologies with a would-be supplier, but decides not to license the technology, and instead go off and design an alternative technology internally, the customer may have contaminated itself with the knowledge of that supplier. A subsequent internal development in a related area by the customer may be challenged by the supplier, who might allege that the customer stole the idea from the supplier without paying anything for it. If the customer is a very large company, and the supplier is a very small company, this David-and-Goliath situation may make a jury very sympathetic to the small company, even if the large company developed its approach in a completely independent manner.

10.4.3 Beyond the Value Chain: Business Networks

The business networks in which a company operates can also be a fruitful source of external possibilities. Informal sharing of information, and knowledge trading, can lead to the discovery of useful ideas that might solve important business problems. Larger communities where public information is exchanged, such as industry conferences and trade shows, also supply a great deal of public knowledge that can lead an alert innovator to useful solutions. These groups exchange substantial amounts of information, but this exchange is considered generally to be in the public domain. The most valuable information here is often where to look and to whom talk that reveals the location of private information, which then would have to be pursued under non-disclosure.

Technical standards bodies comprise another resource for accessing available knowledge about a particular technology, and then forging a shared approach across a number of firms for how to apply that technology. Even here, though, IP issues surface with great frequency. These groups are not purely neutral forums, trying to develop the best technical solution to a particular technology problem. The research of Mark Lemley (2002) shows that technical bodies have a wide range of rules regarding how much IP must be disclosed to others in the technical body. And this variation in rules can be leveraged by alert companies who position themselves to occupy key positions within an emerging standard.

One such example is Rambus, a virtual semiconductor design firm offering a technology to speed up DRAM chips inside computers, which has profited significantly by exploiting loopholes in the rules of its standards setting body. After that body settled on a standard for how to accelerate the speed at which DRAM chips transferred data to the system, Rambus revealed that it had received patents on important elements of that standard. It then began extracting high royalty payments from other standards participants, who had designed products around the emerging standard, and hence, infringed Rambus' patents.

What Rambus did has been found to be entirely legal, in a series of court cases regarding its conduct, and the legal rules around its intellectual property. The company's stock price is something of a "pure play," in that the intellectual property of the company is the only business it has. Therefore, Rambus' daily stock price reflects the market's current assessment of the value of its IP. As it happens, the valuation of the company over the past seven years has experienced wide swings, from more than $100 to below $10, even though the IP itself has been well-publicized for many years now.

10.5 BARRIERS TO EFFICIENT IP MARKETS

10.5.1 The Quality of the IP

While Salomon Brother's bundling of mortgages established a secondary market for mortgages we also know that this subsequent syndication contributed greatly to the credit crunch in 2008. In order to avoid a similar calamity as secondary markets for IP begin to emerge, it will be crucial to assure the quality of the underlying IP assets. Strong patent quality is of primary importance in this respect. At the moment, patent offices around the world spend on average 20–25 hours per case to search for prior art. As a result many of these patents are declared invalid during the course of litigation as they did not meet the criteria for patentability and should not have been granted in the first instance.

10.5.2 Limitations that Impede the Secondary IP Market

While we have sketched the development of intermediate markets for innovation in a few industries, and provided some preliminary evidence that these markets are becoming more widespread, the fact remains that there are many inefficiencies that are limiting the emergence of secondary innovation markets. Understanding some of these inefficiencies allows us the ability to maintain a proper perspective on these markets. They also point the way to some mechanisms through which companies can overcome at least some of these current limitations.

One of the most critical limiting factors is the simple lack of information about the extent and terms of trade in secondary markets for innovations. Markets require information in order to function well, and much of the requisite information needed for coordinating market exchange of innovations is not yet available. For example, while there is an estimated trade of more than $100 billion annually in licensing for technologies, there is no place where this trade is reported and tracked. What we know of the licensing market today comes from occasional surveys of companies (which ask the companies to report their trade in total) or from the occasional IP dispute in court, where the terms of a particular contract become part of the court record, and made available to the public.

This very low level of licensing suggests that patents are by and large seen as a defensive mechanism, a negative right, a right to protect rather than enable. This mindset leads to the under-management of valuable assets (Borod, 2005). Various cases show that treating IP as a mere defensive right is not enough to keep business going. Canadian technology giant Nortel Networks is a prime example of how companies can lose out on value if they use their patents only as defensive tools. Unable to generate sufficient revenue to continue operations, Nortel Networks filed for bankruptcy protection in 2009. Nortel's patent assets were then bought in 2011 during bankruptcy proceedings for $4.5 billion by a consortium led by Apple, Microsoft, Sony, and Research in Motion.[5] Kodak encountered similar problems from its defensive approach to patents. MDB Capital Group estimated in August 2011 that Kodak's digital-imaging patents—which comprise only 10% of its patent portfolio—are worth $3 billion.[6] As part of a turnaround strategy, Kodak attempted to generate revenues through aggressive patent litigation, yet filed for bankruptcy protection in January 2012.

The situation is somewhat analogous to the condition of the mortgage market in the US prior to the advent of Salomon's bundling of mortgages. There is no information standard for technology licensing and associated IP trade. There is no FHA that defines a template or format for such trade. And given the wide range of terms and conditions for trading IP, it would be difficult to aggregate statistics on this trade, unless and until one or more information standards arise.

Without these data, it is hard for companies to know what technology is available in the market. One consultant documented some of her technology scouting work for a client, revealing both the potential and the problems of finding available technology. While she and her colleagues found two highly useful technologies in a short period of time, the client's purchasing organization was unable to find any useful technologies using their normal procedures for soliciting external inputs.[7] This is typical of inefficient markets: you don't know what you don't know, and you don't realize what you may be missing.

It is also very challenging to know how to value available technologies, once they are located. Such value is determined by what a willing buyer would pay to a willing seller. Markets aggregate suppliers and customers, so that any individual technology can go to the highest bidder, and bidders know what similar technologies have sold for in the past, giving them a basis for calculating their bid price.

These conditions are typically not present in IP licensing. There is no systematic reporting of previous prices paid for external technologies and their associated IP. This makes it hard for sellers to know what price to expect to receive, or what price would be reasonable, given similar transactions in the past. So too for the buyers. Both sides to a transaction can have unrealistic expectations in these circumstances, and there is little or no objective data to align those expectations more closely. The poor development of the IP market may also be the result of the hassle related to licensing. Revenues from licensing arrangements between small firms may not offset the costs associated with entering into the licensing arrangements. This can however be overcome through the establishment of intermediaries, such as IP trading platforms, public auctions, and websites that bring licensors and licensees together through standardized arrangements. It is to these that we turn now.

10.6 INSTITUTIONAL RESPONSES TO IP MARKET FAILURES

A host of new monetization mechanisms—such as securitizations, pooled patent portfolios, public auctions, and financial exchanges—have been implemented in order to extract value from IP. Securitizations describe a financial instrument which an issuer creates by combining other financial assets and then marketing different tiers of the repackaged instruments to investors. The process can encompass any type of financial asset and promotes liquidity in the marketplace. By combining assets into one large pool, the issuer can divide the large pool into smaller pieces based on each individual asset's inherent risk of default and then sell those smaller pieces to investors. The process creates liquidity by enabling smaller investors to purchase shares in a larger asset pool.[8] Annie Leibovitz for example securitized the copyright of her photographs and in that way successfully raised substantial sources of funding for her future work. In doing so, she followed the approach taken by David Bowie earlier on, who securitized the copyright to his music very successfully. ("Bowie Bonds") The fashion retailer BCBG securitized its trademarks a couple of years ago and even Dunkin Donuts used its trademark to raise successfully capital through a securitization. Yet, these securitizations of various forms of IP have been the exception to the rule and are not used on a broad scale; primarily because it is hard to identify IP that offers solid future revenue streams.

Against this background IP Exchanges seemed to offer the much needed market place to trade otherwise illiquid assets and thus promote more active markets for technology and other forms of creativity. These intermediaries seek to enable non- or under-utilized patents and to be traded in a transparent marketplace. An exchange is valuable because it makes patents as well as being available to those that are in the best position to monetize them. Certain firms may own valuable patents but have insufficient complementary assets to monetize them (Teece, 1998). An effective exchange mechanism for patents reduces the need for complementary assets to commercialize a product. They can thus be seen as important enablers of open innovation. Exchanges enable innovating firms to monetize their rights without the considerable capital traditionally associated with this (Serrano, 2006; Chesbrough, 2006d). Formal secondary markets for patents are believed to level the competitive playing field by lowering entry barriers and undermining privileged access to technology (Fosfuri & Gambardella, 2001). Again, as with IP securitizations, this type of secondary market is embedded, and relies on a set of financial and regulatory institutions.

The Intellectual Property Exchange International (IPXI) is one intermediary that has attempted to establish such an exchange platform for patent rights. IPXI is modeled after the Chicago based Climate Exchange and offers to trade standardized Unit License Right (ULR) contracts. The Unit License Right contract seeks to turn patent rights into a more transparent and standardized commodity by enabling buyers to utilize a standard setting along with third-party monitoring and enforcement technologies to facilitate exchange. This should enable patent owners to license their technology in a non-discriminatory way to a variety of interested parties. The pricing mechanisms of ULR contracts are rather complicated (Ghafele, Gibert & Malackowski, 2011).

IPXI started its operations in early 2012 and found more interest from patent owners wanting to offer their patents at the exchange than from potential buyers. IPXI found that it was also difficult to put the "right" patents up on the exchange. As many patents are of low value and only a few are worth trading, it has been very difficult to find out which patents to trade and which ones to leave untraded. Finally, many companies that IPXI approached had difficulties understanding the value proposition of IPXI and the concept of IPXI altogether. Very likely more educational work will be needed before IPXI will be able to observe the trading volume it would like to see. The example of IPXI shows that patent exchanges encounter a number of difficulties that the exchange of other commodities does not because of the nature of the rights being traded. Patent rights are by definition a claim to unique and novel technologies. The rights traded are thus extremely heterogeneous. Trading patents is not like trading sacks of rice or bars of gold. A lack of common valuation standards and a multitude of different types of rights complicate the process of turning patents into a standardized and tradable commodity. IP rights cannot be efficiently traded in a transparent market space until there are adequate standards for valuing them (Hagelin, 2002).

Contrary to IPXI's rather complex mechanisms, the TAEUS PatentBooks is less complicated in its approach and has for that very reason found more reflection in business. TAEUS PatentBooks aggregates patents and this enables manufacturers to license all the necessary patents for a specific technology in a single transaction. Royalty income is distributed among patent owners according to the quality of patents submitted. PatentBooks offers significant advantages to both product manufacturers and patent owners. Enabling manufacturers to license hundreds, even thousands, of product-specific patents in a single transaction at a competitive price tag, the PatentBooks eliminates royalty-stacking problems and prolonged bilateral negotiations among multiple stakeholders.

The primary rationale for PatentBooks licenses is based on the fact that the manufacture of technological products requires licensing multiple patents from a variety of actors around the globe. The major value proposition of the TAEUS PatentBooks is that it significantly reduces transaction costs. PatentBooks thus primarily targets two types of firms: product manufacturers that assemble PatentBooks products and firms that own the patents included in the PatentBooks. PatentBooks reduces the total search and information costs of identifying license partners and spreads the remaining costs among multiple users.

This may lead to increased licensing activity. "Demonstration effects" suggest that the more a product becomes prevalent the more it is known, and thus the more likely people are to use it (Arthur, 1989). This phenomenon is evident in social networks like Facebook, MySpace, and Linked-In as well as online marketplaces such as Ebay. As the advantages of the PatentBooks platform are demonstrated through its use, the exchange should attract more participants and thus increase its value. Patent owners should be able to generate greater non-exclusive licensing revenue from manufacturers than they could if licensing their rights in isolation. Economies of scale also permit transaction costs reductions at the enforcement and adjustment stages, where mediation limits disputes and decision costs (Ghafele & Gibert, 2011-A). At present, both IPXI and the TAEUS PatentBooks are too new to offer much evidence of success to date. Sites like IPXI or the TAEUS PatentBooks may see little activity because licensing is still strongly associated with litigation; i.e. licensees have a strong tendency to only take a licence under threat of litigation. To what extent market participants will be ready to take a licence without being threatened of being sued remains to be seen.

ITRI, based in Taiwan, equally seeks to aggregate patents among firms to enable their monetization. The TFT-LCD (thin film transistor—liquid crystal display) alliance is a good example showing the success of this approach. In 1990, ITRI formed a joint venture with the Taiwan TFT-LCD Association to form a pooled patent portfolio of over 200 patents relating to flat-panel displays (Lee et al., 2009). The patent pool enabled local Taiwanese companies to enter the flat-panel display industry quite late. This can be explained by the significant entry barriers of Japanese and South Korean competitors. By facilitating cross-licensing deals with these competitors using the patent pool created through the alliance, ITRI was integral to the development of this now lucrative industry in Taiwan. The activities of ITRI show the multiple strategies that can be implemented to facilitate access to innovation and promote its commercialization. ITRI can be seen as a prototype that may be replicated in many different industries and regions.

Another interesting development following a somewhat similar rationale is the Open Invention Network. Open Invention Network acquires patents and licenses them royalty-free to companies, institutions, or individuals in return for agreements that these actors will not assert patents against the Linux system. This enables companies to invest in Linux infrastructure and related products without fear of infringement liability, fuelling innovation and growth around this technological ecosystem. What is interesting in this approach is that the IP is used to develop an adequate innovation infrastructure and not as a means to litigate against operating firms (Ghafele & Gibert, 2012-A).

10.7 CONCLUSIONS AND IMPLICATIONS FOR FURTHER RESEARCH

The most crucial institutional change needed to promote the management of IP under an open innovation paradigm is the change of established patterns of thought. As long as those responsible for IP strategy see IP primarily as a negative right, a right to exclude, it will be very unlikely that secondary markets for IP will take off. Changing belief systems is, however, a lengthy and sometimes desperate undertaking as history shows. Ignaz Semmelweis, Louis Pasteur, and Gregor Mendel were ground breaking scientists whose research was disregarded by their times in spite of the fact that they were ultimately proven to be right. Perhaps the strong interest in open innovation will help enlighten IP managers on alternative avenues possible for handling IP questions. If IP is to be managed pro-actively and markets for technology are to be developed, then it will be of paramount importance to look at IP as an enabling mechanism promoted under an open innovation paradigm. The TAEUS PatentBooks, ITRI, and Nike's Green Exchange have sought to overcome the existing limitation by providing sophisticated electronic trading platforms. It seems that it will be equally important to educate corporate leaders and help them frame new questions on IP. A new market is not only created by ICT alone.

Outside the domain of IP, online intermediary services have proven to be an important means to bring potential buyers/customers and sellers/service providers together. Successful examples are for example Ebay, Booking.com, Kiva, Zipcar, or Expedia. The success of "infomediation" is based on lower transaction costs as search, coordination, and payment costs decrease (Ng & Yip, 2010). By establishing a space for price comparison and the aggregation of buyers and sellers, it makes a market investable and paves the way for secondary markets for IP. The exchange and pricing of new commodities as offered by IP exchanges bears the potential to stimulate liquidity, transparency and standardization in a manner that positively impacts economic growth. The creation of financial instruments to monetize non- or under-utilized assets encourages greater investment and can even create totally new markets. Standardizing the valuation procedures of the asset in question and rendering price responsive to market fluctuations is still the most significant obstacle to establishing such a market. Various IP Exchanges have created complex financial instruments to try to overcome this limitation, and this commoditization process has helped them tap into previously inaccessible markets. Such innovation fuels a virtuous cycle of productivity growth that underpins a stable increase in GDP.

Open innovation shows how intellectual property protection can be used in a creative way to achieve the goals of the knowledge-based economy. IP bears the potential of a tradable asset that promotes the transfer of technology and the sharing of ideas rather than the opposite. The popular perception of IP as a defensive legal tool stands in the way however. This has a negative impact on the innovation efficiency of an economy. Open Innovation provides a novel rationale for IP that goes beyond the scope to appropriate rents from inventions. It is important to understand that Open Innovation is not promoting a "gift economy," where inventions are freely revealed for no economic gain. Rather, it offers a window of opportunity to a different economic regime, where firms are offered a range of new strategies to generate business from their inventions. This effectively constructs a new appropriability regime (Teece, 1986). Open Innovation is therefore a gateway to a different IP system. A system predicated on the open exchange and diffusion of ideas made possible through clearly codified inventions. If an Open Innovation rationale is applied to the management of IP, the establishment of secondary markets for IP follows by consequence. Open innovation provides the answer to what "Rembrandts in the Attic" leaves unanswered. Firms need to buy and license technology in once they realize that they do not need to have all competences in-house (Ghafele & Gibert, 2012—B).

IP management under an Open Innovation paradigm recognizes the value of knowledge exchange and uses the IP regime to ensure access. In doing so, it is paving the way for a new vision of IP. This paradigm is not founded on the ability to exclude others. It is founded on the ability to use the legislative mechanisms associated with the introduction of private property rights over knowledge. For decades the enabling mechanisms associated with the use of IP remained untapped because of a highly litigious market culture. However, IP is now increasingly accepted in the economic literature as an asset. Analysts argue that intangibles have now fully emerged as a powerful asset class (Millien & Laurie, 2008) while others propose that IP assets can be proactively managed, developed, and nurtured to enhance business value (Reilly & Schweihs, 2004). The shift from an intellectual property—"rights" perspective towards an intellectual property "management" paradigm is a key factor in discovering creative mechanisms to leverage the intellectual property system as a means to fuel Open Innovation. Open Innovation has been at the forefront of this exploration by developing new licensing agreements to ensure openness and thus effectiveness of innovation. Attempts to develop secondary markets for IP have been made before and failed. They failed, not because the idea is wrong, but because an open innovation paradigm has not been sufficiently embraced by market participants. As long as corporations are only willing to take out a license under the threat of litigation and suffer from "not invented here" syndrome, it is not very likely that secondary markets for IP can be established. The absence of buyers/licensees illustrates how little understanding there is about the enabling opportunities associated with IP. This chapter hopes to have made a contribution to change that.

NOTES

1. The literature review for this chapter was undertaken by Robert O'Brian and Eric Motycka. The work on IP intermediaries and IP exchanges draws upon a body of research that Roya Ghafele created jointly with Benjamin Gibert over a couple of years of collaboration.
2. As of June 25, 2012 there were 1397 records available for export from the ISI Web of Knowledge database related to "intellectual property" in journals related to business, economics, and management. HistCite is a powerful tool that helps in reporting this data.
3. Note that licensing can be used to create value from knowledge that is not patented, and may not even be codified. The licensing of know-how, for example, is a frequent aspect of many technology licenses. See Arora et al., (2001a) for examples from the chemicals industry.
4. The category of "neglect" or "do nothing" is usually excluded from academic study, but from our casual observations of research managers, this is a frequent path for many early stage inventions.

5. http://www.guardian.co.uk/technology/2011/jul/01/nortel-patents-sold-apple-sony-microsoft
6. http://www.bloomberg.com/news/2011-08-17/kodak-worth-five-times-more-in-breakup-with-3-billion-patents-real-m-a.html?cmpid=yhoo
7. Citation to Chesbrough and Crowther, *Beyond High Tech*, 2006.
8. Read more: http://www.investopedia.com/terms/s/securitization.asp#ixzz2HP6Fv9cj

11

Managing Inside-Out Open Innovation

The Case of Complex Ventures

Henry Chesbrough and Chris Winter

> The search for innovation needs to be organizationally separate and outside of the ongoing managerial business. Innovative organizations realize that one cannot simultaneously create the new and take care of what one already has. They realize that maintenance of the present business is far too big a task for the people in it to have much time for creating the new, the different business for tomorrow. They also realize that taking care of tomorrow is far too big and difficult a task to be diluted with concern for today. Both tasks have to be done. But they are different. Innovative organizations, therefore, put the new into separate organizational components concerned with the creation of the new.
>
> (Peter Drucker, 1974: 799)

11.1 INTRODUCTION

Open innovation contains two important processes: an outside-in process that seeks external ideas and technologies in one's own business; and an inside-out process that allows and enables unused internal ideas to go outside for use in others' businesses. Both processes can contribute to discovering the "new" that Drucker talks about. In this chapter, we explore the second process, the inside-out process in open innovation.

As we shall see, since Drucker's advice of a generation ago, many companies have tried to separate their new business endeavors from their current business structures, in an attempt to generate additional business growth. These attempts have generally met with only temporary success. For example, in the 1960s and early 1970s, 25% of the Fortune 500 had a corporate venturing program (Fast, 1978). These were largely disbanded, though, during the late 1970s. Then in the early 1980s, as the independent venture capital market grew, corporations renewed their interest in corporate venturing. These initiatives were again discontinued after the market downturn in 1987. Then, as the extended bull market of the 1990s gained momentum, corporations again re-introduced corporate venturing activities (Yost, 1994).[1] The crash of the dot-com era witnessed another retreat from corporate venture activities, while the anemic recovery of the economy from the Great Recession of 2008 has also dampened enthusiasm for venturing inside companies.

Yet what is the alternative to external venturing? Over the last 40 years, industrial research has come under increasing pressure to justify its continued funding. While central research labs have discovered numerous important technologies, their owners have found it increasingly difficult to create new businesses from these discoveries.[2] Other companies, with seemingly less investment in basic research activities, have frequently appropriated much of the gain from new innovations that originated in these central labs. Often, even startup companies have been able to commercialize new discoveries ahead of the companies that invested in the early-stage research that led to those discoveries.

Indeed, at least some trends augur for another revival of corporate venturing. Corporations have enormous amounts of cash as of this writing, and are eager to find new growth opportunities to use that cash. It is unlikely that they will return to funding large internal laboratories to conduct industrial research. Instead, it is more likely that they will opt for an increase in inside-out open innovation.

In this chapter, we will recap the experiences of venturing in Exxon, in Xerox, in Lucent, and in British Telecom's Brightstar, in order to explicate some of the processes involved in inside-out open innovation. These companies nurtured internal R&D labs for many years, but struggled to utilize much of the output from these labs. Venturing becomes an important alternative pathway to market in such instances. Obtaining value from this unutilized R&D output requires inside-out processes that go beyond traditional corporate venturing (where a corporation invests in a startup), to processes that follow a technology from a university into a startup, and from a startup into a larger company and a surrounding ecosystem, with multiple shifts in the business model along the way.

These "pivots" in the business model, and these transitions across different types of organizations, are quite difficult to achieve in a traditional corporate management process, yet must be managed if inside-out processes are to deliver value to the corporation. We further argue that the corporation benefits from a healthy inside-out process, both in its businesses, but also in its relations with internal employees and external ecosystem partners. A particular kind of spin-out venture, which we term a *complex spin-out* venture, offers some natural advantages to corporations in comparison to traditional venture capital firms. We examine the processes that lead to such complex spin-outs, in companies with significant internal R&D investments.

11.2 PREVIOUS RESEARCH ON CORPORATE VENTURING

The first academic evaluations of new venture organizations in corporations were rather cautious in their assessments. Von Hippel (1973, 1977) reported that when the parent firm had significant prior experience in that market, the new venture was much more likely to succeed (vs. having experience with the technology, which was *not* associated with better outcomes). He also noted the problems that venture sponsors faced in building and sustaining internal support for new ventures from the top management of the company.

Fast (1978) conducted another study that attempted to explain the factors that were associated with the success of "new venture divisions" (NVDs). In addition to the issues von Hippel identified above, Fast found a surprising third problem encountered by NVDs inside an organization: the problem of new venture *success*. Fast found that successful NVDs were often viewed as threatening to established businesses in the parent firm. As the venture realized greater success, it required more resources, and these resources were perceived to diminish the amount of corporate resources available to other businesses in the firm.

Rind (1981) further explored the potential inherent conflicts of interest that can arise between the sponsoring firm and the new venture it is trying to cultivate. He noted that if the venture was serving a market already served by the parent firm, that might constrain the venture's marketing options so that they didn't conflict with those of the parent firm. A further issue that Rind identified was the problem of governance: the costs required to manage a new venture successfully would be incurred early in the venture's life under one NVD manager, while the benefits to those investments, if they indeed occurred, would arise later on under another manager. This could create perverse incentives for new venture managers to avoid costly, risky decisions, because they will incur the costs of those decisions, yet may not be around to receive credit for their subsequent benefits.

A study by Siegel, Siegel, and MacMillan (1988) studied the potential conflict between two frequently cited rationales for new venture businesses. One rationale is strategic: to exploit the potential for additional growth latent in the company. A second rationale is financial: to create additional revenue and profit in the new venture itself. Siegel et al. point out that, to maximize the financial return from the new venture, firms are best advised to provide complete autonomy to the new venture's managers. However, if the primary motivation for the venture is strategic, then providing this greater autonomy increases the potential likelihood of conflict with the established businesses of the company. Here, the firm may need to intervene, in order to manage the potential conflicts between the new venture and the established business. Such intervention will likely have the effect of lowering the autonomy and hence reducing the financial performance of new ventures.[3]

Overall, previous studies of corporate venturing activities have reported significant difficulties for the sponsoring companies. There are problems with developing the relevant market experience. There are problems of adverse selection. There are conflicts between the strategic objectives of new ventures, and their financial objectives. There are issues of compensation and internal equity. There are even problems of resource allocation if a new venture actually succeeds.

We turn to four examples of inside-out open innovation, some of which are successes, others of which are failures, to discern some larger patterns in the effective management of the inside-out process.

11.2.1 Exxon's Natural Experiment in Corporate Venturing

As part of its strategic mission to diversify its businesses away from an exclusive reliance on the petroleum industry in the 1970s, Exxon embarked on a two-fold corporate venturing program.[4] One portion of the program was a series of external financial investments alongside private venture capital funds, to be followed by a second program of internal ventures that were to be started and managed in a special unit inside Exxon. The Exxon strategy was to (1) probe and assess new venture opportunities via external investment, then (2) invest in the most promising of these venture opportunities via internally funded venture organizations.

There were 18 such external investments made under the first program, starting around the year 1975. Exxon invested approximately $12 million in these external startup companies. These performed well financially: of the 18 ventures in which Exxon invested alongside other private investors, three of them were sold to other companies at a profit, and five went public via an initial public offering (IPO). By 1982, Exxon's investments in these firms were worth $218 million, for an internal rate of return of approximately 51% per annum (assuming all investments were made in 1975, and making no adjustment for inflation). This was an impressive success in financial terms, whether compared to Exxon's overall rate of return, or to the median return of similar vintage private venture capital funds.

Following through on its strategy, Exxon then initiated 19 internal venture activities, to commercialize the most promising areas identified through its external investment programs. One might have expected the internal programs to fare even better, due to their ability to select areas where significant opportunity had already been demonstrated through the external investment probes.

To the contrary, Exxon's financial results were virtually non-existent from these internal ventures compared to those from its external investments. None of the 19 entities achieved an external liquidity event (such as the sale of the company to an outside firm, or an IPO). None of the 19 ever managed to reach a breakeven point, where their revenues were covering their costs. Exxon terminated and wrote off all of the internal ventures.

Why did this happen at Exxon? Chesbrough (2000) identified four factors:

1) Low-powered incentives to Exxon managers.
2) Misallocation of venture financing.
3) Slow decision-making processes inside the corporation.
4) Myopic evaluations of business potential when the venture's business model differed from that of the parent company.

This last point reflects the influence of the business model of the parent company over the perceived value of a new venture. Exxon's estimates of new venture potential were biased by its own business model (Chesbrough & Rosenbloom, 2002), causing it to fail to invest in potentially disruptive business models (Christensen & Raynor, 2004).

11.2.2 The Challenges of Managing Inside-out Open Innovation at PARC

In reorganizing to capitalize on its technology, Xerox created a new entity in 1989, headed by Robert Adams that effectively established an internal venture capital fund for managing spin-offs. Adams had created a new business for Xerox out of PARC's laser technology, which had fostered a multibillion dollar product line within Xerox. He had the support of Xerox's CEO David Kearns for establishing the new internal structure. It would allow Xerox to exploit Adams's skills in obtaining value from new PARC technologies by allowing him and his team to scan Xerox's technology base and identify investment opportunities that might otherwise have been missed. This structure was called Xerox Technology Ventures (XTV).

XTV was given $30 million of initial capital to manage. It was set up formally as a corporate division within Xerox, but Adams negotiated special terms that mimicked many aspects of independent venture capital firms. Eighty percent of the gains of the fund would go to Xerox, while the XTV principals would share the remaining 20% of the gains among themselves. For investments below $2 million, the XTV principals had sole discretion in whether and when to invest. For investments over $2 million, an oversight group called the Management Board (consisting of the XTV principals, the Xerox CEO, CFO, and a senior staff executive) had final authority. This promised a far shorter deliberation process than was typical of the Exxon experience above.

The intention of XTV was to align Xerox's incentives with those of the spin-offs and to employ venture capital processes to create value for Xerox technologies whenever those technologies did not fit with Xerox's own business model. If there were promising technologies within Xerox that were not being utilized in Xerox's businesses, XTV provided a new path for them to enter the market. If that technology subsequently created significant economic value, XTV's investment enabled Xerox to participate substantially in the resultant value. As Kearns stated in 1993, "XTV is a hedge against the repeated missteps of the past."[5]

By 1996, the XTV pool of $30 million committed in 1989 had returned a total of $219 million to Xerox, after fees and profit sharing with the general partners at XTV, according to one calculation.[6] This represented an internal rate of return exceeding 56%. This compared very favorably with the 13.7% average rate of return from independent venture capital funds that were also started in 1989. By any reasonable standard, XTV had been a tremendous financial success for Xerox. XTV's success occurred not because it created a larger number of spin-offs but because it relied on processes that were more compatible with venture success and bypassed Xerox's internal processes. At the same time, XTV enabled Xerox to own more of the ventures and thus to reap more of the profit when they succeeded.

Yet, despite this performance, Xerox elected to exercise its right to terminate the fund in 1996. Some senior Xerox managers felt (hearkening back to Siegel, Siegel & MacMillan, 1988) that the autonomy given to Adams compromised the ability to develop strategic synergy between the spin-off technologies and Xerox's internal businesses. Once the technologies were diverted into startup companies funded by XTV, the spin-offs treated the Xerox businesses just as they would any other potential customer. There were allegations that some of the success of the most profitable XTV companies, Documentum and Document Sciences, came partially at the expense of Xerox products that customers would have bought instead. The rate of return earned by XTV took no account of this potential lost business to Xerox.

11.2.3 An Improved Inside-out Innovation Model: Lucent's New Ventures Group[7]

Lucent created its New Ventures Group (NVG) in 1997 in order to commercialize technologies out of its Bell Laboratories that did not fit with any of Lucent's established businesses. In addition to capturing value from these technologies, Lucent also wished to speed up the time it took for its technologies to go into its mainstream businesses as well. In contrast to Exxon, Lucent was not interested in unrelated diversification in its growth. In contrast to Xerox, Lucent was not focused on a financial return, but rather focused on additional processes to identify new adjacencies and business opportunities around its core businesses. Chesbrough and Socolof (2000) provide a detailed review of this model.

To manage the cultural change process required for the inside-out model to work, the NVG consciously created what became known internally as "the phantom world." The phantom world was a "half-way house," which would enable people and ideas that weren't ready or able to go out directly to obtain pure venture capital to develop their ideas further within Lucent. By being sensitive about the cultural gaps that had to be bridged, and by being sensible about the right mix of risk and reward to offer, the process created a launching pad for ideas to move out of Bell Labs into markets outside of Lucent's traditional business channels.

Prior to spinning out of Lucent in 2001, the NVG had invested in 28 ventures. Most ventures were in the Internet, networking, software, wireless, and digital broadcast spaces, which were of strategic interest to Lucent. While most investments did not achieve liquidity, five ventures that have reached liquidity have brought in an 80% return on invested capital for NVG's fund.[8] The fund thus was successful financially, as was Exxon's and Xerox's. But it also enabled Lucent to accelerate its entry into promising growth markets like optical networking years ahead of when it otherwise would have been able to do so. This strategic benefit was absent from Exxon's and Xerox's earlier experience.

The NVG process also served as an impetus for Bell Labs technologies to move off the shelf. Once the NVG group identified a promising technology within Bell Labs, the Lucent business units had only a limited amount of time to consider whether or not to take over the technology themselves, and fund its further development. In the past, the business units could wait and see whether a technology would become important, and this often delayed the introduction of new technologies to the market. When the NVG served notice that it was interested in commercializing an internal technology, that effectively became a forcing function, accelerating the speed with which technology is moving out of Bell Labs into the market.

The NVG process also provided more rapid feedback on the value of the technology to Lucent. The three instances where Lucent reacquired an NVG venture arose when it became clear that the technologies were too important to Lucent to have them managed independently of the company. This strategic value would not likely have been visible, had the technologies continued to sit on the shelf. The ability to take them to market through new ventures allowed the market to provide a "second opinion" to the earlier judgment of Lucent's business managers, who judged earlier that the technologies were not yet ready for the market.

The NVG operating model also brought in people from outside to help launch new ventures from within Lucent. Managers hired from outside of Lucent received substantial equity options, a commitment to achieve liquidity for that stock, and a pursuit of financial success no matter what the cost or impact is upon the parent companies' business. This commitment was made more credible by NVG's willingness to syndicate its funding of new internal

11.2.3 BrightStar—Developing a Strategic Venturing model

BrightStar was a corporate incubator formed at the UK-based telecommunications firm BT (formerly British Telecom) in 1999/2000. It was driven by three observations: first, the R&D facility had an international reputation for invention but the business units and investor community failed to see any material benefit from the R&D program in terms of products, services, and value creation; secondly, the company was poor at selecting and driving those technologies out of the labs into the business units, with the latter at times acquiring such products from outside the company; and thirdly, many of its more entrepreneurial managers were leaving BT and creating successful start-ups rather than exploiting those talents and ideas inside the company. This role of attracting and retaining key talent is an under-appreciated benefit of robust inside-out open innovation policies.

Thus Brightstar was conceived to tackle a number of issues simultaneously: monetization of the large pool of under exploited IPR; creation of products and services the new businesses could then sell to and through the parent; and thirdly creating and sustaining a more entrepreneurial process and culture. From its initiation there was as much focus on the strategic goals of supply back into BT and training, education and culture change than simply the creation of value from IP, with these activities being key parts of the strategic story of Brightstar.

In its initial concept Brightstar was conceived as creating new businesses inside BT in common with many earlier BT attempts at stimulating corporate innovation. However, the management team quickly altered the strategy to spinning such businesses out as a better way to ensure they could operate in an entrepreneurial fashion. It was felt that BT processes and decision making would quash the speed and innovation required and that appropriate culture and decision making would only occur if the company was free of the internal processes.

The model developed was similar to that of New Venture Group in Lucent with the intention to retain a significant stake in the businesses as they grew. Projects that were to be spun out were moved into the Brightstar business unit where they were run as though they were separated businesses. Like the Lucent model this enabled external experienced entrepreneurs to be brought in, for the culture and processes to transition to a suitable point for spinning out, and to market test the business model in the relative safety of BT. The intention was to hold the businesses in this transition state for 6–12 months prior to spinning out with a minority but significant stake.

The BT businesses created were a mix of technology exploitation, new products and internal mission-critical systems that could be offered to third parties if spun out from BT. The latter differentiates the Brightstar model from some other corporate venturing activities. Internal systems that were admired by competitors and companies in the sector were spun out with the intention of reducing the cost of the internal supply (as is typically used to justify an outsource arrangement), but also to capitalize on the value of selling to third parties in order to increase volumes, set standards, and spread fixed development and support costs over more customers. These are clear examples of inside-out open innovation (though the expression was not used by BT at the time). Examples include companies such as Vidus, who were responsible for scheduling BT's field workforce, and Azure, who ran BT's interconnect billing system. Both of these companies were successfully exited in due course.

Brightstar also put in place a unique management and governance process in order to avoid accusations that it was operating against BT business unit interests and did not understand the venture world. First, it recruited a blend of managers for the business unit including experienced BT insiders, people who had worked for both BT and start-ups and people employed as contractors with solely start-up and venture experience. Secondly, it created an investment advisory board where all projects were reviewed, which consisted of two external venture capital firms and senior managers from each business unit. This ensured that candidate projects were suitable for venture funding and that early contacts with a champion in the business unit(s) were established.

The Brightstar experience was that spinning out companies that are focused on supply back to the parent still required active and aggressive management of the links back to the business. The business unit often did not share directly in the value created in the start-up and was thus inclined to treat the start-up as an unrelated business rather than part of the BT family. The presence of a champion inside the business unit (via the business unit's representation on the investment advisory board) helped overcome this tendency.

Brightstar also became a training ground for managers with much of the team working on secondment and temporary placements into the unit. Training courses were devised; the analysis tools used in Brightstar became adopted elsewhere in the company; and involvement in Brightstar itself was frequently cited on CVs. Brightstar managers believed this was as important for BT in creating culture change as the value in the companies themselves. A number of BT managers who spun out were welcomed back if they subsequently chose to leave their companies.

While the model performed well on its strategic and financial goals, by 2003 the model nonetheless was under heavy strain due to the virtual shut down in the availability of venture capital in Europe in the period 2001–2. Brightstar had revamped the model from the first two years so that it held companies longer in incubation from the desired 6–12 months; however this had a negative impact on the business unit's financial position. The challenging financial environment of 2001–2 led to a mandate for Brightstar to be EBITDA positive—despite creating start-up businesses. Brightstar looked for an alternative financing structure to enable it to continue its mission. It had always been run off the corporate P&L and not developed a stand-alone fund. As a result it was decided to spin-out Brightstar itself out and supply the service back into BT.[9] The 2001 Lucent deal provided an ideal template, with the enhancement that BT agreed to a five-year deal to supply venturing and incubation services back into BT with an exclusive access to BT's technology base to create new ventures. Brightstar thus was bought out from BT and merged with the former Lucent team to create New Venture Partners.

This progression from the early experiments at Exxon to the more recent ones at Lucent and BT show that inside-out open innovation can play a number of important roles in corporate innovation. One role is to generate new options for future businesses. By spinning out projects that do not fit well with the current business model, inside-out venturing processes enable these projects to search more broadly than would be possible internally for alternative business models. Another role is the possibility of financial benefit, should the venture prove successful. Over time, however, this possibility is given less weight, relative to the other roles played. A third role is the attraction and retention of key talent in the organization.

Yet all of this venturing activity occurs in the shadow of private venture capital (Chesbrough, 2000). If corporate venturing is to be sustained over time, it must generate results that are difficult for traditional venture capital processes to obtain. Below, we examine one such result, the complex spin-out.

11.2.4 The Next Level of Inside-out Open Innovation: Constructing Complex Spin-outs

Start-up companies can be afflicted by a well-known problem for venture capitalists: the founder's death grip. This is the situation that occurs when a founder is unable to let go of control or alter the direction of the company from the founding vision, even in the face of compelling market feedback. In comparison, spin-outs do not seem to suffer this problem. The founding team, although seeing this as their project, have a less proprietary sense of ownership than the pure entrepreneur, since the project was developed in a large organization, with input and approval from multiple sources.

This difference means it is possible to build a spin-out to order rather than launching solely with what was the initial idea. This is particularly true when the spin-out has access to a powerful technology but needs parts from outside the organization to build a complete customer proposition. Through assembling these different parts, corporate venturing can be more than a start-up funded by an existing company. It can be employed as a tool to achieve new business opportunities for the company that it would not likely have been able to build on its own in a similar time frame within an acceptable budget. We call this process of building a spin-out from multiple sources a *complex spin-out*.

British Telecom (BT), after it spun out Brightstar, proceeded to use this approach to build two complex spin-outs with solutions that ideally suited its needs. It had already spun out Vidus (discussed above), which specialized in field service workforce scheduling for the large enterprise and service provider markets. However, BT's sales teams realized there was an opportunity to produce a company that offered field service management products to mid-sized enterprises that would complement the existing workforce scheduling offering BT had already taken to market with Vidus. One of the key differentiators that such a service offering would need is the ability to satisfy customers with a much lower cost support process. Although BT had a strong technical team, which could develop the product, there was little experience inside BT for developing such field service management products. It was felt better to look for a solution outside where a more flexible, low cost route to developing and testing the market could be achieved, and then incorporate that into Vidus.

BT identified a product that was a partial solution in a small UK company. At that time the system being sold was mainly as a product sale and it was felt that a SaaS business model and a broader range of product features would be required. It was also felt that taking this small company inside BT would kill the ability to innovate the new products and services needed to find the combination that worked in the market place. BT thus approached its corporate venture funding partner—New Venture Partners (NVP) – to develop the complex spinout opportunity together. As part of its broader vision and strategy, NVP invested in the UK company and looked to strengthen and widen the product set by extending the UK company's field service management offering into mobile supply chain management products. NVP identified an opportunity to spin-out a non-strategic division from a US company in this sector. This was done and two entities were merged to form Airversent, resulting in an international presence and wider product set for the spin-out. The resulting product and service set was sold through two BT business units in the UK and Europe targeted on its top 21,000 businesses. The company quickly began generating substantial sales across multiple channels of distribution, amounting to tens of millions of pounds annually.[10]

An opportunity emerged to strengthen further the spin-out's offerings to extend to managing, tracking, and storing documents and to gain access to a heavy weight SaaS architecture that had been developed by AirClic. This offered the venture and BT access to further products and extended the size of company that the offerings could address. This interesting combination of spin-outs, VC backed ventures, early stage M&A and venture scouting quickly led to a company with a viable size and market presence, with BT acting as a channel to market. BT gained access to products and services it would never have specified internally and to quick and flexible testing of those businesses in the market place. Best of all, BT did not have to fund all the development internally and solely bear the risks of failure along the way. Instead, the risks and financing were shared with external investors. Thus, the complex spin-out was in one way an R&D project that didn't require internal R&D resources to execute.

A similar complex spin-out model was used to create a governance, compliance, and risk management software for corporate enterprise software customers—Neohapsis. This involved BT identifying an internal need as well as a market opportunity. New Venture Partners, the above-mentioned VC firm, managed the spin-out from a US company of the key product, licensed additional technology from GE, added some BT technology and deployed the solution within BT as its first customer. Subsequently a further merger occurred to add a professional services arm to the product strategy. This is another facet of complex spin-outs: the opportunity to engage with the corporation as a customer for the technology. As a customer BT is able to influence the new venture to produce an offering to its specifications, and is only committed to purchasing the result if those specifications are met. Relatedly, a large corporation could utilize a spin-out to become a supplier to a newly created customer, so that the further work needed to sell into the market is borne by the customer, not BT. Whether as a customer or as a supplier, these complex spin-outs create options for BT, where the company has the right, but not the necessity, to take a further action later in time.

Deals such as these—integrating multiple corporations' technology and executing early stage roll ups—occur more easily in spin-outs than in pure VC-backed start-ups. The venture capitalists are probably more adept at the small scale deals that are needed than corporate M&A departments, and the corporations often possess in their product lines and R&D facilities the missing pieces (that often lack any clear internal path to market on their own) needed to quickly build a winning proposition. New Venture Partners in the last few years has carried a number of such complex spin-outs and it may be that companies "built to order" represent a new wave of innovation strategy. It is worth noting that in both the quoted cases the parent—BT—had strategic sales channel or internal needs that drove the model. The availability of these inside-out ventures allowed BT to increase its utilization of these strategic resources, or what David Teece would term "complementary assets" (Teece, 1986).

11.2.5 The Cultural Benefits of Complex Spin-outs

This chapter has focused on the strategic reasons for using inside-out open innovation processes as a method of innovating. It has also covered many of the operational reasons that such models can fail. However, in the BT model there was as much focus on the wider implications for creating more focus inside the R&D department on commercial exploitation and in creating an entrepreneurial mindset, as there was on financial exploitation of under-utilized assets. Incubators such as those created at Philips also provide this wider cultural environment.

At an operational level it is important in such vehicles to consider the cultural impact of the management style and processes used. For instance, the rate of rejection of ideas for new companies should substantially exceed the number of businesses created, typically in a ratio >100:1, much higher than occurs in many internal project management systems. This means most of the time the venturing team will be turning down proposals. The way they are rejected and the advice given on rejection will be critical in creating an innovative culture. In a similar manner the decision-making process needs to be transparent and fast.

If done well these processes lead to wider benefits. At BT there was a steady stream of managers looking for secondment to the Brightstar incubator as it was perceived as having value in training and career development. People sometimes were seconded into the businesses or supported the venturing team. This led to a wider dissemination of the innovative processes and culture than the simple spin-out of companies would suggest. Unlike scouting technology or innovations outside the parent organization, inside-out innovation has the potential to change the culture and processes internally to support a more innovative environment and make the organization more open to new ideas from both inside and outside the main operating culture.

CONCLUSION

Like other aspects of open innovation, managing inside-out open innovation requires a mindset change in the way innovation is managed, seeing it as part of an ecosystem rather than something controlled solely from inside the corporation. Opening up the venture creation process to other start-up technologies and strategic partners requires different skills and insights to manage innovation beyond those of a traditional VC. It is linked much more deeply to the culture inside the parent organization and thus has much more potential to create, drive and reward business model innovation in the parent than does traditional venture capital. The concept of inside-out open innovation processes, to search for alternative business models, and spinning out complex ventures and technologies as a strategy for creating multiple innovation options, thus deserves to be much more widely used and appreciated.

Freedom from the "entrepreneurs death grip" also allows such complex spin-outs to carry out early stage M&A; with the company built like a jigsaw puzzle out of pieces sourced not just from the parent but other organizations as well. These "built to order" companies need much more operational and market insight from their investors than many venture-backed companies do. These are the conditions best suited to the complex spin-out model.

Other companies have inaugurated such novel venturing processes, where the search for the business model is explicitly part of the mission. Microsoft's IP Ventures operation in Mountain View, CA, for example, places Microsoft technology and IP into external start-up companies and then sees what happens in those firms. In Europe, Microsoft also works with Enterprise Ireland and Sitra in Finland to use its technologies to spur new business models for its technologies in small and medium enterprises (SMEs) (Gutierrez, 2008).

Large corporations are missing an innovation opportunity to use the commercially oriented, large scale R&D facilities they have to generate business opportunities, and then test and qualify them in a venture-like manner by creating multiple spin-out companies. For those spin-outs that succeed, the corporation has the option of re-acquiring the successful new product lines. This would enable them to create flourishing innovation eco-systems around their R&D facilities, speed up the metabolic rate of innovation inside their labs, and be ideally positioned to exploit the resulting disruptive businesses. The key skills will be in portfolio management and looking at the ecosystem for the options that the parent corporation is creating for its future. Equally, vital corporate resources like distribution channels, manufacturing capabilities and support networks can sustain competitive advantage for nascent spin-outs, without burdening the R&D budget.

There is a further, more human, business rationale for enabling greater external use of inside-out open innovation. Companies that forbid inside-out processes are likely to frustrate many of the R&D staff, because many of the ideas these people work on are never deployed in the market. It is reportedly quite common for a pharmaceutical researcher to never see one of her projects ship into the market, over a 30-year career, because the attrition rate of compounds is so high. This is an enormous waste of human talent, and must take a toll on any person's initiative. Companies that embrace inside-out open innovation allow other pathways for internal ideas to get into the market. These other pathways allow the market to provide feedback on those ideas, and lets researchers see their ideas in action in the wider world, even if those ideas do not make it into the company's own products. That also provides new sources of feedback for the researcher on how to improve upon those ideas, and some of those improvements might one day make it into the company's own products.

The converse of this model is the internal control and management prevalent in many corporations, a mentality that probably stops any open innovation occurring by any model. In the future it is likely that companies that cannot or will not spin out complex ventures as part of their strategy will either be crippled in their innovation strategy or find themselves with a company culture that is not supportive of innovation.

Spinning out is not an optional way to dispose of unwanted assets; it is the future of new business creation and business model innovation for truly innovative companies. Complex ventures represent one example of how greater use of inside-out open innovation processes can revitalize corporate innovation.

More research is needed to uncover other ways in which inside-out open innovation can contribute to growth and to innovation. Chesbrough and Garman (2009) make the case that inside-out processes can increase corporate flexibility in economic downturns. Chesbrough and Chen (2013) consider the use of inside-out processes to recover previously abandoned compounds in pharmaceutical drug development. Chesbrough and Ghafele (Chapter 10—this volume) discuss new ways of managing IP to stimulate greater exploitation of technologies not utilized by the company that discovered the technologies originally. But these are just initial inquiries that only hint at the ways in which inside-out processes can contribute to innovation.

NOTES

1. Block and MacMillan (1993: 13) think the cycle historically has run about every ten years.
2. Richard Rosenbloom and William Spencer, *Engines of Innovation: U.S. Industrial Research at the End of an Era*, Cambridge, MA: Harvard Business School Press, 1996.
3. This is a specific instance of a more general problem. See Williamson, 1985, Chapter Six, for a seminal discussion of "the problem of selective intervention," or why a large company cannot do everything a small company can do, and more.
4. This section closely follows Sykes' (1986) first hand account of this experience.
5. Kearns' quote is taken from Larry Armstrong, "Nurturing an Employee's Brainchild," *Business Week*, October 23, 1993, 196.
6. Lerner (Xerox Technology Ventures, Harvard Business School case #295-127, Exhibit 5) provides the general terms of the XTV structure and the calculated financial returns shown here.
7. See Chesbrough and Socolof (2000) for a more complete description of Lucent's New Ventures Group. This section draws heavily from that paper.
8. This is a cash-on-cash return, and excludes markups taken on private companies in subsequent rounds.
9. As we will discuss below, Brightstar itself became a complex spin-out venture of BT as a result of this process.
10. One challenge in complex spin-outs can be quantifying their financial impact after the fact. When we probed for the results of Vidus and Texert, we learned that their financials were co-mingled with those of their partners and distribution channels, making a precise valuation problematic. While a result of tens of millions of pounds in annual sales is brilliant performance for a start-up, it is a very small amount of revenue to a company the size of BT.

12

Patterns of Implementation of OI in MNCs

Letizia Mortara and Tim Minshall

12.1 INTRODUCTION

Open innovation (OI) is an innovation paradigm first described by Chesbrough ten years ago (2003a). Since then it has received substantial interest from practitioners, with firms in many sectors progressively adopting a variety of open innovation practices. Policymakers are also increasingly recognizing OI's potential economic value[1] and are encouraging its adoption. Some academics envision a future where the OI model will be fully integrated in firms' innovation management activities (Huizingh, 2011) even though a recent review (Schroll & Mild, 2012) estimates that OI activities have currently been implemented in only 20–45% of companies (Schroll & Mild, 2012) and very few single determinants behind the adoption of OI practices have been identified. The highest rate of agreement is that turbulent environments (industry, market, and/or technology) encourage the adoption of OI (Schweitzer et al., 2011), whilst there is high uncertainty on other factors.

Underpinning the successful OI implementation is an understanding of *how* firms adopt this model to determine what approaches work and what do not. As OI is for many firms an innovation in itself (Christensen, 2006), studying OI adoption also presents a new opportunity to enhance the theory of innovation implementation.

Only a few scholars to date have committed efforts to researching OI implementation patterns (e.g. Chesbrough, 2003a; Chesbrough, 2006a; Chiaroni et al., 2010; Chiaroni et al., 2011; Christensen et al., 2005; Mortara & Minshall, 2011). This line of enquiry has been popular particularly in European firms (and more specifically, in Italy, e.g. Buganza et al., 2011; Chiaroni et al., 2010; Chiaroni et al., 2011; Di Minin et al., 2010; Pellegrini et al., 2012; Petroni et al., 2012).

In an attempt to consolidate the knowledge to date on OI implementation, this chapter reviews the richest available data, i.e. the evidence on OI adoption in large multinational corporations (MNCs). Based on case study research (the method we consider best suited to revealing the complexities that underpin the adoption of OI), this chapter develops a model of OI implementation and highlights future research needs. In section 12.2, we summarize the relevant theories useful to study this phenomenon and the general trends applicable to OI implementation. In section 12.3 we delve into a review of the evidence published in the literature, highlighting the patterns and key traits of OI implementation that can be used to characterize OI adoption in firms. Finally, we propose a framework which can be used to examine the OI implementation approaches of firms with process research methodologies (Pettigrew, 1990).

12.2 CHESBROUGH'S OI MODEL: A POINT OF DISCONTINUITY IN OI IMPLEMENTATION

Evolutionary theory has been used to analyze OI adoption in firms (Christensen et al., 2005). Accordingly, companies have adopted various approaches to opening their innovation processes but only the "fittest"—i.e. the most successful in that context—survived. However, many of the cases in literature are "static" descriptions of the OI practices and models such as those of Whirlpool (Muller & Hutchins, 2012), Air Products (Tao & Magnotta, 2006) or P&G (Huston & Sakkab, 2006), whilst the abandoning of specific types of innovation activities is not widely recorded. The study of "negative" or "failure" examples (Chesbrough, 2012b; Huizingh, 2011) as well as studies of the disadvantages of adopting OI (Vanhaverbeke et al., 2008) could help clarify what forms of OI did not work and why. It is unfortunate that the memory of negative experiences (i.e. abandoned practice) is hard to identify and document, skewing the evidence to surviving or recently adopted—but not yet tested—approaches.

Since 2003, the widespread diffusion of Chesbrough's OI model (Chesbrough, 2003a) has encouraged firms to experiment with OI practices (Mortara & Minshall, 2011), moving the adoption from a phenomenon of emergent to that of planned change (Livne-Tarandach & Bartunek, 2009). Hence, the OI model represents a new contingency in the implementation of OI. For example, when the head of R&D at FIAT started opening the firm's innovation processes during the early 1990s in response to a sharp performance downturn (Di Minin et al., 2010), there was no OI model to support the implementation of this strategy. In contrast, any firm seeking to do something similar post-2003 had the benefit of a clearly accessible OI model and widely publicized examples to support the design, development, and implementation of activities. Teleological theory would hence seem appropriate to analyze OI implementation after the OI model had become explicit. Adopting this theory, the OI paradigm becomes a goal to pursue and organizations set out to achieve its implementation faster compared to firms who opened up in a more evolutionary manner. In firms where the choice to implement OI came as a conscious decision, several other theories could concur in studying OI such as Bounded Rationality (which assumes that managers who decide to adopt OI are only marginally aware of all the possible consequences of their implementation choices (Simon, 1945)) or Upper Echelon theory (that proposes that firms are direct manifestations of their top managers' personalities (Hambrick & Mason, 1984)). Also, clearly relevant is the change management theory of how companies pursue and maintain the change momentum, and if and when they are ready for change (Jansen, 2000; Jansen, 2004).

It is also evident that the increasing debate surrounding the OI model and its associated terminology has concurred to create "noise" in the ten years since its first publication. The example of other firms, especially within similar innovation systems, provides an extra stimulus to test and develop OI practices. This is evident from the story of OI implementation at Roche (Nakagaki et al., 2012) where managers decided to experiment on OI primarily following the examples of other firms, rather than being pushed by an explicit need. Hence, in the evaluation of recent cases, researchers need to consider the "bandwagon effect" in adopting OI practices (Xu et al., 2012). This is not a new situation in management research (Abrahamson & Fairchild, 1999; Bikhchandani et al., 1992, 1998) which has often looked at issues concerning industry trends of innovation adoption and diffusion (e.g. Webb & Pettigrew, 1999).

Therefore, the study of OI implementation is particularly suited for dynamic (West et al., 2006) and process research methodologies which pursue the "how" question (e.g. Huizingh, 2011; Pettigrew, 1990; Sminia & de Rond, 2012), bringing together the knowledge about contextual (internal and external) and process (evolution of actions). Qualitative and longitudinal methods are hence the most appropriate (Yin, 2009). Although such lines of enquiry are frequently followed by change management and strategy implementation scholars (Sminia & de Rond, 2012), this approach has not been systematically used in evaluating OI implementation.

We focus on the analysis of evidence drawn from published case studies of OI implementation in multinational corporations (MNCs), extracting a set of features which could be used to characterize OI implementation. For each we highlight some key gaps which need further work. We will then show how these characteristics could be used to pursue more developed process studies of OI implementation.

12.3 OI IMPLEMENTATION KEY CHARACTERISTICS

From the current OI literature and particularly from examples of implementation in large companies, we present a comprehensive set of characteristics that can be used to describe OI implementation. These features are summarized at two levels in the following sections:

- Macro characteristics: i.e. characteristics that indicate the high-level, firm-wide tendencies of firms' OI approaches.
- Micro characteristics: i.e. internal characteristics of the process and internal dynamics of OI implementation.

12.3.1 OI Implementation Macro Characteristics

Five key dimensions characterize the macroscopic implementation of OI whose variability is also illustrated:

- Process: Inbound—Coupled—Outbound
- Internal change stimulus: Top-down—Bottom-up
- Coordination: Decentralized—Centralized
- Approach to location: "Go to key places"—"Come to me"
- Networks: Types of partners

12.3.1.1 Processes

To date, research on OI has been based on differentiating between the inbound and the outbound processes of OI, whilst many have pointed also at the existence of coupled or reciprocal processes (Enkel et al., 2009). These three labels help researchers to categorize various OI activities (formal and informal) which companies set up. It has been generally acknowledged (e.g. Bianchi et al., 2011; Enkel et al., 2009; Mortara & Minshall, 2011) that inbound processes are favored, in particular by large companies, and researchers have attempted to measure the impact of these processes on performance. We cannot delve here into the description of each of the instances of inbound, outbound, or coupled processes. However, what is still lacking is a longitudinal analysis to determine when the specific instances of processes get modified or abandoned. From examples where open approaches did not prevent a company from failing we could infer the limitations of these open approaches. For example, Kodak adopted OI approaches when faced with the threats and opportunities of digital imaging. They focused primarily on inbound OI, firstly via strategic alliances (Grant, 2012), then moving towards the model of externalization of the exploration through intelligence operations (Mortara, Thomson, et al., 2010). Although the firm adopted OI activities to support the development of an ecosystem for easy sharing of images (Easyshare) (Grant, 2012) the changes in the technology moved faster than the capability of the firm to adapt to its environment and to develop a sufficiently robust business model. The outcome of this was that the firm filed for a "Chapter 11" (which permits reorganization under the bankruptcy laws of the United States) in early 2012. The adoption of a strong outbound business model has been missing from Kodak's portfolio until very recently, with the firm only now systematically out-licensing non-strategic IP. It could be argued that enabling outbound processes at an earlier stage might have helped the firm to leverage the broad IP portfolio and build more open business models. However, adopting coupled or outbound processes is not a recipe for success as an evolution of approaches could be seen in many firms. Nokia for instance showed a reverse approach to modifying its strategy in response to a change of technology. They followed an exploitation strategy whilst developing the first two generations of mobile telephony and moved to an exploration strategy for the third generation (Dittrich & Duysters, 2007). Nokia also planned to develop an ecosystem around open source technology (see Chapter 4 and West & Wood, 2013) to establish a new stronghold based on a coupled-outbound approach. However, as this model did not match the challenge and the Symbian technology lost the battle for dominancy, they recently decided to abandon the further development of their own ecosystem and instead developed a strategic partnership with Microsoft with the same aim. Some authors (e.g. Chesbrough, 2006a; Buganza et al., 2011; Ferrary, 2011) have attempted this line of enquiry providing some initial insight. Future work should continue to review in greater detail the underlying successes and limitations of each of these business models, linking more strongly the study of OI with that of strategy (Chesbrough, 2012b) and the improved understanding of firms renewal and resilience mechanisms (Hamel & Välikangas, 2003; Vanhaverbeke & Peeters, 2005).

12.3.1.2 Internal Change Stimulus and Coordination of OI Implementation

The OI adoption impetus can be viewed as coming from two opposite directions. In many firms top management are the stimuli for the implementation of open practices such as in P&G (Cloyd & Euchner, 2012) or LG (Ryu, 2011). As might be expected, these firms have adopted coordinated and centralized approaches to OI implementation often relying on teams of OI managers (Mortara & Minshall, 2011). We will discuss the role of individuals, teams, and top managers in the next section. However, it should be highlighted that it is not uncommon for the adoption of OI to originate from lower levels in the business hierarchy and then move upwards. The stimulus may come from middle management who seek to develop innovation procedures starting by experimenting with different practices in isolated pockets (Mortara & Minshall, 2011). When these managers realize the disaggregation of their activities, they may attempt to coordinate their efforts through communities of practices with the ambition to demonstrate validity of OI and then get buy-in from the top. This is shown by the case of Roche (Nakagaki et al., 2012) which is currently relying on OI champions for the implementation and is seeking both coordination of OI implementation through an open innovation "Network of Practice" and a "eureka moment"—a demonstrator of the benefits of OI—prior to sharing the changes with top management. In this and other cases, open practices started to appear organically, distributed throughout the firm (i.e. decentralized OI). At a later stage, when the firm recognizes that OI could be used to explain the changes that have already happened, it may use OI as a language to frame and coordinate the various activities. For example, BT started many OI activities in the 1990s and early 2000s (e.g. it experimented with a corporate incubator activity known as "BT Brightstar" (Ford & Probert, 2010) for the commercialization of new discontinuous technologies internally developed. A coordination team for innovation was formed in 2006 with the name of Innovation Central that later became BT Innovate and Design. This function was set up "[...] *to deliver strategic innovation and technology vision for BT through effective and coherent engagement with other parts of the business as well through powerful relationships with third party organisations—including some of the world's most famous academic institutions.*" The OI terminology started appearing soon after in presentations and articles (e.g. Bross, 2009).

Recent observations have indicated that some companies tend to move from decentralized activities to centrally coordinating OI approaches (Mortara & Minshall, 2011) but more data is required to evaluate if reverse cycles also exist. Tirpak et al. (2006) observed that there are on occasion centralization–decentralization cycles in R&D. What is not known is whether the centralization of coordination is always the successful approach in implementation (Linton, 2002) as some recent evidence shows that both structures have benefits (Hollenbeck et al., 2011). Linked to this is the need for an investigation of the R&D structures and how they change because of the advent of OI (Petroni et al., 2012).

12.3.1.3 Approach to Location

OI implementation could also be evaluated according to the infrastructure deployed and the choice of location for different OI activities. OI infrastructure can range from the "hard" (e.g. on-campus research labs and new venture incubators) to the "soft" (e.g. on-line platforms for innovation competitions and technology scouting activities) (Minshall et al., 2014). Some infrastructure is location-independent (e.g. use of innovation competitions or website portals to attract ideas) while some geographically embedded traits are fundamental for their functioning (e.g. development of open innovation campuses around corporate R&D labs). Some infrastructure requires substantial and long-term investment (e.g. establishment of embedded labs within universities) whereas others require more modest, short-term commitments (e.g. sponsoring student projects).

There are examples of OI infrastructure aimed at attracting partners to a specific location. We brand these the "come-to-me model." Examples are the OI campuses developed on corporate R&D sites to foster an array of OI activities as well as to generate returns from high cost laboratory facilities. Among the most famous OI campus examples is that of the High Tech Campus at Eindhoven developed around the former Philips' R&D infrastructure (Doppen, 2008; Tödtling et al., 2011; Torkkeli et al., 2009). Another example is that of Colworth Science Park, formally housing exclusively Unilever's corporate R&D facilities that has now been developed and managed under a joint venture with a property development firm, Goodman. This joint venture enabled the raising of additional funding from a regional economic development agency that wanted to stimulate the economy in that part of the East of England and saw the development of OI infrastructure as an effective means to achieve that goal (Minshall et al., 2014).

The opposite approach to OI location is the "go-to-key places." For instance, the location and OI organizational structure links are important for Italcementi. It created an office in Brindisi specifically to operate in a particular European Union region with priority access to public funding (Chiaroni et al., 2011). Also Nokia (along with many other technology intensive multinational corporations) decided to set up research centers worldwide in partnership with key institutions (mainly universities) to develop long-term future options.[2]

A categorization of OI infrastructure and analysis of the suitability of different approaches in specific contexts would seem to be a useful path for exploration. This analysis would need to draw together diverse literature ranging from, among others, regional innovation systems (Asheim & Gertler, 2005), the changing role of universities within national innovation systems (Etzkowitz, 2003), use of innovation competitions (Lampel et al., 2012), business incubation (Hackett & Dilts, 2004), technology intelligence (Kerr et al., 2006) and corporate venturing (Markham et al., 2005). Such work would also complement the issues discussed in Chapter 8, where the established literature on global R&D location is merged with OI literature to provide insight into the role of geography in OI implementation.

12.3.1.4 Networks

Collaboration with external partners is the core of OI implementation and it has been widely used as a measurement of OI since the publication of Laursen and Salter's 2006 paper (Laursen & Salter, 2006; Schroll & Mild, 2012), mainly utilizing two indices: the breadth of partners (i.e. how many different types of partners) and the depth of interaction with partners (i.e. how strong is the collaboration with the partners). Some researchers have verified that openness towards customers, suppliers, and universities has a significant positive impact on the different innovation performance measures (Inauen & Schenker-Wicki,

2011). However, different and sometimes contrasting results appear in the literature relative to the merit of collaborating with each single partner type (see for example Bianchi et al., 2011). This may be because, even when the relationship is with only one type of partner, there is the potential for a variety of different types of relations to be established (e.g. with the whole university, one of its departments, or an individual academic) (Melese et al., 2009). For example, an increasing body of literature looks in particular into universities as partners. These are being encouraged to place greater emphasis on the commercialization of research results, but have variable levels of maturity with respect to commercialization activities (Philpott et al., 2011). Regulations regarding ownership of research results between firms and university partners vary with location and time (Rhoten & Powell, 2007). In particular, the Bayh-Dole Act (Kenney & Patton, 2009) has drastically modified the commercial position of US academia, and its indirect effects are also noticeable on universities in many other national innovation systems (Etzkowitz, 2003).

As for the analysis of the other OI implementation features, the understanding of the evolution of the partnerships over time requires further attention. Although network analysis has been advocated by scholars researching implementation theory (Linton, 2002) and can be pivotal in understanding how companies have evolved their OI strategy over time (Bianchi et al., 2011; Dittrich & Duysters, 2007; Vanhaverbeke & Cloodt, 2006; West et al., 2006), there is still a lack of understanding about how partners and networks impact on the process of adoption of OI in large firms and which ones best fit each context.

12.3.2 OI Implementation Micro Characteristics

The dynamics of implementation internal to the firm can be characterized by observing:

- Implementers: Top management, champions, and OI implementation teams, departments and functions.
- Divisibility: Incremental—"single stroke" implementation.
- Social interactions: Training, recruitment and retention processes and communication and PR implications of OI.

12.3.2.1 Implementers

Many scholars highlight that the role of top management is fundamental in the implementation of OI (Elmquist et al., 2009; Giannopoulou et al., 2010) but literature to date has not yet seriously approached this theme. The role of the leadership is advocated to create a sense of urgency and disruption (Buganza et al., 2011), and as the swaying element (Slowinski et al., 2009) for the establishment of OI. For instance, in P&G, A. G. Lafley's leadership for OI was reported as being key. He put himself personally at the forefront of the OI adoption and publicly set the target for 50% of innovation to be achieved using external sources. His commitment and credibility were pivotal in P&G's transformation to OI, according to Cloyd (2012). The leadership role is important also if OI is not directly the target. For instance, at BP, Lord Browne's clear strategic interests in developing sustainability as part of corporate strategy and branding led in 2005 to the creation of an Alternative Energy business that used OI principles. This new business was set up to create an ecosystem of partners through investments in technology and firms operating alternative energy businesses, in particular biofuels and wind power. Similarly, the appointment of a new CEO, convinced of the importance of developing more sustainable vehicles, started the experiments with OI in a motorbike manufacturer, and in a cement manufacturer the CEO's original commitment to the idea of applying photocatalytic elements to cement advanced their first OI project (Boscherini et al., 2010).

Therefore, even if the study of leadership's responsibility in the implementation of change is prone to biases (Linton, 2002), several aspects related to OI and leadership should be investigated in more detail. There is for instance uncertainty on what particular characteristics of leaders could facilitate the implementation of OI. The entrance of top management with experience outside the industry or from firms with very different cultures could enable change because they "help [...] to challenge existing routines and practices" (Di Minin et al., 2010). Managers from other backgrounds bring personal expertise from other domains and sometimes have been purposefully recruited, such as the IP manager at Italcementi who had previously worked in pharmaceuticals (Chiaroni et al., 2011). However, the long-term experience within a firm could facilitate the establishment of trust and the recognition of the authority in directing changes. For example, it was the head of R&D at FIAT, Gian Carlo Michellone, who had been with the firm many years, who successfully implemented a long-plan transformation towards open practices (Di Minin et al., 2010). This could be a particularly important factor for companies where the influence of the leader is potentially greater, such as SMEs and family-owned and family-run firms, situations that are more common in certain countries than in others (e.g. South Korea (Ahn & Minshall, 2012)). For this particular type of study, Upper Echelon Theory could provide helpful foundations (Hambrick & Mason, 1984).

Other influential figures in implementation are the managers who, at an operational level, design and lead the activities that drive OI implementation. Many firms who wanted to adopt OI soon after the publication of Chesbrough's OI model in 2003 relied on OI implementation teams, often lead by senior R&D managers tasked with establishing new practices and changing the culture (Mortara & Minshall, 2011; Mortara, Slacik, et al., 2010). As the role of

OI in firms became gradually more strategic, OI functions and roles have been progressively formalized (the "Institutionalizing" phase of change (Lewin, 1947)). For instance Unilever started committing to OI in the mid-2000s and since then it has progressively acquired strategic importance that has been signalled by the appointment in 2008 of a Vice-President for OI. The experience of Unilever is not uncommon, as illustrated by the examples of OI Directors at Crown Packaging[3] and Philips (Manceau et al., 2011). This analysis should be extended to consider a) the maintenance and further development of OI activities and b) the impact that taking on this OI championing role has on managers' long-term careers.

12.3.2.2 Divisibility

Small-scale implementation activities for OI have been reported in the literature (Boscherini et al., 2010; Mortara & Minshall, 2011) and Chesbrough indicates that incremental approaches can be a successful recipe for implementing new open business models (Chesbrough, 2006a). In contrast, many firms across the FMCG domain such as P&G decided to plan for a big overhaul (e.g. Cloyd & Euchner, 2012). Perhaps it is still early to evaluate the relative success of these two strategies, but this remains a variable worth monitoring.

12.3.2.3 Social Interactions

There is a range of social dynamics linked to OI implementation. In particular, we would like to highlight two areas where we feel further work is needed: 1) the Human Resource (HR) management issues regarding the ways in which firms hire, progress, and integrate employees in the age of OI; and 2) the connections between communication/public relations (PR) and OI.

HR: There is evidence that firms adopting OI are changing the way in which they recruit new staff and the skills they are seeking. FIAT, for instance, changed the way it recruited personnel by including "entrepreneurial attitude" in the assessment of prospective staff (Di Minin et al., 2010). The changes brought in by OI relate also to the development and support of staff key capabilities. Petroni at al. argue that the Anglophone model of dual career ladder, with those with technical competence confined to R&D activities, does not provide the "T-men" required for OI (Chesbrough, 2012b). The required model, they argue, is instead that of an "open dual ladder" whereby firms promote mobility of R&D employees across the firm (Petroni et al., 2012). Several case studies also mention the need for stimulating entrepreneurial behavior in their R&D workforce (e.g. Cloyd & Euchner, 2012; Di Minin et al., 2010; Dodgson et al., 2006; Huston & Sakkab, 2006). Case studies have shown how firms pursued this aim using different means. For example Qualcomm encouraged and rewarded entrepreneurship through an internal Idea Competition, the "VentureFest" (dos Santos & Spann, 2011). FIAT organized a program called "researchers with the briefcase" not only to train researchers to become effective gatekeepers, but also to explain where the boundary between open versus closed should be drawn (Di Minin et al., 2010). Similarly, Unilever[4] attempted to roll-out a training program to develop professional OI managers (Duff, 2011), where the practices and skills of OI are linked to the Want/Find/Get/Manage process outlined by Slowinski (2010). Philips instead chose to setup an incubator for spinning out internally developed ideas that did not immediately fit the strategic pipelines of the core businesses. Like Xerox (Chesbrough, 2002), they encouraged their workforce to create new businesses by transferring the required management skills to new opportunity areas (Ford et al., 2010).

Other aspects deserving further attention concern the management of innovation systems, the integration of internal and external knowledge (Wallin & Von Krogh, 2010) and the knowledge management procedures adopted to do so. There is some evidence that shows that there are different ways in which employees integrate with the main business unit (Broring & Herzog, 2008). However, there is not conclusive understanding of the effectiveness of these aspects.

PR: It is now becoming increasingly clear that the OI implementation and the firm's image are linked. In particular, the aspect of corporate communication for building and maintaining reputation, once the domain of marketing strategy, is becoming more relevant for innovation activities in an OI context. However, with few exceptions (e.g. Pfeffermann, 2011a, 2011b), studies to date have not yet concentrated on communication and its impact on OI implementation. Four initial themes emerge from our exploratory research on this topic. Firstly, to succeed in OI, a firm is required to be perceived as a potentially attractive partner. This can be shown by the experience of scouts at Kodak: *"[...]We knew that contacts would be doing their "due diligence" on us, both as KER (Kodak European Research), and us as individuals—we would be "Googled"! We wanted to make sure that they found accurate and "open" information that would help facilitate interaction"* (Ruth Thomson, Innovations Leader for KER 2006-2009, cited in (Dang et al., 2011)). Communication can be seen as one of the competences that employees need to possess and which could impact on how they are recruited and promoted within an OI environment.

Secondly, given the open nature of OI, political and social pressures seem to influence the adoption of OI innovation practices. Several examples point towards the need of understanding whether and how communication and PR issues mediate in the adoption of OI. For instance, as a response to the environmental crisis in the Gulf of Mexico in 2010, BP's management set up an open portal to collect suggestions and solutions to the emergency. The studies on idea competitions and crowdsourcing highlight how very few of the ideas submitted are sufficiently ready or feasible to be accepted (Alexy et al., 2012; Poetz & Schreier, 2012). However BP, already in the news spotlight, was highly criticized and negative PR issues quickly emerged.[5] In particular some contributors felt that their ideas were not given sufficient consideration. Similarly, in idea competitions, exposing the public to firms' innovation processes could also lead to biased outputs (Mortara et al., 2013). Whilst firms in a closed innovation model do not have to give public account on how they select certain ideas rather than others, in these public contests ideas might be selected with criteria other than those of immediate self-interests. For example they need to aim to produce public goods or take into account what the public thinks (Lampel et al., 2012). An example is shown by the case of GE in its Ecomagination challenge, whereby GE invited comments on submissions from the public. The ideas which found favors with the public were not necessarily those which the firm would have acquired, but were nevertheless awarded prizes (Chesbrough, 2012a).

Thirdly, the impacts of adopting OI on the public image of the firm could also be included in the evaluation of the success and performance of OI practices. For example, idea competitions help to develop the public profile and branding of participating firms (Cornelissen et al., 2012). For example GE used this method to support the "Ecomagination" brand (Chesbrough, 2012a). Reputation building is mentioned by many firms as an additional success factor considered whilst planning for this type of activity (Mortara et al., 2013). For this purpose, idea competitions are often linked to a message of firms' social responsibility, such as sustainability (e.g. Shell's "Springboard" idea competition targeted at identifying low carbon business ideas from UK SMEs).[6] Positive messages have the effect of facilitating the "obtaining" of innovation at the start of the inbound process (West & Bogers, 2014) by providing intrinsic motivations to contributors (West & Gallagher, 2006). On the other hand, this same factor could be a limitation for the implementation of these OI methods for firms in industries of potentially lesser public appeal such as defense (Mortara et al., 2013).

Fourthly, an as yet underexplored link between OI and PR is the use of OI as a term to represent the firms' innovation activity. In some cases the term "open innovation" is not part of the language used by firms to communicate their innovation activities, whilst in others OI is used publicly as a defining image of the firm. For instance since the 1980s Rolls-Royce has developed a network of research centers integrated in universities (University Technology Centres or UTCs) which, starting from the UK, were gradually extended across the globe. These UTCs focus on many engineering domains and the relationship with the experts in the partner universities is clearly strategic for the firm. However, the term "open innovation" does not appear at all on their website.[7] In contrast, Nokia has decided the opposite and has clearly used the label "Open Innovation" for its Nokia Research Centers worldwide: "*Nokia Research Center is actively engaging in Open Innovation through selective and deep research collaborations with world-leading institutions. [...].*"[8] So which strategy might be more effective? And more importantly what is the PR value of OI? Firms such as P&G used OI as a message to promote its path to renewed growth. Having become such strong symbols for OI, the success or failure of these single firms, independently from their specific choices in OI, might impact on the OI implementation phenomenon. This theme would seem to be one that merits further research.

12.3.3 Moderating Factors: Enablers and Barriers

A number of elements have been described as potential enablers and obstacles for the implementation of OI. The most widely acknowledged is culture, but to date this theme has not been explored in depth (Elmquist et al., 2009; Giannopoulou et al., 2010). Some qualitative studies have attempted to tackle this subject but the only quantitative study in this area is one that reviewed employees' attitudes and cultural metrics in open and closed innovation at Evonik Industries' Creavis (Herzog & Leker, 2010). Culture is seen most often as a barrier in the adoption of OI in large companies but has also been signaled as an enabling factor (Mortara & Minshall, 2011). Lichtenthaler and Ernst defined six different attitudes which could distort, act as barriers or overplay the importance of OI (Lichtenthaler & Ernst, 2006), of which the "Not-Invented-Here" (NIH) syndrome (Katz & Allen, 1982) is the most frequently cited. From the examples of Roche (Nakagaki et al., 2012) and the firms studied in Italy (Boscherini et al., 2010; Chiaroni et al., 2011) it emerges that "demonstrators" can be important enablers for the acceptance of OI, particularly to institutionalize the change. Westergen and Holmström use the case of LKAB (a traditionally very closed company at the start of the process of opening up) to show that the building of trust with external partners supported the implementation of the initial OI projects, and thus provided a demonstrator for further OI activities (Westergren & Holmström, 2012). Mortara et al. (2010) found that to support the establishment of an OI culture, the OI champions supported different groups in R&D, providing specific types of motivators in accordance with the underlying subculture. Particular emphasis was placed on the delivery of skills and the creation of boundary spanning objects. Control mechanisms such as incentives were found to have a positive impact on external (outbound) (Persson, 2006) and internal (Minbaeva, 2005) knowledge transfer and on the search performance (Salge et al., 2012).

Whilst the company's OI culture is clearly important, national (Savitskaya et al., 2010; Chesbrough & Crowther, 2006) and regional (Tödtling et al., 2011) culture may also impact on OI implementation. Culture is a subject as yet understudied in the context of OI. This theme could be particularly suited for more qualitative analysis, for instance by adopting a close analysis of the "social dramas" that may be a feature of the adoption of OI.

Other internal firm characteristics, primarily concerning the presence of internal R&D capacity (e.g. Berchicci, 2013; Cassiman & Veugelers, 2006) and that of a set of OI capabilities (Lichtenthaler & Lichtenthaler, 2009) such as absorptive capacity (Bogers & Lhuillery, 2011) have been suggested as enablers or barriers. Certain management practices have been observed to support the implementation of OI (Salge et al., 2012). For example, in Creavis, three types of projects enable the firm to access and integrate external knowledge (Broring & Herzog, 2008). The experience of P&G (Dodgson et al., 2006) and Italcementi (Chiaroni et al., 2011) shows also that IT systems can support the shift towards OI. IT infrastructure is seen as an enabler of communication across boundaries (Boscherini et al., 2010) and as an element of control (Kuschel et al., 2011). Management tools (Griffiths et al., 1998), taxonomies (Di Minin et al., 2010), or "watch lists" (Mortara, Thomson et al., 2010; Tao & Magnotta, 2006) can also be used to find the balance between what to do openly or internally. The impact of the adoption of "virtual" platforms for carrying out specific innovation activities such as the interaction with users has also been explored (Bughin et al., 2008). In addition, as a change of strategy is often linked to leadership, the political climate and the internal dynamics of power need to be viewed as mediators in the adoption of OI (Pye & Pettigrew, 2006).

12.4 THE OI IMPLEMENTATION FRAMEWORK

The external environments in which organizations are embedded provide the context and the underpinning drivers for these transformations. The literature reveals that turbulence is associated with the implementation of OI (Schroll & Mild, 2012; Schweitzer et al., 2011). Technology characteristics (e.g. its uncertainty, appropriability and clockspeed (Buganza et al., 2011)), and industrial systems of innovation and ecosystems dynamics (Christensen et al., 2005) provide the background factors influencing the opening of innovation processes. The dynamics of the transition from closed to open, documented via case studies (Buganza et al., 2011), has also been confirmed quantitatively in a sample of Dutch firms, indicating that the timing of the shifts from closed to open differs from industry to industry (Poot et al., 2009) and that the discontinuities in OI adoption may be particularly linked with crises (e.g. FIAT (Di Minin et al., 2010), ENI (Pellegrini et al., 2012) and related changes of management (Mortara & Minshall, 2011)). Other exogenous contextual factors such as proximity have a less clear implication for OI adoption (Schroll & Mild, 2012). Schroll and Mild conclude that OI adoption is certainly on the rise but that more data is required to understand its determinants (Schroll & Mild, 2012).

We feel that OI implementation should hence be evaluated contingently to understand OI implementation traits in relation to external determinants. The next problem is that of judging the "successful" implementation. According to prior research (Linton, 2002) there are four different levels to assess implementation, not only related to economic performance:

- Implementation, Integration, and Institutionalization: Has the innovation changed how work is performed? Has the change been institutionalized (become a routine) or has it been abandoned before routinization? This is the approach of works such as Charoni et al. (2011).
- Human Partnership Dynamics: Does the innovation change the nature of the organization's structure or employee interaction with work?
- Economic Performance: How does the innovation perform in quantifiable economic terms? (this is the favorite approach of quantitative studies).
- Operational Effectiveness: Does the innovation improve the operation in a manner which cannot be easily quantified?

Other methods of evaluation of what constitutes "success," such as goal attainment or management satisfaction, are more value-laden and can be more difficult to use (Linton, 2002).

Figure 12.1 attempts to bring all the elements related to OI implementation described above into a coherent framework which can be used to evaluate the patterns of implementation of OI in large firms, linking external contextual to internal characteristics of the implementation and monitoring their evolution. We believe this analysis could bring increased understanding of OI and its potential, and better explain why companies in similar circumstances (i.e.

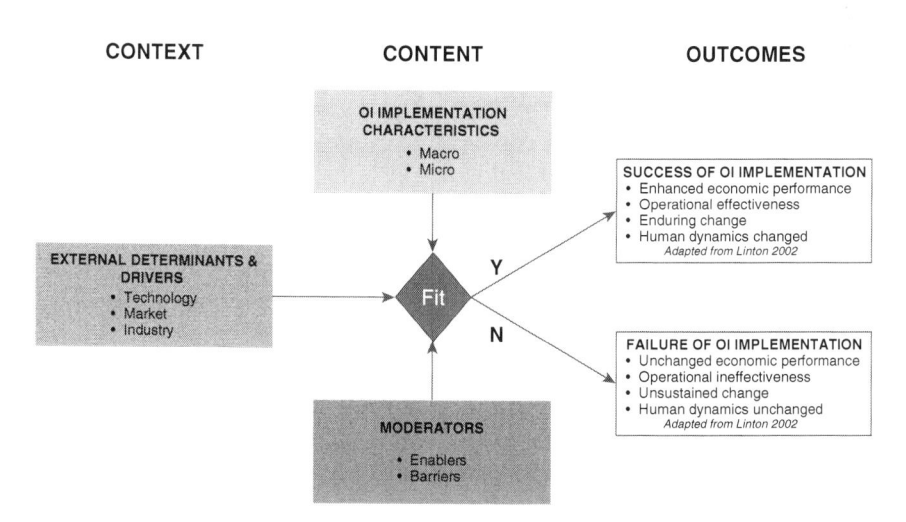

Figure 12.1 An open innovation implementation framework

facing similar turbulence) have adopted OI in different ways with variable rates of success (e.g. Cisco & Lucent, (Ferrary, 2011)).

12.5 CONCLUSIONS

This chapter reviewed the current knowledge on how companies, specifically large multinationals, implement OI, drawing upon—and highlighting the benefits of—longitudinal case studies to shed light on this complex multi-dimensional issue.

Firstly, as a result of the 2003 discontinuity in the adoption of OI, we have emphasized that the study of OI implementation needs to use different theoretical bases. Evolution theory provides a strong foundation, already used to analyze OI emergence. However, the high visibility of OI in recent years means that some firms may view openness as an end in itself (i.e. the 2003a Chesbrough OI model has been a disruption in the implementation trend) rather than a mechanism by which firms respond to some other internal or external disruption. The patterns of OI implementation evolution hence do need a broader theoretical basis which includes Teleology, Upper Echelon, Bounded Rationality and the understanding of "bandwagon" effects in the adoption of innovations. We feel that a process research approach is the best suited to study OI implementation in that it is capable of linking context to content and action. The nature of the discontinuities that firms are facing needs to be taken into account when analyzing OI implementation, pointing to a rich area of potential research that makes this connection much more explicit, and delivers insight at a finer level of granularity, bringing OI research more strongly linked with strategy research. Longitudinal analyses of databases and the use of ethnographic studies are most suited for this purpose.

Hence, from an analysis of the current state of the art in OI implementation research we proposed a framework which not only aids by expanding on the theory of implementation (Linton, 2002), but also comprises the elements needed for a process analysis. In particular it lists the key implementation traits in firms along which OI implementation should be characterized over time. These are divided into: (1) Macro-characteristics illustrating high-level traits of OI implementation and (2) Micro-characteristics which concentrate on internal factors and social dynamics in OI implementation. In addition the framework lists moderating factors (enablers and barriers) and suggests a method of evaluation for OI implementation successes and failures.

The template in Table 12.1 summarizes all the dimensions in the framework and the key specific areas for future research we have highlighted along the way. This schema could also be used to capture the evolution of approaches for each individual firm.

Table 12.1 Key areas for future research.

This template summarizes the key OI implementation features in MNCs (in the gray boxes). The template provides also the chance to recapitulate the key areas for future research in OI implementation discussed in the chapter (in the white boxes).

Time-->

				Pre 2003	2003	Post 2003
External Contingencies	Market					
	Industry					
	Technology	Uncertainty Appropriability Clockspeed				
OI Implementation traits	Macro	Processes	*Inbound Outbound Coupled*	How did the Macro OI implementation features change over time? How did they match the external contingencies? Which configuration of OI implementation is more successful in what circumstance?		
		Implementation stimulus	*Top-down Bottom-up*			
		Coordination	*Centralised Decentralised*			
		Approach to location	*Come-to-me Go-to-key places*			
		Networks	*Depth Breadth*			

(continued)

Time -->

OI Implementation traits				Pre 2003 — 2003 — Post 2003
Micro	Implementers	Leadership		What characteristics of leadership facilitates the implementation and the maintainance of OI practices?
		OI Managers-Teams Champions		How are the OI champions continuing their career, after OI has been adopted?
		Functions		How do individual functions interpret and implement OI?
	Divisibility	Single-stroke Incremental		What is the impact of OI implementation strategy on success outcomes?
	Social Interactions	HR		How are the HR practices changing because of the implementation of OI?
		PR		What is the role of OI for PR (and vice versa)?
Enablers and barriers	Culture			Is culture a barrier or an enabler of OI implementation? What are the detailed consequences of OI implementation on firms' culture?
	Politics			How is OI influenced by political tensions?
	Internal R&D			To what extent do internal R&D capabilities facilitate or hinder the adoption of OI?
	Innovation and Knowledge management tools and procedures			How/which management practices help or hinder the implementation of OI?

NOTES

1. http://www.whitehouse.gov/open/toolkit
2. http://research.nokia.com/open_innovation. Accessed February 20, 2013.
3. http://www.crowncork.com/innovation/open.php
4. http://bit.ly/hu6TWQ
5. See, for example, http://www.guardian.co.uk/environment/2011/jul/12/bp-deepwater-horizon-oil-spill-crowdsourcing
   http://blog.crowdspring.com/2010/06/crowdsourcing-bp-oil-spill-innocentive/
6. http://www.shellspringboard.org/
7. http://www.rolls-royce.com/about/technology/uni_research_centres/index.jsp
8. http://research.nokia.com/open_innovation

13

Getting Help From Innomediaries

What Can Innovators do to Increase Value in External Knowledge Searches?

Nadine Roijakkers, Andy Zynga, and Caroline Bishop

The authors would like to acknowledge the help of H. Lopez in collecting the data; the willingness of many Ninesigma clients to share their views with the authors; and the much appreciated observations of R. Wielens and F. Tropper of Ninesigma in relation to some of the topics covered in this chapter.

13.1 INTRODUCTION

This chapter aims to show what actions innovating companies can take to increase value when they make use of intermediated OI services in different phases of their external knowledge searching and hence improve their chances of ultimately establishing a successful tech-transfer agreement with a solution provider. Over the past decade more than 25 articles have been written on intermediated services indicating their increasing importance for both academia and practitioners. Some of these publications have been descriptive in nature highlighting important trends in the intermediated OI service industry and describing its main characteristics (see Fosfuri & Gambardella, 2001; Enkel, Gassmann, & Chesbrough, 2009). Other studies have focused on the value innovating companies can potentially derive from using intermediated services and have identified the factors (usually outside firms' sphere of influence) that affect the potential value accruing to innovating companies when interacting with innomediaries/solution providers (Dushnitsky & Klueter, 2011) and the motivations behind solution providers' involvement in markets for solutions (Boudreau & Lakhani, 2009; Che & Gale, 2003). Despite this growing body of literature on intermediated services relatively little attention has been paid so far to what innovating companies themselves can do to increase the value they derive from working with innomediaries throughout all phases of external knowledge searches. Some companies are more successful at deriving value from their interactions with innomediaries than others. These differences can partly be linked to the actions innovators take internally to add to their value capturing potential. The objective of this chapter is to identify these internal actions and thus help innovating companies to increase the value they generate when using intermediated OI services.

We can distinguish between two types of innomediaries in the intermediated OI service industry:

1. Innomediaries that offer their intermediated services on the basis of interaction between their staff and the clients they serve and thus rely heavily on experienced personnel. Three subtypes are currently in existence:
   - Innomediaries that support innovating companies in their external knowledge searching and find technical solutions that are integrated in the products/services of their clients (e.g. Ninesigma, Innocentive, IXC)
   - Innomediaries that help innovating companies make use of their unused Intellectual Property (IP) (e.g. Yet2.com, Innovaro)
   - Staff-augmentation companies that provide staff to help clients solve OI problems (e.g. IXC, YourEncore)
2. Innomediaries that offer their services on the basis of interaction between innovating companies and technology and thus rely on software programs and search engines. We can distinguish the following three subtypes:
   - Platform providers that offer platforms where innovating companies can post their technological needs/offerings (e.g. Hypios, IdeaConnection)
   - Software companies that create platforms for ideation/searches (e.g. Inno360, Spigit)
   - Crowd sourcing companies that provide access to consumers (e.g. IdeaScale, Threadless)

In this chapter we study how innovating companies can increase value when interacting with the first type of innomediaries in their external knowledge searches. Innovating companies that make use of intermediated services in external knowledge searching are usually active in highly innovative fields of industry (e.g. automotives, chemicals, consumer packaged goods, food and beverages, pharmaceuticals and medical devices, communications and defense, energy and utilities, electronics) and have a broad technology portfolio that is simultaneously covered by both internal sourcing and external knowledge searches (Cassiman & Veugelers, 2006). Examples of such companies include Philips, Siemens, Glaxo-Smithkline, Kraft, Jaguar Land Rover, Res Med Crown Packaging, PepsiCo, etc. Our exploratory research entails a set of 21 interviews conducted at Ninesigma clients, which were recorded, transcribed, and thematically analyzed resulting in quote sheets categorized by theme. Furthermore, we administered an online survey to 260 innovating companies working with Ninesigma using Survey Monkey. Fifty-two managers at innovating companies (i.e. a response rate of 20%) provided information on their interactions with Ninesigma.

When innovating companies engage in external knowledge searches whilst seeking the help of innomediaries they typically go through four phases: Orientation, exploration, selection, and engagement. In orientation innovating companies join forces with innomediaries to formulate their technological needs and translate these needs into Requests For Proposals (RFPs). In exploration innovating companies rely on innomediaries to retrieve interesting solution proposals that meet their needs. In the selection phase companies and innomediaries jointly determine the value of submitted proposals and decide which solution provider(s) (if any) to engage with. In engagement innomediaries help their clients to set up meetings with solution providers and sign agreements with these parties (non-disclosure agreements (NDAs) or tech-transfer agreements). While each phase can result in valuable outcomes for innovating companies the process of external knowledge searching is successfully completed once a beneficial agreement with a solution provider is set up. The quality of the intermediated services offered by innomediaries partly influences the likelihood of a signed agreement between their clients and solution providers. However, this represents only one side of the story. Some innovating companies are better equipped to make effective use of intermediated services in their external knowledge searches than others. These variations in success rates can partly be linked to the actions innovating companies take internally to add to their ability to benefit from their interactions with innomediaries. For each of the phases of external knowledge searching we identify a set of actions that innovating companies can take to increase the likelihood of successfully engaging with innomediaries and solution providers. Figure 13.1 visualizes the relations between the services offered by innomediaries in each of the phases of external knowledge searching, the value-adding actions by innovating companies, and the key success factors linked to each phase.

On the basis of Figure 13.1 the chapter is structured as follows. In the next section we first outline the context of the chapter. We start off by providing an overview of the important literature that has come into existence to date and that focuses on relevant aspects of innomediaries. The second part of this first section deals with the most important trends we currently observe in

Getting Help From Innomediaries 245

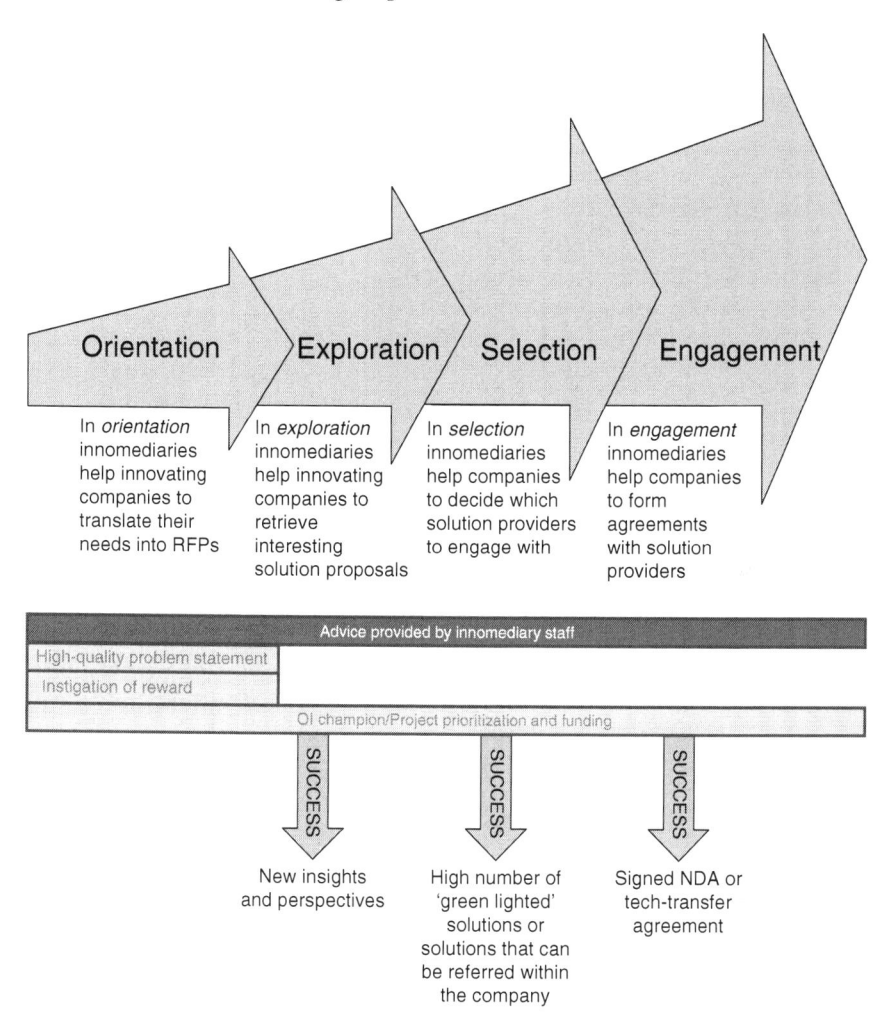

Figure 13.1 Potential value added by innomediaries in different stages of external knowledge searches, value-adding actions by innovating companies, and key success factors related to each phase

Orientation — In *orientation* innomediaries help innovating companies to translate their needs into RFPs

Exploration — In *exploration* innomediaries help innovating companies to retrieve interesting solution proposals

Selection — In *selection* innomediaries help companies to decide which solution providers to engage with

Engagement — In *engagement* innomediaries help companies to form agreements with solution providers

Advice provided by innomediary staff
- High-quality problem statement
- Instigation of reward
- OI champion/Project prioritization and funding

SUCCESS:
- New insights and perspectives
- High number of 'green lighted' solutions or solutions that can be referred within the company
- Signed NDA or tech-transfer agreement

the intermediated OI service industry. Next, we describe the services offered by innomediaries in each of the phases of external knowledge searching and we explain how Ninesigma clients define success in relation to the help they receive from the innomediary in all phases. In the following, we pinpoint the actions that innovating companies can take for themselves to stimulate value creation in their relations with innomediaries. Finally, we draw conclusions from our exploratory analysis and identify the main venues for future OI research on innomediaries.

13.2 BACKGROUND ON INNOMEDIARIES

13.2.1 Theory Development

The increasing importance of the role of innomediaries in helping innovating companies to stimulate their external knowledge searches is witnessed by the growth of academic study in this field. Over 25 publications have arisen in recent years that shed light on innomediaries. Several studies have been published that describe trends occurring in the intermediated OI service industry (Arora et al., 2001a; Arora & Gambardella, 2010; Athreye & Cantwell, 2007; Dushnitsky & Klueter, 2011). In 2001 Arora et al. were among the first to describe the market for solutions as an effective medium for technology transfer where innovating companies can establish contact with solution providers that offer them resolutions to their internal technical problems. In recent years an increasing number of innovating companies have searched for external knowledge through innomediaries adding significantly to the growth of this marketplace. Specifically, in the mid-1990s the size of the global market for solutions was estimated to comprise around 55–60 billion US dollars in royalty and licensing revenues while this figure grew to around 90–100 billion US dollars in the year 2000 (Arora et al., 2001a; Athreye & Cantwell, 2007). Other studies focus on the characteristics and dynamics of the intermediated OI service industry (Enkel et al., 2009; Huston & Sakkab, 2006). Specifically, depending on the number of innovating companies (one or multiple) involved in external knowledge searches markets for solutions have been classified as being either internal to the firm or external (Huston & Sakkab, 2006). As an increasing number of innovating companies have begun to search for knowledge outside their boundaries several innomediaries have come into existence that facilitate the match between innovating companies and solution providers and offer intermediated OI services to firms through their staff. Enkel et al. (2009) point out that innomediaries promote external knowledge searches by creating effective bridges between innovating companies and solution providers.

Other authors have described the role played by innomediaries in helping companies to search for external knowledge (Arora & Fosfuri, 2003; Nambisan & Sawhney, 2007; Tapscott & Williams, 2006). Innomediaries help their clients to formulate effective problem statements in RFPs thus enhancing the likelihood of obtaining high-quality solutions from solution providers (Sieg, Wallin, & von Krogh, 2010). Others focus on the factors affecting the value that innovating companies can potentially derive from interacting with innomediaries and solution providers, such as the not-invented-here syndrome (NIH), a low level of absorptive capacity, a lack of complementary knowledge/assets, and tacit knowledge components (Arora & Gambardella, 2010; Bresnahan & Trajtenberg, 1995; Ceccagnoli, Graham, Higgins, & Lee, 2010; Dushnitsky & Klueter, 2011; Gans & Stern, 2010). The NIH syndrome refers to the unwillingness in innovating companies to engage in external knowledge searches and a clear preference for internal knowledge development. Typically this opposition is grounded in the organizational culture of innovating companies (Arora & Gambardella, 2010). A low level of absorptive capacity (Cohen & Levinthal, 1990) in innovating firms is related to the underdeveloped capacity in some companies to make use of external knowledge, which discourages their efforts to engage with innomediaries. Some solution proposals tend to be more valuable when innovating companies are able to combine these technical resolutions with internal complementary assets and knowledge components. Not all innovating companies possess these important complementary skills, which leads them to derive less value from their external knowledge searches (Bresnahan & Trajtenberg, 1995; Ceccagnoli et al., 2010). If solution proposals involve tacit knowledge components it is very difficult for innovating companies to appreciate the value of these proposals or apply this knowledge within their business (Dushnitsky & Klueter, 2011).

A final stream of research is focused on issues related to solution providers. Some authors point out that a high number of solution providers responding to RFPs may decrease the quality of solutions as solution providers are less likely to invest large amounts of resources if the chance of winning is low (Che & Gale, 2003; Taylor, 1995). Others argue that a higher number of solution providers adds to the diversity of solutions thus offsetting possible negative effects (Pisano & Verganti, 2008; Terwiesch & Ulrich, 2009; Terwiesch & Xu, 2008). Silveira and Wright (2010) examine the ambiguous role of IP in the intermediated OI service industry where innovating companies need access to full information to rightfully assess the value of solution proposals while solution providers look for ways to protect their knowledge through IP (Laursen & Salter, 2012). Few studies examine the drivers behind solution providers' involvement in markets for solutions and conclude that monetary rewards are important motivators besides softer motivations such as personal enjoyment (Boudreau, Lacetera, & Lakhani, 2011; Boudreau & Lakhani, 2009; Frey, Lüthje, & Haag, 2011; Lakhani & Jeppesen, 2007).

13.2.2 Trends in Practice

There is increased acceptance in large companies that OI is an essential part of their business. Early adopters of OI such as Philips, Kraft, and PepsiCo have learned from their experiences and have consequently refined their use of OI (Cassiman & Veugelers, 2002; Chesbrough, 2003a, 2006b; Laursen & Salter, 2006). They have moved from exploring technological solutions through traditional partnerships to multi-party collaborations. This is particularly the case where companies recognize the need to work with universities and a multitude of other partners at the same time. An example of this is the San Diego Zoo Global Bioinspiration program where biologists, engineers, and chemists join forces to solve problems in the areas of technology, transportation, and renewable energy. The more refined OI practice of large companies also becomes clear from their current use of crowd sourcing, which has evolved from writing RFPs based on a few predefined search criteria to setting up competitions in areas where they need to rapidly acquire new knowledge and build partnerships (e.g. Siemens Smart Grid Contest 2010). Furthermore, most early adopters of OI no longer view crowd sourcing and technology scouting as OI actions in themselves but more so as part of the day-to-day job of all members of their R&D teams. A final piece of evidence with respect to the increased refinement of large companies in OI relates to their objectives for external knowledge searches that have broadened out from in-licensing to acquisitions, joint ventures, joint development agreements, and collaborations to secure government funding of development programs. As a result of their increased sophistication in OI and their increased reliance on internal staff for technology scouting early adopters typically rely on innomediaries for external knowledge searches outside their normal field of view or to provide staff as an interim resource to handle specialist OI projects or peaks in workload.

In small innovators we witness increasing recognition of the importance of OI and strong efforts to engage in OI. Particularly, the more the small innovator makes use of complex technologies the more it is likely to use OI for product improvement, cost reductions, solving technology challenges, new product and service development, and opening up new markets. However, a lack of skills and resources as well as a short-term focus seem to restrict the scope for OI action in small companies (Chesbrough, 2011; Dahlander & Gann, 2010; Freel, 2000; Gans & Stern, 2003; Laursen & Salter, 2006; Lee, Park, Yoon, & Park, 2010; Narula, 2004; Spithoven, Vanhaverbeke, & Roijakkers, 2012; Van de Vrande, De Jong, Vanhaverbeke, & De Rochemont, 2009b). In order to spur their innovation activities small innovating companies seem to continue relying on interaction with universities. When they do make use of intermediated OI services small firms are more likely to engage with individual consultants rather than innomediaries.

Increased interest in OI among all types of companies and a heightened refinement in the use of OI in large companies have expanded the market for innomediaries and encouraged new entrants (Diener & Piller, 2010). OI is a people intensive process and whilst techniques such as crowd sourcing now make it much easier to find data, people are still needed to make sense of what is found. Hence, many innomediaries have increased their focus on supplying hard to replicate analytical and soft people skills for OI to ward off competition. Ninesigma, Innocentive, IXC, Yet2.com, and YourEncore have expanded their services to now cover advice not only in the orientation and exploration phases of external knowledge searching, which were traditionally the phases where innomediaries would offer their services, but also in selection and engagement (see Figure

13.1). Innomediaries nowadays provide OI consultancy through their staff throughout all phases of external knowledge searching and stimulate companies to take actions for increasing the value they can derive from using intermediated OI services (e.g. recruiting, training, and appointing OI champions).

13.3 EXTERNAL KNOWLEDGE SEARCHES: WHAT INNOVATING COMPANIES PERCEIVE AS SUCCESSFUL OUTCOMES OF INTERMEDIATED OI SERVICES

A number of studies describe the intermediated OI services offered by innomediaries in external knowledge searches (Mortara, 2010a). Few researchers have paid attention to what constitutes a successful outcome of interactions between innomediaries and their clients. Most observations in this respect are of a general nature where researchers mention that value pertains to lower internal R&D costs, reduced R&D risks, shorter time-to-market, and access to new ideas (Arora et al., 2001a; Enkel et al., 2009). In this section we describe the intermediated OI services offered by innomediaries and we link these intermediated services to the various phases that innovating companies typically go through when searching for technological knowledge outside their boundaries. Furthermore, we specify what constitutes a successful interaction with an innomediary in each phase as perceived by Ninesigma clients (see Figure 13.1).

In orientation the intermediated services of innomediaries are focused on helping their clients to understand and formulate their technological needs as well as coaching them to write RFPs that contain high-quality problem statements. One of the most difficult facets of writing RFPs is associated with formulating problem statements that adequately cover the innovating company's technological needs. The quality of the problem statement in an RFP (in terms of pinpointing the precise technological problem) determines the quality of the solutions that are offered by solution providers (in terms of the extent to which the proposed solution meets the technical needs of the innovating company) (Sieg et al., 2010). Interviewees mention the following in relation to intermediated OI services in this phase: "(...) Ninesigma asks questions that force us to think carefully about the problem we are trying to solve (...) they work with us on defining the problem (...)."

In the exploration phase innomediaries facilitate the retrieval of new and unexpected technology solutions. Based on their network of technology companies, universities, research institutes, etc. they are able to invite solution providers to respond to their clients' specific technological needs. As such, innomediaries help their clients obtain proposals from parties that they might not have considered as relevant technology sources or that they might not have been able to reach otherwise. Managers at innovating companies phrase this as follows: "(…) Ninesigma can use its platform to identify thousands of engineers working on particular research problems and within a couple of weeks we know whether to continue or terminate a project (…) they provide us with information regarding possible solutions and solution providers that we could not find on the internet or in the scientific literature. We could have been searching forever and never found them (…)." The exploration phase is considered to be successful by Ninesigma clients when they are able to gain access to new insights and perspectives with respect to their internal problem-solving initiatives through their use of intermediated OI services. An interviewee described this successful outcome in the following manner: "(…) through Ninesigma we get exposed to companies with different views (…) different perspectives or angles that we did not consider before and sometimes this leads us to re-evaluate our projects (…)."

In the selection phase innomediaries offer services that aid innovating companies in selecting valuable solution proposals. When solution providers submit possible technical answers to problem statements that were worded in RFPs, innovating companies have to determine the value of these solutions. Furthermore, they need to decide whether or not to engage in further interactions with solution providers. Innomediaries capture and deliver information that their clients need to make such difficult decisions. In addition, they provide methods and tools to assist in the decision-making process. When working with innovating companies Ninesigma makes use of a traffic light system to recommend and prioritize solution proposals meeting companies' original objectives where "green lighted" solutions represent the most valuable solutions in terms of meeting technical specifications. As one client put it: "(…) Ninesigma ranks solutions based on relevant criteria that allow us to prioritize the top responses (…)." In selection Ninesigma clients perceive a high number of "green lighted" solutions as a successful outcome of their interaction with the innomediary. One interviewee states that: "(…) it is not just about the number of solutions we receive (…) it is about the number of solutions we find useful (…)." Another successful outcome in selection is the number of solution proposals that can be referred to other departments within the client organization. Sometimes innovating companies receive solution proposals that are not directly relevant for the team/department that got involved with the innomediary in the first place. In these cases proposals may get transferred within the company to other teams/departments working on different sets of technological problems benefitting more directly from the proposed solutions. As one manager put it: "(…) it was not directly of use to us (…) so we knew of some scientists within our company working on such issues (…) we sent it to them and they could use it (…)."

In engagement innomediaries help to bring about agreements between their clients and solution providers that were selected in the previous phase.

Innomediaries assist their clients in conducting negotiations with solution providers and help to set up NDAs to stimulate further technological discussions and more definitive arrangements pertaining to the transfer of technological knowledge/IP or joint knowledge creation. Ninesigma clients describe this service as follows: "(…) Ninesigma facilitated interaction with a solution provider, which allowed us to communicate through face-to-face talks and clarify those nuances that came out of telephone conversations and were still unsolved (…)." Innovating companies perceive this phase to be successful when they sign a beneficial NDA or a tech-transfer agreement with a solution provider. One interviewee mentioned the following in this respect: "(…) for me a successful project is the one that ends in a signed agreement for further cooperation (…)."

Throughout all phases of external knowledge searching innomediaries rely on their staff to provide clients with technical advice and OI consulting where necessary. One of the interviewees referred to this service aspect in the following manner: "(…) we have had discussions with Ninesigma project managers to describe our needs (…) they capture what we need and then translate that need to make it work for their network of solution providers (…)."

13.4 EXTERNAL KNOWLEDGE SEARCHES: WHAT INNOVATORS CAN DO TO INCREASE VALUE FROM USING INTERMEDIATED OI SERVICES

Clients' perceptions of successful outcomes of using intermediated OI services are linked to different phases of external knowledge searching. The successful completion of this search process is evidenced by a signed agreement between innovating companies and solution providers. This valuable end result is by no means an isolated outcome as it builds on the results obtained in previous phases: A high-quality RFP with a well-defined problem statement is likely to attract a high number of interesting solution providers resulting in a high number of "green lighted" solution proposals. A high number of valuable technical solutions positively affect the chances of successful negotiations between innovating companies and solution providers. Being successful in each of the phases of external knowledge searching whilst joining forces with an innomediary partly hinges on the quality of the intermediated OI services offered by the innomediary; it is also partly dependent on the ability of the client to make effective use of these intermediated services. Some innovating companies have a higher likelihood of signing an NDA or a tech-transfer agreement with solution providers than others following their interaction with innomediaries. Broadly we can say that for less successful innovating companies, which have a signed agreement as one of their main targets, around 40% of their RFPs lead to a signed agreement whereas more successful companies, targeting a signed agreement, are able to turn 60–70% of all their external knowledge searches through RFPs into successful contracts. These differences in success rates can be linked to the actions innovating companies take internally to add to their ability to make effective use of intermediated services in each phase thus increasing their chances of signing an agreement with a solution provider in engagement (Ihl, Piller, & Wagner, 2012). In this section we identify the actions innovating companies can take in different phases of their external knowledge searches to increase their chances of success (see Figure 13.1).

To increase value from using intermediated services innovating companies need to develop skills with respect to rightly formulating problem statements and writing up high-quality RFPs. In orientation, one of the main challenges is defining a high-quality problem statement (Sieg et al., 2010). If the problem statement contains more than one technical issue, there is a high likelihood that very few solution providers will respond to the RFP. If the problem statement is too descriptive in terms of applications solution providers from industries other than the innovating company's may fail to address the problem or offer their solutions. Innovating companies with the skills and experience necessary for translating their technical needs into specific problem statements tend to be more effective in teaming up with an innomediary and jointly developing an RFP. Ninesigma clients mention the following in this respect: "(…) it is all about how you craft the RFP (…) formulating a problem statement that is free of industry language is of eminent importance (…)." Another important action that innovating companies can take to be more effective in their interactions with innomediaries is to instigate a reward for successful solutions in the orientation phase. Most of the successful innovating companies make available a financial reward for solution providers that present them with valuable solutions. The amount of the reward available for a solution provider should be proportional to the potential value created through the successful solution. There is a high correlation between the reward offered by an innovating company and the likelihood that this company will ultimately sign a mutually beneficial agreement with a solution provider either to engage in further negotiations or to transfer knowledge from the solution provider to the innovating company (Boudreau et al., 2011; Boudreau & Lakhani, 2009; Frey et al., 2011; Lakhani & Jeppesen, 2007). As one interviewee put it: "(…) the financial incentive shows that you are serious and willing to spend money to solve a problem (…)."

To stimulate more effective interactions with innomediaries throughout all phases of external knowledge searching innovating companies need to appoint an OI champion to support the project as well as label the project as "strategic" and fund it as such. Innovating companies that effectively make use of intermediated OI services experience very strong support for OI initiatives and intermediated services from the very top levels of their organization.

Most of these companies have an OI champion in place when interacting with innomediaries. An OI champion is typically an experienced, high-status individual who is well-connected within the innovating firm and is empowered to manage all phases of external knowledge searching. Innovating companies interacting with Ninesigma describe the role of an OI champion in the following manner: "(...) the champion is an early point of contact for outside companies (...) the champion preaches the OI mentality (...) the champion is cheerleader, coach, strategist, organizer, maintainer of the OI philosophy (...)." When all phases of external knowledge searching are driven by a champion, the success rate doubles and triples compared to companies that do not have such executive support. As one interviewee stated: "(...) this whole process has been supported by senior management; otherwise it simply would not happen (...)." Furthermore, innovating companies that are most effective in signing agreements with solution providers typically designate their OI projects and their engagements with innomediaries as "strategic." Strategic projects are a corporate priority and are not likely to be cancelled due to changing corporate agendas. When an innovating company uses the most strategic projects for external knowledge searches that have a certain urgency about them, chances for success go up dramatically. In the interview phase we found that "(...) the main reason why projects do not succeed is because they are not strategically relevant to the company (...)." Another Ninesigma client mentioned that "(...) senior management has to provide long-term financial support (...) you have to have full engagement of internal resources (...) for it to be successful (...)."

13.5 CONCLUSIONS AND VENUES FOR FURTHER RESEARCH

On the basis of interviews and survey research among Ninesigma clients this chapter has shown that there are several actions innovating companies can themselves take to increase value from interacting with innomediaries in all phases of their external knowledge searching. More and more innovating companies with broad technology portfolios in automotives, pharmaceuticals and medical devices, communications and defense, and other sectors of industry make use of intermediated OI services to stimulate their external knowledge searches. While these typically large companies (early adopters such as Siemens, Glaxo-Smithkline, and Jaguar Land Rover) have gained substantial experience in OI and the use of sophisticated OI tools, innomediaries have intensified their service offerings in the initial phases of external knowledge searching and expanded their activities to also cover intermediated services in the final phases of external searches and OI consultancy in all phases. In recent years several interesting publications have emerged describing the role played by innomediaries in the external knowledge searches of large innovating companies (Enkel et al., 2009). While some authors (Ceccagnoli et al., 2010) have identified a number of factors influencing the potential value innovating companies can derive from using intermediated OI services these factors are mostly difficult to change or beyond the direct sphere of influence of most companies. Building on a few other studies (Boudreau & Lakhani, 2009; Lakhani & Jeppesen, 2007; Ihl et al., 2012) in this chapter we focus on what innovating companies can do internally to increase the likelihood of success when engaging with innomediaries.

Our exploratory research identifies several routes to improving the chances of success when engaging in external knowledge searches: Optimizing the RFP writing process; rewarding solution providers that add value to internal problem-solving; appointing an experienced, high-profile OI champion to stimulate external knowledge searches; prioritizing OI projects and allocating long-term funding to these initiatives. What is interesting about these results is that there is a specific set of actions that innovating companies can take at the very beginning of their external knowledge searching, that is before even putting out an RFP (defining a high-quality problem statement and making mention in the RFP of a reward coupled to receiving valuable solutions), which significantly increase the likelihood of engaging with solution providers in the final stage of external knowledge searching.

On its website Ninesigma hosts an OI scorecard diagnostic tool that captures companies' OI adoption rates or maturity level. Of the companies visiting this website (not necessarily clients of Ninesigma) and providing input to the tool 34% are not involved in OI at the moment; 37% of all respondents report to be in the early stages of OI; 23% of companies are currently optimizing an existing OI program; 6% are re-launching an OI program. From these figures we can conclude that although there are several large innovating companies with a fairly high level of experience in OI and highly sophisticated OI programs in place, there are also many (small) companies that are only just beginning to learn about the potential benefits related to engaging in external knowledge searches. For all of these innovating companies (both experienced and inexperienced at OI) aiming to learn about new technologies through using intermediated OI services it is of utmost importance to realize that there are several actions they can take themselves, such as building skills in the RFP writing process and appointing an OI champion, to significantly add to the value they can potentially derive from using these intermediated services (Enkel, Bell, & Hogenkamp, 2011; Ihl et al., 2012).

We discern several venues for further research by OI scholars targeted towards increasing the efficiency and effectiveness of intermediated OI services in all phases of external knowledge searching leading to more benefits for all parties involved. Several case studies in intermediated OI services have emerged over time. However, they have lagged behind practice due to reluctance in innovating companies to publish how they gained competitive advantage and due to the fact that product/service launches and resulting revenue generation occur years after their interactions with innomediaries. There is thus still a need for case study research as well as large-scale data research to identify best practices in intermediated OI services. Detailed research into the effects of engaging in intermediated OI services on commercial success and other innovation-related Key Performance Indicators (KPIs) may help innovating companies to better assess the value of these intermediated services for their business and choose the venue that is right for them. While early adopters of OI become increasingly knowledgeable with respect to external knowledge searches and innomediaries intensify their service offerings to address more refined needs in these large companies it seems that smaller companies are mostly in need of more basic OI services that help them to build up experience in external knowledge searching. With their lack of skills and resources the question comes to mind of how to make intermediated OI services accessible and affordable for small companies. As an increasing number of innomediaries intensify their service offerings to include OI consultancy, the question arises which business model is most appropriate to commercialize these new intermediated services. As large companies become more experienced at external knowledge searches they develop their own competitions to attract solution providers. How does this development affect innomediaries? Other interesting questions relate to the topic of group-based problem solving: Are RFPs best written in isolation or with the help of outside parties? Are solution providers best off responding to RFPs by themselves or should they team up with other knowledge parties to provide higher-quality solutions? Although both theory and practice point to the importance of monetary rewards for solution providers in stimulating their participation in markets for solutions more research with respect to the drivers behind solution providers' involvement may shed light on how to attract the most knowledgeable ones.

14

Theories of the Firm and Open Innovation

Wim Vanhaverbeke and Myriam Cloodt

14.1 INTRODUCTION

The concept of open innovation primarily stems from observing changing innovation management practices in companies (Chesbrough, 2003a, 2006a). This practice-based approach explains to some extent why open innovation research has not been grounded systematically in prior management research. The failure to connect the phenomenon of open innovation to existing theories of the firm represents a weakness in the open innovation literature and, therefore, there is an urgent need to overcome this theoretical deficit as: "…a better theoretical foundation of open innovation research is needed. […] In particular, open innovation studies need to be sufficiently grounded in prior research into both open innovation and related fields. […] A cumulative development of open innovation research that integrates earlier findings is essential to arrive at a coherent body of knowledge about open innovation" (Lichtenthaler, 2011: 87).

The growing interest for open innovation management in companies provides several opportunities to shed new light on existing theoretical frameworks on innovation. To develop a consistent body of knowledge about open innovation, we argue that existing management theories should be combined as none of them can fully explain how companies benefit from open innovation. We will shed light on open innovation from multiple perspectives and bring theories together in an attempt to develop a better theoretical grounding of open innovation. In particular, we explore the need to link open innovation to the strategy literature and to different theories of the firm, such as transaction cost economics, the resource-based view, the resource dependence theory, the relational view, and real options theory. We also relate open innovation to theoretical concepts such as absorptive capacity and dynamic capabilities. Thus far, only a few open innovation researchers have attempted to connect open innovation concepts to the existing innovation management literature and its underlying theories of the firm.[1]

To connect open innovation to existing theories, we choose the open innovation funnel as a starting point. Chesbrough (2003a) explains open innovation using the innovation funnel as a central visualization tool in his first book to discuss differences between closed and open innovation. We use the open innovation funnel (Grönlund et al., 2010) as a starting point for this chapter because a quick examination of its constituent dimensions already highlights the need to delve into strategy, business modeling, transactions or collaborations with external partners, internal and external resources, etc.[2]

In the following we first describe the open innovation funnel and its most important dimensions in detail. In this part, we will explain which literature streams and theoretical concepts underlie the open funnel concept. We will then proceed with a discussion of these literature streams in an effort to enhance our understanding of the different dimensions of the funnel. Interestingly, although open innovation requires connecting different theories with each other, it also implies that particular interpretations of these theories are compatible with open innovation and others are not. In a final discussion we will shed light on the lessons learned through our venture to build the contours of a theory that has the potential to frame the open innovation phenomenon.

14.2 THE OPEN INNOVATION FUNNEL AS A STARTING POINT

The innovation funnel has a long history in the innovation management literature. It has been used within the Chandlerian, "closed innovation" framework where firms organize research, development, and commercialization processes within the corporate boundaries. Chesbrough (2003b, 2006a) uses the (open) innovation funnel as a central concept to develop several key insights about open innovation. The funnel is an interesting concept, not only summarizing and visualizing key lessons of open innovation, but it has also the potential to connect open innovation to existing management and theories. In this chapter, we use particular constituent parts of the open innovation funnel (Grönlund et al., 2010) as a starting point to connect open innovation to existing management theories and theoretical concepts. We explain these potential links referring to the open innovation funnel represented below in Figure 14.1.

Firstly, "new market" and "current market" at the right side of the innovation funnel refer to the business model of a company. Business model thinking is at

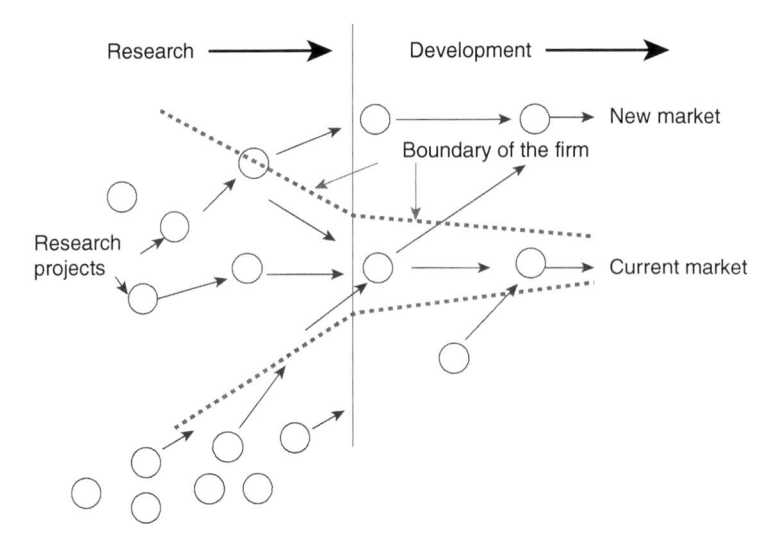

Figure 14.1 The open innovation funnel
Source: Chesbrough (2003b: 37)

the heart of open innovation: internal knowledge not supporting the firm's business model will be out-licensed or sold and external knowledge complementing its business model will be insourced to develop new products or businesses. Accordingly, open innovation can be only correctly understood when it is integrated into firms' strategy. Yet, few publications have examined how the overall strategy, the innovation strategy, and open innovation in companies interconnect with each other (a notable exception is Chesbrough & Appleyard, 2007).

Secondly, arrows crossing the organizational boundary in the figure represent different types of inter-organizational agreements to pull ideas and technologies into the funnel or to monetize on unused technologies. Insourcing knowledge takes place through research agreements, co-development deals, corporate venturing, in-licensing agreements or outright-acquisitions. In other cases, companies tap into communities of users or experts, or they rely on the services of specialized innomediaries.[3] Firms outsource their own knowledge via alliances, licensing agreements, and spin-offs. Firms practicing open innovation are continuously making decisions about which knowledge they have to develop internally, which to buy or to co-develop, and which to sell or license. The choice between internal and external technology developments within the open innovation framework is closely related to the make-buy-ally decision-making process (Geykens et al., 2006; Jacobides & Billinger, 2006; Gulati & Nickerson, 2008; Mudambi & Tallman, 2010). Consequently, future research should examine how the make-buy-ally and the open innovation literature streams relate to each other and how they can enrich each other.

In this respect, open innovation research has certainly to pay more attention to the question how collaboration with external innovation partners redraws the boundaries (and the organization) of (the innovation activities in) the firm.

Choosing appropriate sourcing modes is crucial to deploy open innovation successfully. The choice between them depends on the uncertainty both in the technology and market (and will thus change in different stages in the funnel) (Van de Vrande et al., 2006, 2009a) and the type of knowledge involved. The arrows in the figure can be considered as some type of inter-organizational transactions. In section 14.4 we investigate whether the arrows in Figure 14.1 should be examined through a transaction cost minimization lens or not (Williamson, 1975, 1985). In open innovation, firms can trade technology through market like transactions, but in most cases they also invest in complex, long-term agreements with their innovation partners.

Thirdly, we can focus on the R&D projects (the dots inside and outside the funnel). They start as ideas and develop over time into new business products. This process requires resources and capabilities to nurture the projects. Some of them are available internally, but others have to be sourced from other organizations. Getting access to these resources or capabilities is a major driver of open innovation. Open innovation should therefore be linked to the resource-based view (Barney, 1986, 1991; Wernerfelt, 1984). We also take the knowledge-based view into account, considering knowledge as a specific case of the resource-based view (RBV) (Grant, 1996). The RBV and knowledge based view are primarily focused on the internal development of capabilities. To account for the growing use of external competencies, several authors have developed frameworks to incorporate the sourcing of capabilities and knowledge from external sources: The relation-based view of the firm has focused on how companies can tap into external resources (Dyer & Singh, 1998). Similarly, the knowledge-based view has been applied to inter-organizational relations such as strategic alliances (Grant & Baden-Fuller, 2004), external corporate venturing (Keil, 2004; Schildt et al., 2005) and acquisitions of start-ups (Wagner, 2011).

Finally, the open innovation funnel can be managed as a Stage-Gate process. Stage-Gate methodologies are widely used in managing closed innovation processes (Cooper, 1999), but they can also be used—in a different way—in open New Product development (NPD) processes (Grönlund et al., 2010). Innovation is a risk-laden activity that requires a sequence of investments, first with small reversible steps followed by investments with increasing financial commitment. The Stage-Gate process is an approach that aims to reduce technological and market uncertainties early on in the process and it disciplines companies to postpone major investments until uncertainty is low enough. In this way, the open innovation funnel can also be considered as a staged decision-making process that can be analyzed from a real options theory perspective (Vanhaverbeke et al., 2008). Collaboration with external partners in early stages of the funnel can be considered as option-creating decisions, which offer companies an opportunity to make better-informed decisions about more costly investments in external technology during later stages of the funnel (Van de Vrande et al., 2009a).

The (open) innovation funnel provides us the different intersection points between open innovation and prior management literature. We start by exploring the relationship between open innovation and strategy.

14.3 LINKING OPEN INNOVATION TO STRATEGY

Careful observation of companies with a good track record in open innovation indicates that open innovation practices have to be embedded in firms' strategy. We touch upon three aspects of this (potential) link between open innovation and strategy.

Firstly, business model thinking as part of firms' strategy is at the heart of open innovation. Discussing Figure 14.1, we mentioned already that internal knowledge, which cannot be aligned to the firm's business model, would be licensed or sold. In contrast, external knowledge complementing a firm's business model would be insourced to develop new products or businesses. Yet, few publications have examined how strategy and open innovation interconnect with each other (some exceptions are Chesbrough & Appleyard, 2007; Dittrich & Duysters, 2007). This lack of attention is surprising as practitioners have called attention to the fact that open innovation is pointless if it is not integrated with strategy (Slowinsky & Sahal, 2010; Kirschbaum, 2005). This nexus between strategy and open innovation deserves a central place in new open innovation studies.

Secondly, there is a connection between corporate strategy and open innovation, which has not been explored in the open innovation literature. Most examples of open innovation illustrate how firms can benefit using external knowledge sources to develop new products in existing businesses. This overemphasis on the use of open innovation in existing businesses eclipses other strategic uses of open innovation. At DSM, a Dutch specialties chemical company, management has established the EBAs (Emerging Business Areas), developing and incubating complete new divisions at the corporate Innovation Center to drive future (in contrast with current) growth of the company by establishing new businesses that do not exist yet in the company (Vanhaverbeke & Peters, 2005; Wijen et al., 2011; Vanhaverbeke & Roijakkers, 2013). These businesses are developed in collaboration with a variety of external (technology) partners, which are different from the partners that are involved in open innovation initiatives to grow existing businesses. To generate incremental growth in current businesses requires a different form of internal organization compared to the case when the company intends to develop completely new businesses in the long run. Different strategic growth targets lead to different ways to organize open innovation in a company, different departments in the organization will be responsible to lead the projects, and other types of partners are required to collaborate with. Consequently, open innovation should be explicitly linked to corporate (growth) strategy. Sourcing knowledge from partners is not only useful to spur growth in existing businesses, but also to incubate early stage ventures in business areas that are targeted by top management as corporate growth areas (beyond the existing divisions in the company). Likewise, a firm may use open innovation to realize corporate renewal. Organizing open innovation for reasons of corporate growth and renewal implies that a firm has to develop new (technological) competencies. How companies develop new competencies collaborating with different external knowledge partners is a topic that has not received the attention it deserves in the innovation management literature. This, in turn, can be linked to interesting new developments in our understanding of dynamic capabilities (Helfat et al., 2007; Teece et al., 1997; Teece, 2007).

Thirdly, making use of innovation communities, corporate venturing, ecosystems, licensing deals, and venture acquisitions requires new approaches in strategy. The main literature streams in strategy—growth and diversification strategy (Chandler, 1962), positioning strategy (Porter, 1980,1985) and extensions (Brandenburger & Nalebuff, 1996), resource based strategies (Wernerfelt, 1984; Barney, 1986,1991) and control of key complementary assets (Teece, 1986)—are all focusing on internal assets, ownership and control as key sources of strategic success. Each of these directions has proven to be fruitful for understanding business strategy when companies relied mainly on internal technological capabilities to develop new products. None, however, adequately accounts for collaborative or open innovation as an empirical phenomenon emerging in many industries. There is an imperative to find new or adapted strategic approaches that can fully account for collaboration and cooperative strategies. In this respect, one approach to take into account is the recent work on the management of innovation ecosystems (Nambisan & Shawney, 2011; Adner, 2012; Borgh van der et al. 2012; Leten et al., 2013).

14.4 TRANSACTIONS WITH SINGLE AND MULTIPLE ACTORS

Open innovation is about setting up relations with external innovation partners. Those inter-firm relations are shaped into transactions between different legal entities. Companies can collaborate in different ways with their innovation partners and the choice between these different types of collaboration lies at the core of the *transaction costs theory* (Williamson, 1975, 1985). As inter-organizational transactions are essential in open innovation, we have to examine the potential of the transaction costs theory as a theoretical framework for open innovation. The productivity of the collaboration between innovation partners increases through cooperative specialization (Alchian & Demsetz, 1972). However, when innovation partners make transaction-specific investments, transaction costs arise because of fear of opportunism (Williamson, 1985). Although co-specialization increases productivity, firms' incentive to make transaction specific investments is alleviated because specialized resources have lower value in alternative uses (Dyer, 1997). Owners of specialized resources are exposed to greater risks compared to owners of general resources (Klein et al., 1978). The transaction cost theory argues that companies will choose a specific governance mode (arm's length transactions, hierarchical control or intermediate forms of governance) to *minimize transactions costs* given the threat of opportunism in a competitive setting. Market transactions are highly efficient as a transaction governance mode, but when opportunistic behavior is likely to occur, companies can mitigate the threat of opportunism by choosing for an intermediate (e.g. strategic alliances; joint ventures) or hierarchical governance mode (e.g. acquiring the economic exchange partner) (Oxley, 1997; Van de Vrande et al., 2006).

The central question remains whether minimizing the transaction costs is a useful approach for open innovation? One assumption underlying the transaction cost theory is that transaction costs increase as asset specificity increases. However, Dyer (1997: 539) shows that "while transaction costs may increase with asset specificity, they will also vary independently of asset specificity" due to the use of appropriate ways to govern the relationship. Consequently, partners can simultaneously achieve the twin benefits of asset specialization and lower transaction costs (through optimal governance choice). In this way, inter-organizational transactions should not be structured only to economize on transaction costs but also to maximize the transaction value (Dyer, 1997). Careful observation of open innovation deals reveals that companies choose a particular type of collaboration governance to jointly maximize the value of a transaction rather than to minimize transaction costs.[4] As such, transaction value optimization offers a more interesting theoretical framework for open innovation deals. First, open innovation takes shape through different types of long-term relationships between a set of partners to execute complex tasks related to the development and commercialization of innovations. This requires a high level of understanding and confidence among the partners to learn from each other and to realize the full strategic potential of the cooperation. Within such a context, the value of cooperation among firms is not only based on transaction costs objectives but also *on strategic value and learning opportunities* (Kogut, 1988). Accordingly, partners are not interested in transaction cost minimization, but in the pursuit of transactional value. They may choose cooperative modes with higher transaction costs, as long as "[…] the expected joint gains outweigh transaction cost considerations" (Zajac & Olson, 1993: 138).

Secondly, although the threat of opportunistic behavior is obviously present within an open innovation context, companies are often involved in a long-term relationship to develop and/or commercialize a new product offering. In such a context, the transaction or relationship between the partners becomes a value-bearing asset in its own right. Transaction specific expenditures are no longer considered as costs but as an investment in future value (Madhok & Tallman, 1998). Consequently, the inclination to act opportunistically is dominated "[…] by the firm's estimate of the negative impact that the opportunistic behavior will have on the value of the future exchange with its partner" (Zajac & Olson, 1993: 137).

Thirdly, transaction costs theory focuses explicitly on transaction specific uncertainty—or uncertainty *within* the relationship (endogenous uncertainty). However, research examining the choice between different government modes in an open innovation context has shown that this choice depends on both exogenous and endogenous uncertainty. Both types of uncertainty will differ depending on the stage in the innovation funnel, and as such have an effect on the preference for a specific governance mode. For example, Van de Vrande et al. (2006) argue that under conditions of high exogenous uncertainty (front-end stages of the funnel) companies prefer to undertake flexible governance modes that are highly reversible and involve a low level of commitment. When external uncertainty decreases (back-end stages of the funnel) firms prefer to undertake more hierarchical governance modes, which involve a larger amount of investment and commitment.

Based on the above-mentioned argumentation we conclude that open innovation seems to be more in line with the *transaction value theory*. As transactions among partners in an open innovation context can be considered primarily in terms of maximizing transactional value, it is useful to further analyze open innovation activities from a transaction value theoretical perspective.[5]

Another topic related to the inter-organizational transactions (arrows in Figure 14.1) that we like to address is the fact that reaching out to external innovation partners can be done in a direct or indirect way. Companies can set up relationships indirectly with partners through innovation intermediaries—or innomediaries. Over the last decade, companies made increasingly use of services offered by innovation intermediaries such as Innocentive, Ninesigma, Yet2com, Ocean Tomo, and many others (Chesbrough, 2006a; Huston & Sakkab, 2006; Lopez & Vanhaverbeke, 2009). An innovation intermediary can be defined as a platform provider in two-sided innovation markets created to coordinate the flow of innovation requests and solutions across distinct, distant and previously unknown innovation actors.[6, 7] The mechanics and dynamics of two-sided markets have been analyzed in detail by industrial economists (Rochet & Tirole, 2003, 2006; Parker & van Alstyne, 2005), and these theoretical insights may help open innovation scholars to better understand the drivers of success of innomediaries. Embedding innovation intermediaries into the two-sided (innovation) markets framework will provide a double benefit: First, open innovation has been focusing traditionally on direct transactions while a two-sided market approach brings in the role of indirect transactions and describes the conditions under which a platform creates value. Second, transactions become *more efficient* as *technology intermediaries* develop and shape the technology market. The platforms that innomediaries create increase the value of external searching of technology (as the reach is much broader than in direct searches) and the transparency and neutrality of innomediaries decreases transaction costs related to external knowledge and IP acquisition. This, in turn, makes open innovation an even more interesting strategy to follow.

Next, we should connect innomediaries to the ubiquity of knowledge and markets for technology. Innovation intermediaries become more interesting when knowledge is both plentiful and widely distributed across the globe (Chesbrough et al., 2006). The growing globalization of R&D and the increasing diversity of knowledge sources increase the value of working via innomediaries. Innomediaries can also be linked to the market for technology literature (Arora et al., 2001a; Arora & Gambardella, 2010). The market for technology denotes trade in technology disembodied from physical goods. The focus is mainly on the efficiency of market transactions in technology and the resulting division of innovative labor between technology specialists licensing their technology to firms that integrate that externally developed technology to create new products and businesses. This literature has, however, a strong focus on bilateral technology transactions such as R&D contracting and licensing between a technology specialist and a technology buyer. The role of innovation intermediaries, which ties together technology suppliers and technology buyers in a triangular set of relations, is to our knowledge not yet discussed within this framework.

Finally, the choice between firms' use of innomediaries or own portals is an interesting but unexplored topic. Large companies like Procter & Gamble, Unilever, Starbucks, Kraft, Pfizer, Lego, and Dell make use of the services of innomediaries but they also have their own portal connecting them directly with thousands of external solution providers. Future research should determine when it is advantageous to work with innomediaries and when it pays to have one's own portal.

14.5 RESOURCES

A core feature of the *resource-based view* of the firm is that a firm requires a unique collection of difficult-to-imitate resources, competencies and capabilities to be competitive (Barney, 1986, 1991; Grant, 1996; Wernerfelt, 1984). To create a competitive advantage and capture above-normal rates of returns (i.e. rents) these resources must, by definition be scarce, valuable, and reasonably durable (Barney, 1991). The resource-based view argues that firms can create and capture value according to the unique bundle of resources they possess and the differences between these resources are held responsible for the differences in performance between firms (Bierly & Chakrabarti, 1996). In other words, proponents of the resources-based view emphasize the fact that a sustainable competitive advantage is based on those resources and capabilities that are owned and controlled within the boundaries of a single firm (Dyer & Singh, 1998). This view about how companies develop and sustain a competitive advantage is typically in line with the closed innovation framework.

In open innovation, firms rely for their new products on both internal and external resources, and internal resources can be deployed using inside as well as outside paths to market (Chesbrough, 2003b). Companies get access to external knowledge and integrate it into the development of their new offerings. Resources and capabilities of different organizations are brought together in an effort to offer value to the targeted customers. Firms—even the largest ones—cannot develop the required resources internally and have to team up with innovation partners enabling resource flows between firms. Such an increase in permeability of firms' boundaries will enhance the match between market opportunities and capabilities as well as a more efficient use of resources (Elmquist et al., 2009; Arora et al. 2001a, 2010). Subsequently, whereas the resource-based view stresses such issues as *independence* and the crucial role of *competition* between autonomous companies based on the unique set of resources and capabilities they possess, open innovation emphasizes the *interdependence* of *complementary* resources of firms to develop and launch the innovation in the marketplace (Vanhaverbeke et al., 2008). However, given this difference both the resource-based view of the firm and open innovation underline the importance of resources and competencies to generate a sustainable competitive advantage.

Given the similarities and differences, there is a need to align the resource-based view with the theoretical premise underlying open innovation. One option is *the relational view* of the firm. The relational view emphasizes that critical resources can and should also be found outside the firm's boundaries (Dyer & Singh, 1998). Collaborating firms that combine resources in unique ways may realize a competitive advantage over others that compete on the basis of a stand-alone strategy. To do so, firms have to combine resources with partners in unique ways and be willing to make relationship-specific investments. These relationship-specific investments generate relational rents that are a property of the dyad or network and cannot be enjoyed by a firm in isolation (Dyer & Singh, 1998). The relational view identifies complementary resources or capabilities of firms as a potential source of inter-organizational competitive advantage: this is in line with a major premise of open innovation to consider the sourcing of knowledge from external partners a source of competitive advantage. In particular, the relational view of the firm considers the dyad/network as the unit of analysis and, as a consequence, the complementary resources that create the relational rents are essentially beyond the control of the individual firm (Dyer & Singh, 1998).

The latter also implies that organizations become dependent on their environment (i.e. other organizations) for the provision of vital resources, which they need to survive and prosper.[8] Consequently, there is a risk that organizations lose their freedom of action as they are constrained by external interdependencies with other organizations (Pfeffer, 1987). However, they can try to reduce their environmental dependence based on the concept of power (Hillman et al., 2009), which is defined as the control over vital resources (Ulrich & Barney, 1984). According to the *resource dependence theory* the ambition of an organization is to minimize the dependence on other organizations and to find ways to influence these organizations to secure the resources that are needed (Pfeffer & Salancik, 1978). To manage this resource dependency, organizations engage in different inter-organizational arrangements, such as co-optation, contractual alliances, equity alliances (e.g. minority participations or joint ventures), and mergers and acquisitions (Pfeffer & Salancik, 1978; Drees & Heugens, 2013). The open innovation literature has not paid attention to the power structure and the interdependence between partners. On the one hand, it would be interesting to investigate when power structure and interdependence becomes important in innovation networks. On the other hand, the resource dependency theory is static in nature. Most industries nowadays experience strong dynamics and face the challenges of new technologies and business models. Power ebbs and flows in networks and ecosystems over time. Look at IBM with Intel and Microsoft in the 1980s, or Nokia and any of its ecosystem partners in the past few years. We know Philips as a consumer electronics company, but it currently redefines itself as a lighting, healthcare, and lifestyle company. Although we do not exclude power and interdependence issues in some industries, coping with the rapidly changing dynamics in an industry is a more important topic to understand open innovation. This is the subject of the following section.

14.6 DYNAMIC CAPABILITIES

A large number of companies currently operate in an environment in which the knowledge landscape is much more diverse and globally distributed than in the past (Chesbrough, 2003a). In such an environment, core competences based on scarce, unique, and difficult-to-imitate resources can turn into competence traps as companies get stuck with resources that become increasingly irrelevant when technologies and the competitive environment change. As a result, obtaining a sustainable competitive advantage goes beyond the ownership of difficult-to-imitate resources and requires difficult-to-replicate dynamic capabilities (Teece, 2007).

Teece (2007: 3019) classifies dynamic capabilities into the following three categories (1) the capacity to sense and shape opportunities and threats, (2) the capacity to seize opportunities and (3) the capacity to maintain competitiveness through enhancing, combining, protecting, and, when necessary, reconfiguring the business enterprise's intangibles. We argue that the three classes of dynamic capabilities are closely related to several important elements of open innovation.[9] In addition, we argue that the underlying micro foundations of these dynamic capabilities offer interesting insights into *organizational* and managerial processes, procedures, systems, and structures to implement open innovation in the firm.

The first class (sensing and shaping opportunities and threats) deals with the sensing of both technological and commercialization opportunities. To identify and shape these opportunities Teece (2007) argues that companies must overcome a narrow search horizon by combining internal and external knowledge that originates in the core as well as the periphery of their business ecosystem. This is closely related to the outside-in approach of open innovation. To start, firms can develop strong knowledge bases internally by investing heavily in R&D activities (Park, 2002). These R&D activities will enable the firm to develop its own core capabilities in new technological fields that form the basis of possible "first mover advantages" (Chesbrough & Teece, 1996). However, building up an internal knowledge base is a long and costly process and it is also very difficult to change the profile of technological foundations over time (Breschi et al., 1998; Giuri, Hagedoorn & Mariani, 2004; Granstrand et al., 1997). Therefore, relying on internal capabilities only is likely to lead to "core rigidities" (Leonard-Barton, 1992, 1995) or the so-called familiarity trap (Ahuja & Lampert, 2001) reducing the chance that companies can benefit from new technological opportunities (Jaffe, 1986; Lunn & Martin, 1986; Levin, Cohen & Mowery, 1985). By tapping into external knowledge a firm can overcome the limits of internal learning (Capron & Mitchell, 2000; Karim & Mitchell, 2000).

This reliance on external knowledge is closely related to the imperative of open innovation, where external knowledge is equally important to internal knowledge. The same balance between internal and external knowledge sources is advocated by Teece (2007). To establish this balance, microfoundations consisting of processes to direct internal R&D, processes to learn about changing customer needs as well as processes to tap into exogenous science and technology, supplier, complementor, and customer innovation, need to be established within the firm (Teece, 2007). The open innovation literature has not paid sufficient attention to these microfoundations and clearly can benefit from integrating them into the open innovation research.

The second class of capability (seizing opportunities) deals with the choices one has to make after the opportunity is sensed. This is particularly related to making the right decisions regarding the investments in development and commercialization activities, as multiple investment paths are possible. The selection or creation of the proper business model is fundamental as it defines a firm's commercialization strategy and investment priorities (Teece, 2007). Consequently, the creation, adjustment, or replacement of business models is a very important microfoundation of the second class of dynamic capabilities (Teece, 2007). However, business models can create strong inertial forces, once they have become well established (Chesbrough & Rosenbloom, 2002), making it difficult to adjust or replace them. As we have been arguing before, the business model also plays a central role in open innovation (Chesbrough, 2006a). However, Teece is focusing on internal paths to market while open innovation is also emphasizing external paths to market.

The third dynamic capability class is focused on the reconfiguration of assets and organizational structures as a result of changing technologies and customer needs. This change is needed to escape from unfavorable path dependencies (as discussed earlier) and to maintain evolutionary fitness (Teece, 2007; Helfat et al., 2007). One of the important microfoundations underlying this dynamic capability class is the achievement of decentralization and near decomposability. On the one hand, growing enterprises must decentralize to remain flexible and be able to respond to changing technological and customer needs. On the other hand, organizations should be able to achieve integration as well to benefit from potential economies of scale and scope. Obtaining this subtle balance is called "near decomposability" (Simon, 2002; Teece, 2007). Teece (2007) explicitly links the concepts of decentralization and near decomposability to open innovation as they rely on a distributed model of innovation to access and integrate external knowledge.

In sum, dynamic capabilities explicitly address some of the important elements of open innovation. Especially, the balance between internal and external knowledge, the important role of the business model and embracing open innovation as a way to access and integrate external technology, play a role in Teece's in-depth analysis of dynamic capabilities. However, we can also observe some differences between the dynamic capabilities theory and open innovation. Most importantly, the inside-out approach of the open innovation model is not addressed in the former. According to Teece (2007: 1343), the sensing and seizing of opportunities as well as reconfiguration are all closely related to an outside-in approach of the open innovation model, where the company integrates internal and external knowledge to create new products or services that are commercialized via internal paths to markets. However, the inside-out approach of the open innovation model clearly diverts from the above mentioned definition of dynamic capabilities as a firm can also make money by leveraging external paths to market. Companies can only go for external paths to market when they have the right dynamic capabilities in place. Selling or spinning off technology requires the development of particular practices in the company. Philips' IP ventures are an example in case: when the company decides not to use a technology internally, the related IP might not be interesting enough to sell to other firms. In this case, Philips will develop the technology till there is a proof of concept or a prototype in which case it is much easier to sell or license the technology.

14.7 ABSORPTIVE CAPACITY AND BEYOND

Absorptive capacity was first defined by Cohen and Levinthal (1989, 1990) as a firm's ability to recognize the value of new information, assimilate it, and apply it to commercial ends. Both the literature about open innovation (Chesbrough, 2003a, 2006a; Chesbrough et al., 2006; Christensen et al., 2005) and the one about absorptive capacity (Lenox & King, 2004, Arora & Gambardella, 1990; Ireland, Hitt & Vaidyanath, 2002) focus on how innovating companies can benefit from these external sources of technology. More in particular, since absorptive capacity focuses on acquiring and utilizing external knowledge inside the firm (Lichtenthaler & Lichtenthaler, 2009) it is a concept at the heart of the outside-in side of open innovation. To gain a better understanding of the relationship between absorptive capacity and open innovation we will first explain how the two concepts are linked to each other. Thereafter we will also explain the differences between them.

Both open innovation and absorptive capacity stress that there should be a proper balance between internal and external knowledge. In order to access and assimilate external knowledge firms need prior related knowledge to understand the knowledge that is absorbed (Cohen & Levinthal, 1990; Jansen et al., 2005). Firms thus require internal R&D capabilities to recognize and monitor interesting technologies that are developed elsewhere. On the other hand, internal research capabilities are indispensable to effectively exploit external know-how (Arora & Gambardella, 1994; Rosenberg, 1990; Cohen & Levinthal, 1989). In-house R&D activities remain crucial to develop technological know-how, to increase the firm's learning capacity and to improve its absorptive capacity. This is also echoed in the open innovation literature; internal R&D improves the effectiveness in monitoring and using external knowledge resources (Rigby & Zook, 2002; Chesbrough, 2003b; 2006a). In line with Cohen and Levinthal (1990), open innovation scholars stress the need to balance the ability to profit from external knowledge sources and the ability to develop and exploit internal knowledge (Chesbrough, 2003a, 2006a;

Gassmann & Enkel, 2004). Accordingly, the growing focus on external knowledge sources does not diminish the need to understand how companies can generate and manage internal knowledge (Gambardella & Giarratana, 2004). Therefore, to understand the relationship between a firm's absorptive capacity and open innovation, one should focus on the internal organization of the innovation process, which determines the diffusion and exploitation of technological knowledge within the organization (Nooteboom et al., 2007; Levinthal & March, 1993; Argyres & Silverman, 2004). The presence of valuable external sources of knowledge does not imply that the inflow of new ideas into the organization is an automatic or easy process. External knowledge can only be recognized, accessed, and assimilated when firms develop routines and change their organizational structure and culture to facilitate open innovation processes (Dahlander & Gann, 2007). Consequently, open innovation scholars have to take advantage of the development in the absorptive capacity literature to understand how firms develop new organizational routines to tap more effectively into external knowledge.

There are, however, also noteworthy differences between absorptive capacity and open innovation. Absorptive capacity is related to the assimilation and integration of external knowledge and is therefore limited to the outside-in perspective of open innovation. This also implies that other aspects of the open innovation process are neglected. For example, the purposive outbound flows of knowledge and technology through licensing and spin-offs, has been accorded no recognition in the absorptive capacity literature. Absorptive capacity cannot explain all dimensions of open innovation in terms of capabilities. Therefore, we need to complement the concept of absorptive capacity with new theoretical developments. The study of Lichtenthaler and Lichtenthaler (2009) provides an important first step in this direction. They built a capability-based framework for open innovation by distinguishing internal and external knowledge exploration, retention and exploitation. We restrict our attention here solely to the external dimension. Firstly, external knowledge exploration describes the assimilation of knowledge from external sources (Lane et al., 2006) which matches with the absorptive capacity concept. Secondly, external knowledge retention refers to the knowledge that is embedded in firms' interorganizational relations such as research agreements, technology alliances, corporate venturing investments, technology acquisitions, etc. They identify a firm's *connective capacity* as its ability to maintain and subsequently reactivate knowledge in interorganizational relationships (Lichtenthaler & Lichtenthaler, 2009: 1320). Contrary to absorptive capacity, here the focus lies on maintaining and managing knowledge externally instead of inward knowledge transfer. Gaining access to external knowledge without immediate transfer of the partner's knowledge is an important but often neglected aspect of the first stage(s) of the open innovation funnel. We hope that this concept will be applied more in the future partially as a consequence of the growing popularity of innovation ecosystems. Thirdly, they introduce external knowledge exploitation as the monetization of knowledge through external paths to market (Chesbrough, 2003b). It is not straightforward to monetize effectively on unused technology. A firm's *desorptive capacity* is its capability to generate revenues through external knowledge exploitation (Lichtenthaler & Lichtenthaler, 2009: 1321), which is complementary to internal knowledge application in a firm's own product-markets (Lichtenthaler, 2007). It consists of identifying external knowledge exploitation opportunities based on strategic and monetary motives and the subsequent transfer of the knowledge to the partner firm(s) (Lichtenthaler & Lichtenthaler, 2009). Given that external knowledge exploitation refers to outward knowledge transfer and external paths to market, it is a capability related to the inside-out dimension of the open innovation model.

The extension to three capabilities to explain open innovation practices is a remarkable extension of the absorptive capacity framework. Future contributions should identify how companies build these capabilities and to which extent they are different from each other. In any case, it is obvious that an extension is needed and that the distinction between different types of external knowledge capabilities is a fruitful step to understand the complex reality of open innovation.

14.8 REAL OPTIONS

The alleged benefits of open innovation can be partly explained by the real option approach. The creation of new products and businesses involves a high level of uncertainty. One way for firms to cope with the technological and market uncertainty associated with new business development is by making small investments in multiple options on technology. These small, initial investments can be regarded as a real option, that is "the right, but not the obligation, to take an action in the future" (Amram & Kulatilaka, 1999: 5). Real options reasoning is a tool for uncertainty reduction—making a small, initial investment under high levels of uncertainty to create an option while waiting until the uncertainty about the opportunity has decreased. When the uncertainty has decreased, the investing firm can decide whether to make a follow-on investment or whether to terminate the project (Adner & Levinthal, 2004; McGrath & Nerkar, 2004). Vanhaverbeke et al. (2008) show how real options reasoning can be applied to open innovation. The open innovation funnel is composed of different stages where uncertainty decreases, going from the idea generation stage till market launch. Compared to closed innovation, open innovation practices have four advantages in terms of real options; 1. Firms benefit from early involvement in new technologies or business opportunities; 2. They may benefit from delayed entry or delayed financial commitment to a particular technology; 3. Open innovation offers firms the advantage of an early exit, and the ability to realize some value from projects that do not go forward internally; 4. Finally, open innovation allows firms to benefit from delaying an exit.

Firstly, firms may benefit from open innovation through early involvement in new technologies or business opportunities. External partners help the firm to get access to a broad range of externally developed technologies and market opportunities. Partners can trace new technologies or market developments earlier and more accurately than a single company. The latter can take advantage from partners' knowledge through early stage investments such as minority stakes in start-ups, participation in VCFs, educational investments or crowdsourcing. These are minor investments for the company enabling them to scout technologies (through its partners) in a more effective way. Secondly, companies can also benefit from a delayed internalization of new technologies or ideas. Internalizing research and development projects implies that the company is paying all the related costs. Delaying these expenditures is beneficial when technologies are in an early stage and when therefore investments are still very risky. With open innovation firms have more flexibility in deciding when to start the internal portion of the innovation process: Firms start exploring the commercial possibilities of a technology outside initially, via relationships with universities, start-ups, research-labs, suppliers, and other innovation sources. Delaying investments in internal innovation enables the firm to take fewer risks in starting the innovation inside the company. Thirdly, open innovation offers firms the advantage of an early exit for R&D projects that do not live up to expectations. In a closed innovation setting, an R&D project can only turn a success (a new product or service) or a failure (abandoning the project). Because of this dichotomy, firms tend to invest too long in unpromising projects. In contrast, firms have additional options with open innovation: they can license out technologies or spin-off ventures that are not promising enough or do not fit with their business model. Finally, open innovation allows firms to benefit from extended control until a full exit of a project. Licensing out technologies or spinning out ventures frees cash for other, more interesting projects. Yet, the firm may still want to monitor "externalized" knowledge for strategic reasons while delaying the (full)exit decision. While the licensed technology or the spin-off matures and the value for the company becomes apparent over time, it can delay the decision whether to spin-in the venture or sell its remaining interests to the other capital providers.

These four advantages are just a simple exercise in reinventing open innovation in terms of real options. We are convinced that real options reasoning can be applied in different open innovation contexts. Collaborating with external partners should not only be considered in terms of competencies that a firm requires but also in terms of strategies to reduce uncertainty during research projects.

14.9 CONCLUSION: BRINGING THE PIECES TOGETHER

Ten years after Henry Chesbrough (2003a) coined the term open innovation we see a continuing increase in open innovation practices. This growing openness of firms is illustrated in a recent survey among large companies, where 78% of responding firms reported using open innovation processes (Chesbrough & Brunswicker, 2013). Van de Vrande (2009b) observed a similar trend among SMEs. Yet, despite the popularity of open innovation, few scholars have tried to link and integrate it into existing theories. This research deficit shows a strong need for further theoretical and empirical work. While the open innovation framework includes interesting conceptual insights and some theoretical extensions to theories of economic spillovers from R&D (Chesbrough & Bogers, Chapter 1 in this volume), its connections to existing theories of strategy are not yet well developed (Van de Ven, 1989). Therefore, we started in this chapter with a simple exercise linking the open innovation framework with existing theories and theoretical concepts. We limited our attention to the constituent parts of the open innovation funnel to narrow the focus and to keep the analysis tractable.

In our attempt to connect open innovation to existing theories and theoretical concepts, we found that: (1) open innovation largely neglected the link with firms' business and corporate strategy; (2) some theories are more in line with the phenomenon of open innovation than others; (3) most theories that can be aligned with open innovation still have to be adapted to grasp a particular dimension of open innovation; (4) open innovation is a complex, multi-dimensional phenomenon which compels us to combine different perspectives into a broader, dynamic (or stepwise) framework. We will discuss these findings in the following paragraphs.

Firstly, decisions to engage in open innovation should always be related to a firm's strategy. In particular the link between open innovation and corporate strategy needs to be highlighted in future research. Nowadays, benefits of reaching out to innovation partners are considered independently from a firm's business and corporate strategy. Embedding open innovation decisions in firms' strategies will offer us a more accurate picture of the diversity of goals companies envision with the opening of their innovation activities. Nowadays, open innovation activities are lumped together in "inside-out" and "outside-in" modes of open innovation. In this way, the diversity of strategic goals and the multitude of outcomes that can be achieved through open innovation are hidden by these encompassing concepts.

Secondly, some theories are difficult to align with the underlying assumptions of the open innovation framework. Minimization of transaction costs (TCE) is not reflecting the rationale of transactions between organizations that are practicing open innovation. Maximizing the joint value between innovation partners represents this rationale much better as it explains why companies engage in relationships with high transaction specific costs (Zajac & Olsen, 1993; Dyer, 1998). Similarly, we discussed the possibilities to consider open innovation through the lenses of the relational view (Dyer & Singh, 1998) and the resource dependency theory (Pfeffer & Salancik, 1978). In the latter, attention is paid to the power structure and the dependence between partners through inter-organizational agreements. In the relational view, the focus is on the complementarity of the resources of the partners and how partners can achieve strategic targets that they cannot achieve on their own. The open innovation literature has been focusing mainly on the benefits of inter-firm collaboration—that is, how partners can create jointly value. Therefore, open innovation can benefit from a closer examination through the lens of the relational view of the firm. The resource dependency theory seems to be less promising as a theoretical framework to understand the open innovation phenomenon. However, although the relational view is very useful to theoretically back open innovation, the theory is focused mainly on dyadic collaboration (Dyer & Singh, 1998). More attention should be paid to innovation networks, where the role of a network orchestrator or systems architect in the ecosystem becomes crucial to understand the benefits and dynamics of open innovation (Adner, 2012; Nambisan & Sawhney, 2011).

Theories that can be aligned with open innovation still have to be modified to grasp open innovation. The resources-based view—and knowledge-based view (KBV) as a special case—is a valuable theoretical background for open innovation in which firms try to have the right (external and internal) resources in place to create new products and services. Yet, the RBV (and KBV) are focusing on internal resources only—in line with closed innovation—not on a balance between internal and external resources. Similarly, "theories such as transaction cost economics, resource-based view and resource dependence theory are *static* or nearly so" (Eisenhardt & Tabrizi, 1995: 108). Open innovation is an answer given by innovating companies to cope with increasing technological complexity and competitive dynamics. Moving quickly, fast product innovation and seizing new market opportunities are core capabilities. Therefore, in order to explain open innovation, we are in need of *dynamic theories*. Dynamic capabilities and absorptive capacity are *dynamic theories* and thus appropriate for understanding open innovation in fast changing settings. However, both are only linked to the outside-in perspective of OI and they neglect the inside-out perspective. It would be interesting to see more research explicitly linking dynamic capabilities to open innovation (Teece, 2007 was a seminal work in that respect). Similarly, we need a more detailed investigation of new capabilities—such as the connective and desorptive capacities to understand which internal capabilities firms have to develop to engage successfully in open innovation.

There is definitively a need to integrate different theories. The different theoretical perspectives can only explain a particular dimension of open innovation. Therefore, we have to bring the pieces together to draw the contours of what could become a theory of open innovation.

Step 1: We argued that open innovation is inextricably related to strategy. Strategy has always been present in the background of the open innovation literature through the continuous emphasis on business models and their role in determining what external knowledge need to be in-sourced and which internal knowledge can be validated through external paths to market. Yet, in many cases practitioners and researchers start focusing on the benefits of open innovation projects without framing the projects within the broader strategic objectives the company has in mind. Framing open innovation within both business strategies and corporate strategies is crucial to understand with whom a firm needs to collaborate, when and how it will collaborate and what type of governance mode is needed to reach the targeted objectives. More importantly, strategic objectives also determine which capabilities have to be developed internally, which in many cases requires external partners to source or to co-create the requisite capabilities. Combining for example, the resource-based view of the firm with the resource dependence theory and the relational view provides us with a better insight in the combination of an internal perspective (specification of resource needs for a firm's strategy) and an external perspective (outside-in and inside-out innovation opportunities leading to the strategic renewal of an existing business) within the broader strategic context of the firm.

Step 2: Once the strategic objectives are fixed, managers have to determine where the interesting sources of the required external technologies are located. It is important to realize that this is a stage before a firm actually develops relationships. This stage is strongly related to the sensing (dynamic) capability developed by Teece (2007). When open innovation projects are following existing routines, firms can rely on existing networks of partners. However, if they engage in new technologies or markets, existing networks are of little help and firms have to search for new external sources of knowledge. This capability has been underemphasized in the open innovation literature. Similarly, looking for new ways of valorization of external paths to market for internal knowledge is an elusive capability that most companies find difficult to master. In this stage, firms should already focus on several determinants that might affect the final outcome. Examples are the fit of the competencies of the partner with the new product or service (KBV, RBV, and relation-based view); the risk profile of the project, the appropriability of IP, the irreversibility of the investments (real options theory), the potential value of the transactions, and the power structure between the firm and particular partners (resource dependency theory), etc.

Step 3: Once a firm has determined who are the best partners to team up with, it still has to start and execute the partnership. In contrast with step 1 and 2, this step has received a lot of attention in the prior literature, although there were few attempts to clarify this stage using existing theories. In this step, reaching out to external innovation partners can be done in a direct way or indirectly via innovation intermediaries—or innomediaries. Furthermore, success is determined by designing and crafting the right relation with the partner. In this stage it is important to determine how to choose for the most appropriate governance for the collaboration and how to guarantee that each partner captures enough value from their joint activities. The choice for a particular governance mode is a function of the joint value partners can realize net of the transaction specific costs they incur (transaction value theory), the risks of a power game and strategic interdependence (resource dependence theory), the absorptive capacity in the firm, the learning potential of the partnership, and the irreversibility of the investment (real option theory). For example, when an innovating firm intends to source emerging technologies, uncertainty about the future business potential of the technology is very high. Hence, under conditions of high uncertainty (environmental turbulence or technological newness), firms will prefer to maximize their flexibility and make small investments, which facilitate the reversibility of actions (Van de Vrande et al., 2009a). In that case, for example, a minority investment in an alliance can be used to achieve more certainty before executing the option of a merger or acquisition (Hillman et al., 2009). However, we have to keep in mind that we described so far only the outside-in mode of open innovation. Choosing the right governance modes when a company wants to exploit internally developed technologies through external paths to market also has to be considered.

Step 4: Open innovation is a competence companies have to learn over time. Open innovation is not only about insourcing and outsourcing of resources. Companies learn over time to work effectively with open innovation. Managing internal innovation is not trivial, but managing open innovation is even more challenging. Recently, scholars have been focusing on the need to organize and manage open innovation internally. Only firms that have the right structures and processes in place can work effectively with external partners. Examples are Enkel, Bell and Hogenkamp (2011) and Chiaroni et al. (2010, 2011). These publications indicate that open innovation is essentially an exercise in becoming a *learning organization*, which brings us back to the need to link up open innovation with another broad literature stream (Argyris, 1999; Senge, 1990). In addition, Teece (2007) discusses several important underlying micro foundations of *dynamic capabilities* that offer interesting insights into the *organizational* and *managerial* processes, procedures, systems, and structures needed to implement open innovation in the firm.

NOTES

1. To the best of our knowledge, Laursen & Salter (2006), Christensen et al. (2005), Gassmann & Enkel (2004), Lichtenthaler & Lichtenthaler (2009), Dahlander & Gann (2010), and Vanhaverbeke et al. (2008) are a few notable exceptions.
2. Focusing on the different dimensions of the open innovation funnel is a deliberate choice and there are certainly other potential inroads to link open innovation practices to the existing management literature.
3. An innomediary or innovation intermediary can be defined as "an organization or body that acts as an agent or broker on any aspect of the innovation process between two or more parties" Howells (2006). They have been described in detail in Chesbrough (2006, chapter 6).
4. Several authors (Madhok, 1997; Ring & van de Ven, 1992; Zajac & Olson, 1993; Dyer, 1997) have argued that firms choose to optimize the *net transaction value*.
5. Given the critics with respect to the transaction cost theory, one could also argue that the transaction is not the proper unit of analysis, but rather the relationship between innovation partners. Pierre Azoulay has some nice work on this topic (see for instance Azoulay et al. 2010; Azoulay, 2004).
6. We could also define an innovation intermediary in a broader way as "an organization or body that acts as an agent or broker on any aspect of the innovation process between two or more parties. Such intermediary activities include: helping to provide information about potential collaborators, brokering transactions between two or more parties; acting as mediator, or go-between, bodies or organizations that are already collaborating; and helping find advice, funding and support for the innovation outcomes of such collaborations" (Howells, 2006: 720). However, we prefer to stick with a narrow definition more in line with the innomediaries that have been described extensively in the open innovation literature.
7. Boudreau and Lakhani (2009) discuss the choice between competitively structured innomediary approaches versus cooperatively structured approaches. This topic is important in determining the efficiency of the intermediation process and the incentive structure of the different actors.
8. Dyer and Singh (1998) argue that there is an important relationship between the relational view and the resource dependence theory (Dyer & Singh, 1998: 675, footnote 7: "This is based on their expectation that the distribution of the relational rents is consistent with a resource-dependence perspective as partners that

bring the more critical (i.e. scarce) resources to the relationship will be able to appropriate a higher percentage of the rents."

9. Chesbrough et al. (2006) mention several points of differentiation for open innovation relative to prior theories of innovation (see Table 1.1). Several of the micro foundations of the dynamic capabilities confirm the new insights of open innovation.

Part V

Conclusions

15

Surfing the New Wave of Open Innovation Research

Wim Vanhaverbeke, Henry Chesbrough, and Joel West

The concept of open innovation has become increasingly popular in the management literature on technology and innovation. Open innovation is a decade old and alive and kicking. Chesbrough and Brunswicker (2013) recently conducted the first large sample survey of the adoption of open innovation by large firms (sales in excess of USD 250 million). None of the respondents report abandoning their practice of open innovation and 82% report that open innovation is practiced more intensively today than three years earlier. These survey results suggest that open innovation is a practice that will continue for years, and is widely viewed as important for improving performance.

Despite the popularity of open innovation, many firms still struggle to properly manage open innovation. While open innovation poses many new challenges to firms when adopted, it is the internal organizational challenges that are perceived as most difficult to manage. Managing the journey from closed to open innovation implies several organizational changes at various levels of the firm. Making these changes happen is difficult according to the respondents. Similarly, respondents to the survey were not satisfied with their metrics for measuring open innovation.

The growing penetration of open innovation in companies is reflected in the astonishing growth of the number of open innovation related publications (see Chapter 1 for a detailed overview). However, despite the large volume of publications about open innovation, there are still many important research topics that have not been fully addressed. Open innovation has been mainly studied at the firm level, while other levels of analysis that could enrich our understanding have not been touched upon. Despite the mushrooming volume of empirical work, many hypotheses still have to be tested properly through empirical studies based on fine-grained data and more advanced empirical work. Prescriptions being proposed are often fairly general and not specific to particular contexts and contingencies.

The current state of open innovation in practice and research calls for extended research in the next decade. This was one of the major drivers for publishing this volume: each of the 14 preceding chapters is a contribution in a research area that has been understudied and deserves more attention from the research community in the next decade. One of the objectives of this volume is to shift from a debate between proponents and opponents about the benefits of open innovation towards an analysis that identifies the mediators and moderators of such benefits. Research has shown that patterns of innovation differ fundamentally by sector, firm, and strategy and thus we would expect that the mechanisms and outcomes of open innovation models would also be sensitive to the context in which they are analyzed. Instead of providing general prescriptions, the different chapters offer conceptual and empirical insights into the precise mechanisms underlying the successful implementation of open innovation research and management practice.

In the remainder of this chapter, we offer specific suggestions for the research community about how open innovation research could and should evolve in the next decade. Some are a direct consequence of ideas developed in previous chapters. Here, we integrate these ideas and crosslink them to each other. We also highlight topics that, while not covered in this book, we believe deserve more attention in future research.

The chapter is structured as follows: firstly, we discuss the need to connect (and integrate) open innovation research into mainstream management theories. Secondly, we look for possible extensions of open innovation research into new application fields, such as SMEs, low-tech firms, and non-profit organizations. Open innovation also has several implications for public policy and multinational companies. Thirdly, we elaborate on the need to examine open innovation at different levels of analysis. Fourthly, we identify the need to develop frameworks to understand how companies must change internally to successfully apply open innovation. Fifthly, we highlight how open innovation has implications for functions beyond R&D that have not traditionally been involved in implementing open innovation: such as HRM, PR, and legal. Finally, we reflect on the ongoing transformation of open innovation research and practice, and the implications for how to benefit from open innovation.

15.1 LINKING OPEN INNOVATION TO OTHER RESEARCH

The open innovation literature originated with reflections on observations about changing innovation management practices in companies (Chesbrough, 2003a, 2006a). Literature about open innovation has grown rapidly, and there is a growing need to relate or integrate it into existing innovation management research.

Firstly, we need to integrate open innovation into the broader stream of innovation literature focusing on collaboration with external partners. Such external collaboration is increasingly important to firms, and other researchers (before and since Chesbrough) have examined such collaboration from perspectives other than open innovation. Such a proliferation of perspectives and terminology has the potential to create ambiguity and confusion, while contrasting and integrating open innovation with other research should improve the precision and predictive value of both streams of research.

In our earlier book (*Open Innovation: Researching a New Paradigm*), West and Gallagher (2006b) examined the distinction between open innovation and open source. In this volume, Piller and West (Chapter 2) unravel the distinction between open innovation and user innovation, while Chesbrough and Bogers (Chapter 1) consider more broadly how open innovation relates to other forms of openness. Others have sought to integrate open innovation with prior research, as Perkmann and Walsh (2007) did for university-industry collaboration. We welcome further research on distinguishing and combining open innovation with other research on external collaboration, including user innovation, co-creation, cooperative R&D, technology sourcing, and related topics.

Another way to extend open innovation research is illustrated by Vanhaverbeke and Chesbrough (Chapter 3), who explain the differences in firm strategies by classifying them into two dimensions: open vs. closed innovation and open vs. closed business models. In the case where open innovation is combined with open business models, new strategies can be developed in which the focal company is not involved in new product development but builds an ecosystem where others are delivering technical solutions (eventually in an open innovation style) to develop the business model of the focal firm. This expanded view on open innovation offers an interesting inroad to link open innovation to innovation ecosystem thinking (Nambisan & Sawhney, 2011). One suggestion for a further expansion of the logic of open innovation is to consider new product development as a particular case of a strategic driver. Using the innovation capabilities of others to leverage the strategic drivers of its business can open opportunities even for companies selling commodities.

Open innovation between organizations fosters the emergence and growth of innovation ecosystems. The open innovation literature has been focusing mainly on bilateral relations with innovation partners viewed from a single company's (and usually technology taker's) point of view. However, more and more companies use platforms and multi-partner networks as the basis of their business model. Consequently, the open innovation literature should shift its attention beyond the one-on-one relations between innovation partners to more complex settings of partnerships to create new open business models. For example, after explicating the differences between networks, ecosystems, and platforms, West (Chapter 4 in this volume) examines the challenges that a startup company faces in managing and funding an open innovation platform strategy.

Finally, Vanhaverbeke and Cloodt (Chapter 14) dig into the underlying assumptions of open innovation and try to clarify the phenomenon of open innovation, through use of existing management theories and theories of the firm. Although the goal is to obtain a better theoretical understanding of open innovation, the results of this chapter suggest broader applications. It is a relative straightforward exercise to investigate which theoretical assumptions are consistent or not with open innovation. The authors also suggest that open innovation challenges existing theories to rethink some of their theoretical assumptions. We invite scholars to take up this challenge, and provide a solid theoretical underpinning to open innovation that will lead to more substantive insights and conclusions. Such research will also help open innovation play a more central role in innovation studies.

15.2 EXTENDING OPEN INNOVATION TO NEW APPLICATION AREAS

Open innovation was originally conceived as a paradigm shift for large manufacturing companies. These companies were also among the first to deliberately adopt open innovation as part of their innovation strategy. Beyond companies such as IBM and Intel that were profiled in Chesbrough's (2003a) seminal book, there are numerous examples of large firms that have adopted open innovation as a direct consequence of the publication of that book. P&G, General Mills, Philips, Siemens, Lego, Natura, and DSM are good examples of companies that have embraced open innovation and systematically improved open innovation management over time.

Chesbrough himself expanded the scope of open innovation during the last decade. Firstly, he extended the focus from innovation to business models, arguing that companies could create and capture more value through open business models (Chesbrough, 2006a). Later, he introduced open innovation into services, showing how companies in a wide range of service industries can benefit from applying open innovation practices (Chesbrough, 2011). It is beyond doubt that major changes in sectors such as banking, insurance, publishing, retailing, logistics, and telecommunications will be seen as service companies start to adopt game-changing open innovation and open business models.

Most chapters in this volume provide ammunition to further broaden the scope of open innovation research. As it is impossible to mention all the possible extensions, we describe here just a few of the more interesting examples.

Open innovation in SMEs: an initially neglected research area was how open innovation and open business models can be applied in small and medium-sized enterprises (SMEs). Only later did researchers investigate the relevance and specific nature of open innovation in SMEs (see van de Vrande et al., 2009; Lee et al., 2010; Wynarczyk et al., 2013; Spithoven et al., 2013) and there is still much that we do not know about open innovation in SMEs.

In Chapter 7 of this volume, Van de Vrande and Brunswicker set the stage for future research on open innovation in SMEs. They discuss the specific nature of this context and map the field of existing research. Against this background, they propose a future research agenda and discuss four key areas of open innovation research in SMEs that have not yet received sufficient attention from researchers: IT-enabled crowdsourcing in SMEs for involving a large number of "outsiders"; the importance of different kinds of networks (personal, R&D, and value networks) when SMEs engage in open innovation; the interplay of IP management and open innovation in SMEs and the internal dimensions of managing open innovation in SMEs.

The limited research thus far suggests that SMEs can successfully embrace open innovation and that openness contributes substantially to their innovative and financial performance. Yet, it also indicates that open innovation management in SMEs substantially differs from open innovation management in large (manufacturing) companies. We strongly encourage scholars to develop a conceptual understanding of why open innovation in SMEs differs materially from open innovation in large firms, and thus why adopting the open innovation practices of large firms is not appropriate for SMEs. Based on in depth interviews at ten SMEs that embrace open innovation, Vanhaverbeke (2012) concludes that open innovation management in (traditional) SMEs is so different from open innovation in large companies that lessons learned from good practices in large firms cannot be transferred to SMEs.

Here are some possible research topics for how open innovation might be re-conceptualized when applied to SMEs. Firstly, open innovation is inextricably linked with the strategy or business model of the firm, and so its role can only be understood within a broad strategic setting. Thus an analysis of the business model innovation would logically come first, and the usefulness of open innovation hinges on the role it plays in achieving broader strategic goals. A second research topic that should be explored in greater detail is the link between open innovation in SMEs and the role of the founder or manager of the firm. In SMEs the entrepreneur plays a crucial role in shaping the entire innovation process. He perceives and explores new business opportunities, and his personal commitment and conviction help determine the success and development of the innovation network. This suggests an opportunity to link open innovation to the entrepreneurship literature, particularly that on the attitudes and behaviors of founders. Thirdly, one approach for SMEs to utilize open innovation might be to apply the principles of discovery driven growth theory to maximize opportunities and minimize risks (McGrath & MacMillan, 2009). Finally, we need to integrate different disciplines such as innovation management, entrepreneurship, and strategy, bridging these independent perspectives to understand the complexity of open innovation in SMEs.

Open innovation in high-tech and low-tech industries: open innovation has been mainly associated with high-tech contexts where companies develop new business opportunities based upon technology sourced from other organizations. This is not surprising because such technological breakthroughs are an important form of innovation. However, new technologies are not the only way in which firms can develop new offerings and generate competitive advantage. Product design, new market insights, customer intimacy, and business model innovation are a few examples of how firms may realize the benefits of open innovation from non-technological factors, and we suspect these drivers will be particularly important in low-tech environments.

Although not an explicit focus of this volume, we would like to see more research on open innovation in low-tech industries and on open innovation beyond technological innovation.[1] Consumer product companies such as Procter & Gamble and Matsushita were early adopters of inbound open innovation (Dodgson, Gann & Salter, 2006; Christensen, 2006), and many of the SMEs (in the studies mentioned earlier) were in low-tech industries. However, only a few papers have explicitly focused on open innovation in low-tech settings (Chesbrough & Crowther, 2006; Spithoven, Clarysse & Knockaert, 2010; Vanhaverbeke, 2012) and there is a need to systematically analyze how open innovation functions in such settings.

The open innovation practices that are effective in high-tech settings may not work in low-tech industries: with little or no internal R&D capabilities, firms in low-tech industries might not work on technological innovations themselves (including firms with poorly developed internal R&D capabilities), but they can still thrive using innovations developed by firms in other (high-tech) industries. Laursen and Salter (2006) found that low-tech firms in industries such as paper and printing had a relatively narrow search for external innovations; instead, firms in these industries tend to rely on suppliers (of capital equipment or key inputs) to provide innovation (Dosi, 1988). The strategic logic behind successful applications of open innovation in low-tech industries should be analyzed in detail and compared to open innovation practices in companies in high-tech industries. Such insights would lead to a more nuanced and predictive theory of open innovation, if the comparison leads to the conclusion that open innovation practices in low-tech industries are significantly different from those in high-tech industries.

Open innovation and not-for-profit organizations: Open innovation has largely been studied in the private sector of the economy. In recent years, the not-for-profit sector has increasingly realized that open innovation can generate considerable potential benefits for organizations such as charities, NGOs, or government agencies. Chesbrough and Di Minin (Chapter 9 of this volume) examined case studies of three such organizations that successfully applied open innovation at the core of their strategy. They conclude that open innovation is also relevant for social entrepreneurs, non-profits, and public agencies, to support their efforts to grow their operations and establish partnerships to achieve social change. Chesbrough and Di Minin explain how the principles of open innovation can be applied in this context through a process they term Open Social Innovation.

Other examples exist of NGOs and other non-profit organizations being centrally involved in open innovation initiatives. For example, the World Wildlife Fund (WWF) established an incubator where individuals or groups outside WWF can pitch ideas that would be financed and guided by WWF. It is also interesting to follow the increasing number of collaborations between multinational corporations and NGOs using this process to develop completely new innovation ecosystems. In May 2007, Unilever became the first company to commit to sourcing all its tea in a sustainable manner. Working with the Rainforest Alliance, an international environmental NGO, Lipton and its parent company, Unilever, announced that all Lipton Yellow Label tea bags sold in Western Europe would be certified by 2010. In this case, the Rainforest Alliance certifies Lipton's tea farms in Africa. The collaboration with the NGO should guarantee consumers that the tea they are drinking contributes to incomes and livelihoods of nearly one million Africans and to the protection of the environment. The alliance provides Unilever a unique way to differentiate the Lipton brand from those of its competitors.

Yet, not only social entrepreneurs and NGOs can profit from innovations. There is a diversity of non-profit organizations that can be analyzed through the open innovation lens: universities, research labs, libraries, trusts, museums, and even regulatory agencies (cf. Chesbrough, 2003b). Each faces considerable strategic challenges for growth and renewal, and—as with companies—open innovation may provide a way forward. As with companies, non-profit organizations are increasingly recognizing the crucial role of partnerships in implementing a successful strategy. Therefore, we strongly encourage scholars to further explore the potential of open innovation applications in the non-profit sector.

Open innovation and public policy. Open innovation has major implications for public policy. However, we did not address this topic in this volume, and thus far, only a few publications have focused on this area. In a study about the policy implications of open innovation in the European Union, Chesbrough and Vanhaverbeke (2012) argue that public policy should follow the evolution of the private sector towards open innovation strategies. This report offers a set of broad guidelines on how, by starting from an understanding of open innovation, public policy can facilitate open innovation in the European Union and create more economic growth and jobs. Their recommendations range from action points in education and human capital development, over ideas about how to finance open innovation, to a new approach to intellectual property in government funded organizations. Other policy recommendations would facilitate open innovation by supporting startups and SMEs, which bring new ideas to market and stimulate competition by established firms. Finally, the report calls on governments to expand open government.

In sum, supporting open innovation policy means going beyond the traditional innovation policies, with new approaches that cut across different policy areas to advance and support innovation. We believe that many policy measures in developed countries were created for an era of closed innovation and have remained largely unchanged since then. As economies have entered the era of open innovation, innovation policy needs to change accordingly to remain effective. Given the limited attention thus far, we strongly encourage researchers and policymakers to develop policy frameworks that facilitate open innovation.

The geographic reach of open innovation. A recent extension is broadening the geographical dimension of open innovation. In the past, open innovation literature focused on why external knowledge is important, but rarely considered where that knowledge should be sourced. Asakawa, Song and Kim (Chapter 8 of this volume) remark that as open innovation is becoming increasingly global and global R&D is becoming increasingly open, cross-fertilization between open innovation and global R&D management literature streams is a natural direction. Connecting open innovation to R&D globalization as currently practiced by a growing number of MNEs will make open innovation more interesting and relevant for MNEs struggling with the geographical dimension of open innovation. The organization of open innovation becomes more complex because management has to source knowledge in different parts of the world, while organizing and coordinating knowledge flows internally to extract the most out of the insourced or co-created knowledge.

However, how open innovation can be coupled to global R&D management in MNEs is only one of the potential topics when we introduce the geographical dimension in open innovation. Researchers have considered the impact of differences in corporate culture upon open innovation, but not the impact of differences in national culture. Earlier innovation researchers have examined how the latter differences affect firm innovation processes, as when Shane, Venkataraman, and MacMillan (1995) found differences in innovation leadership styles between countries in a study of 30 national cultures. Thus, exploring the link between differences in culture with differences in the incidence and success of open innovation would help identify moderators and limits of open innovation.

One possible approach would be for researchers to analyze the impact of open innovation in Asian economies and compare the results with Western economies. Most cultures in the Far East are considered to favor relational links rather than the transactional approach used in most Western economies. How do different cultures change the approach to building collaborations with outsiders? How do they impact the speed of building and dissolving relations? Are relationships in Asian economies more stable? What does this imply for the concept of trust in Asian economies? Thus far, there is preliminary evidence that open innovation works differently in northern and southern Europe, while there may also be a link between the penetration of venture capitalists in a country and the perception of open innovation.

Finally, we should also analyze the role of local clusters in determining the effectiveness of open innovation. The literature on open innovation highlights the relevance of networks of interrelated firms as a key determinant of the ability to successfully innovate (Chesbrough, 2003b; van de Vrande et al., 2010). Regional clusters can be defined as "geographical concentrations of interconnected companies and institutions in a particular field" (Porter, 1998: 78). In our earlier book, Simard and West (2006) concluded that the benefits of open innovation may be better realized in regional clusters, which provide an improved environment for information exchange, and because collaboration among firms (and research organizations) in a cluster is a major factor in the success of such clusters. They noted the potential linkage between the relational theories of social network analysis, and the collaborative processes that lie at the heart of open innovation. While Lee et al. (2010) used such measures in their study of open innovation among Korean SMEs, we are unaware of this approach being used in studies of regional clusters. We welcome further research linking the open innovation literature to the regional cluster theory. Of particular interest is the evolving knowledge-based cluster theory, as it highlights the role of interactive learning processes and the development of relational capital as a primary source of positive cluster effects (Bahlmann & Huysman, 2008; Bathelt, 2008; Cooke, 2007; Staber, 2007; Mesquita, 2007).

15.3 ANALYZING OPEN INNOVATION AT DIFFERENT LEVELS OF ANALYSIS

West, Vanhaverbeke and Chesbrough (2006) identified the need to study the phenomenon of open innovation at different levels of analysis. We outlined five levels of analysis for future research in open innovation: individuals and teams (groups), organizations, inter-organizational (networks, industry, or sector), and regional and national innovation systems. Extensive research has been done at the organizational level, but the other levels of analysis were and are still heavily underrepresented. The overemphasis of the firm level has some unsavory effects: first, open innovation is studied with a narrow, managerial perspective that is focused on topics mainly relevant to top management. Second, a firm perspective prevents us from taking an unbiased view on collaboration between different innovation partners. A dyadic or innovation network perspective is required to understand the objective incentives of all the partners involved. Third, an analysis at the organizational level does not provide detailed information on the mechanism driving open innovation within an organization. A lack of understanding of these mechanisms will prevent us from reaching a better understanding of how open innovation should be managed and organized. Finally, the firm-level analysis may reflect the comfort zone of management scholars, pointing to the need to engage a broader range of research perspectives.

Analyzing open innovation at other levels requires inter-disciplinary research which may be harder to conduct and publish. As we discuss below, understanding open innovation management at the levels of individuals requires some understanding of HRM, IP management, team dynamics, or new product development. A good understanding of strategic alliances, the role and legal understanding of contracts, or trust development are likely to be key elements for studying innovation networks in depth. Finally, describing the role of open innovation in regional innovation systems and public policy (as recommended above) requires collaboration with specialists in policy decision-making.

Three chapters in this volume suggest some of the benefits of analyzing open innovation at different levels of analysis. Chapter 4 suggests the difficulties that young innovative firms have in building a sustainable business model when funding and other key resources are provided by ecosystem partners with divergent interests. In Chapter 5, Christensen demonstrates the increasing importance of collaboration between a firm and its environment and the opportunities that firms practicing open innovation have to enact and shape that environment. In a similar vein, there are opportunities to connect open innovation to earlier research on the emergence of technology markets (cf. Arora et al., 2001a, 2001b; Arora & Gambardella, 2010).

In Chapter 6, Vanhaverbeke and his colleagues examine a specific case of sub-firm unit level of analysis, identifying the potential benefits of researching (and managing) open innovation at the level of the R&D project. Open innovation can speed up or (slow down) R&D projects, it can lead to more technology transfers or innovations with a larger financial impact, while firms can utilize openness during part or all of the R&D project. Researchers can also consider the impact of different types of partners in an R&D project, or the contrasting impacts between openness in radical or incremental innovations. Studies of R&D projects would be even more fruitful when utilizing systematic data about the composition of (internal or external) teams and the traits of individuals working in open innovation projects.

15.4 ORGANIZING AND MANAGING OPEN INNOVATION

While the first decade of open innovation research focused on the adoption and notable successes of open innovation, less attention has been given to measuring the net benefits of open innovation. During this same period, many companies adopted open innovation on an ad-hoc basis or in a more systematic way. However, a recent survey found that most companies fail in the internal organization of open innovation (Chesbrough & Brunswicker, 2013). Examining the adoption of open innovation in large firms, the study found that while 78% of those surveyed use open innovation, their level of satisfaction with the metrics they employ to manage open innovation is low. The authors conclude that much more knowledge is needed about how to manage and organize open innovation.

Open innovation does not automatically lead to improved innovation or financial performance. There are many ways in which an open innovation strategy can go wrong. For example, firms utilizing inbound open innovation may be unable to find suitable external innovations, may fail to integrate the technology into their organization or bring it to market, or may find that the costs of external sourcing exceed the benefits (West & Bogers, 2014). Companies need the right internal organization to gain from open innovation. This includes efficiently organizing and managing collaboration with external innovation partners, as well as following through with commercialization and measuring results. If researchers look within companies to study the internal organization and management of open innovation, they may produce results that are relevant for managers, as well as help shift the direction of open innovation research towards new and interesting directions.

Mortara and Minshall (Chapter 12 in this volume) provide a potentially valuable framework for implementing open innovation that suggests how large companies organize to profit from open innovation. They distinguish between characteristics describing high-level, firm-wide tendencies of open innovation configurations on the one hand; and internal characteristics and dynamics of open innovation implementation processes on the other hand. These "macro" and "micro" characteristics of open innovation configurations by firms should be aligned with the requirements of the environment (market, industry, technology) and can be moderated (in a positive or negative way) by corporate culture, politics, internal technological capabilities, and corporate knowledge management tools and procedures. Although this framework is a promising way to structure research about open innovation implementation, we still have a long way to go and therefore invite more researchers to examine this area. We are only aware of limited publications thus far that have considered these topics (e.g., Chiaroni et al., 2010, 2011).

In other chapters of this volume, the authors focus on specific implementation issues. For that purpose, we tapped into the experience and knowledge of practitioners or experts who work closely together with open innovation practicing companies. Topics covered include: 1. How should firms be organized to work effectively with innomediaries? (Chapter 13 in this volume); 2. How should firms organize to work effectively in the inside-out mode of open innovation, specifically in establishing spin-offs? (Chapter 11 in this volume); 3. How to manage IP in the company for optimal use in open innovation? (Chapter 10 in this volume). These topics cover only a small set of the possible research topics about how to organize and manage open innovation in medium-sized and large companies.

Other questions require more investigation. For example, what is the role of top management in supporting open innovation? How should firms set up organizational, management, and communication structures supporting open innovation projects? How should they recruit, select, train, etc. for open innovation? What skills, attitudes, and personalities are needed? (see also Chapter 12 in this volume) How can firms create a corporate culture where open innovation can thrive? How can IP be used strategically to accommodate open innovation? How should the R&D department be changed to facilitate open innovation? How should employees work with an open innovation implementation team? What is the best way to evaluate the success of open innovation? What does it take to move from closed to open innovation? Many companies are struggling with such questions: in-depth investigations of these topics would help companies organize and manage open innovation more successfully.

15.5 CONNECTING OPEN INNOVATION TO FUNCTIONS BEYOND R&D

The focus of open innovation has been on the challenges it creates for the R&D department or new business development units; but the potential impact extends to many other functions within the corporation. For example, human resource management (HRM) has received scant attention in open innovation literature, even though promoting open innovation actually requires specific management practices and an appropriate organizational culture. Efforts to explain (or enable) the success of open innovation may fail without considering a firm's HR practices. For example, such practices often focus on encouraging and reinforcing individual performance and development, forming potential hurdles for an effective implementation of open innovation. Such incentives mean that employees are not encouraged to innovate outside the bounds of their company or business unit, or are even actively discouraged from doing so.

When we began this book project in early 2012, there was virtually no literature on the relationship between HRM practices and open innovation performance. As we finish this chapter two years later, there are only a handful of publications that have addressed this topic in any depth. Based on a survey of 158 companies in innovative regions of Russia, Podmetina, Volchek, Dabrowka, and Fiegenbaum (2013) find a positive relationship between HR learning and training practices, human capital value and employee motivation on external technology sourcing—and of the first two factors on external cooperation. Van Steerthem, Delcour, and De Stobbeleir (2013) interviewed managers of Belgian organizations successfully practicing open innovation. Their study is a first attempt to develop a systematic approach on the HRM implications of open innovation, considering recruitment and selection, training and development, appraisals and assessment, and the required change in corporate culture.

Although some scholars have been pioneering the implications of open innovation for HRM, it is fair to say that the human aspect of implementing open innovation remains largely overlooked. Success in open innovation depends on the people involved and thus HRM practices and culture that support those people and the process of open innovation. Traditional HR practices that form a roadblock to such efforts must be moved aside for new, innovative approaches to HR. We encourage scholars in innovation management to team with HRM specialists to study the implications of open innovation for HRM and develop HRM practices that facilitate open innovation.

HRM is not the only corporate function that is impacted by (and impacts) a firm's approach to open innovation. Others include the legal department (as considered by Chesbrough and Ghafele in Chapter 10), public relations (studied by Mortara and Minshall in Chapter 12) and manufacturing (Bogers & Lhuillery, 2011). Procurement, quality control, support, and IT are among the functions that might also be studied by open innovation scholars.

15.6 THE TRANSFORMATION OF OPEN INNOVATION PRACTICE AND RESEARCH

Open innovation and open innovation research are not only expanding rapidly, but they also are transforming over time. Originally, open innovation was explicitly developed as a management practice for large manufacturing companies. It is now applied in smaller companies, service companies, universities, research labs, and even government agencies. In parallel, the focus of research has been shifting and evolving, as we note in this chapter (and reviewed in detail in Chapter 1).

Several chapters nicely indicate that the practice of open innovation is still in full development and transforming over time, as indicated by how the topics have changed in the decade since open innovation was launched. Ways of managing open innovation are different, and companies have become mature in managing open innovation (see Enkel, Bell & Hogenkamp, 2011; Chiaroni et al., 2010, 2011). Open innovation practices are gradually professionalizing as illustrated by the examples of intellectual property management in Chapter 10, staged spin-outs in Chapter 11, open innovation implementation mechanisms in Chapter 12, and the professionalization of the use of innomediaries in Chapter 13. These examples show how open innovation practices are becoming more complex and that standard open innovation management will not automatically lead to a competitive advantage. Differences in the processes, structure, and people used to implement open innovation will increasingly predict differences in the ability of firms to realize sustained benefits from open innovation.

Similarly, research in open innovation is changing rapidly. Not only the number of publications is growing exponentially, but also themes are shifting. Open innovation research is no longer only about the advantages (and possible disadvantages) of open innovation. Topics now include how to balance costs and benefits, when open innovation is (or is not) beneficial, how to align and implement open innovation to gain maximum benefit, and how to measure results.

In conclusion, open innovation practices are changing continuously, and therefore open innovation research will follow this evolution. We encourage open innovation researchers to maintain the alignment of their research to open innovation practices, so that managers can benefit from academic research and so that the academic community can continue to offer theoretical and empirical insights relevant to practice.

NOTES

1. Some managers and researchers have sought to expand "open innovation" beyond innovation to any form of external collaboration (Bogers & West, 2012). We do not endorse such expansive views, and consider as an innovation any "idea, practice, or object that is perceived as new" (Rogers, 1995: 11) that leads to improved outcomes for an organization.

Reference List

Abernathy, William J. and Utterback, James, M. (1978). "Patterns of industrial innovation," *Technology Review*, 80(7): 40–47.

Abrahamson, Eric and Fairchild, Gregory. (1999). "Management fashion: lifecycles, triggers, and collective learning processes." *Administrative Science Quarterly*, 44: 708–740.

Acs, Zoltan J. and Audretsch, David B. (1987). "Innovation in large and small firms," *Economics Letters*, 23(1): 109–112.

Acs, Zoltan J. and Audretsch, David B. (1988). "Innovation and firm size in manufacturing," *Technovation*, 7(3): 197–210.

Adamczyk, Sabrina, Bullinger, Angelika C. and Möslein, Kathrin M. (2012). "Innovation contests: a review. classification and outlook," *Creativity and Innovation Management* 21(4): 335–360.

Adams, Richard, Bessant, John and Phelps, Robert. (2006). "Innovation management measurement: A review," *International Journal of Management Reviews*, 8(1): 21–47.

Adner, Ron. (2006). "Match your innovation strategy to your innovation ecosystem," *Harvard Business Review*, 84(4): 98–107.

Adner, Ron. (2012). *The Wide Lens: A New Strategy for Innovation*, New York: Portfolio/Penguin.

Adner, Ron and Levinthal, Daniel A. (2004). "What is not a real option: considering boundaries for the application of real options to business strategy," *Academy of Management Review*, 29: 74–85.

Adner, Ron and Kapoor, Rahul. (2010). "Value creation in innovation ecosystems: How the structure of technological interdependence affects firm performance in new technology generations," *Strategic Management Journal*, 31(3): 306–333.

Afuah, Allan and Tucci, Christopher L. (2001). *Internet Business Models and Strategies: Text and Cases*. Boston: McGraw-Hill.

Afuah, Allan and Tucci, Christopher L. (2012). "Crowdsourcing as a solution to distant search," *Academy of Management Review*, 37(3): 355–375.

Ahn, Joonmo and Minshall, Tim, (2012). "The influence of the characteristics of CEOs on open innovation performance in SMEs: the case of Korea." The DRUID Academy Conference, January 19–21, Cambridge, UK.

Ahuja, Gautam and Lampert, Curba Morris. (2001). "Entrepreneurship in the large corporation: a longitudinal study of how established firms create breakthrough inventions," *Strategic Management Journal*, 22(6–7): 521–543.

Alchian, Armen A. and Demsetz, Harold. (1972). "Production, information costs, and economic organization," *American Economic Association*, 62(5): 777–795.

Alexy, Oliver, Criscuolo, P., and Salter, A. (2009). "Does IP strategy have to cripple open innovation?," *MIT Sloan Management Review*, 51(1): 71–77.

Alexy, Oliver, Criscuolo, Paola and Salter, Ammon. (2012). "Managing unsolicited ideas for R&D," *California Management Review* 54(3): 116–139.

Almeida, Paul and Phene, Anupama. (2004). "Subsidiaries and knowledge creation: The influence of the MNC and host country on innovation," *Strategic Management Journal*, 25(8–9): 847–864.

Ambos, Björn and Schlegelmilch, Brodo B. (2004). "The use of international R&D teams: An empirical investigation of selected contingency factors," *Journal of World Business*, 39(1): 37–48.

Amit, Rafael and Zott, Christoph (2001). "Value creation in e-business," *Strategic Management Journal*, 22(6–7): 493–520.

Amram, Martha and Kulatilaka, Nalin. (1999). *Real Options: Managing Strategic Investment in an Uncertain World*. Boston: Harvard Business School Press.

Anderson, Philip and Tushman, Michael L. (1990). "Technological discontinuities and dominant designs: A cyclical model of technological change," *Administrative Science Quarterly*, 35(4): 604–633.

Andersson, Ulf, Forsgren, Mats and Holm, Ulf. (2002). "The strategic impact of external networks: Subsidiary performance and competence development in the multinational corporation," *Strategic Management Journal*, 23(11): 979–996.

Antikainen, Maria, Mäkipää, Marko and Ahonen, Mikko. (2010). "Motivating and supporting collaboration in open innovation." *European Journal of Innovation Management*, 13(1): 100–119.

Appiah-Adu, Kwaku and Ranchhod, Ashok. (1998). "Market orientation and performance in the biotechnology," *Technology Analysis & Strategic Management*, 10(2): 197–210.

Argyres, Nicholas S. and Silverman, Brian S. (2004). "R&D, organization structure, and the development of corporate technological knowledge," *Strategic Management Journal*, 25(8–9): 929–958.

Argyris, Chris. (1999). *On Organizational Learning.* 2nd ed., Malden, Mass.: Blackwell.

Arora, Ashish. (1995). "Licensing tacit knowledge: intellectual property rights and the market for know-how," *Economics of Innovation and New Technology* 4(1): 41–60.

Arora, Ashish and Fosfuri, Andrea. (2003). "Licensing the market for technology," *Journal of Economic Behavior and Organization*, 52(2): 277–295.

Arora, Ashish, Fosfuri, Andrea and Gambardella, Alfonso. (2001a). *Markets for Technology: The Economics of Innovation and Corporate Strategy*. Cambridge, Mass.: MIT Press.

Arora, Ashish, Fosfuri, Andrea and Gambardella, Alfonso. (2001b). "Markets for technology and their implications for corporate strategy." *Industrial and Corporate Change* 10(2): 419–451.

Arora, Ashish and Gambardella, Alfonso. (1990). "Complementarity and external linkages: the strategies of the large firms in biotechnology," *Journal of Industrial Economics*, 38 (4): 361–379.

Arora, Ashish and Gambardella, Alfonso. (1994). "Evaluating technological information and utilizing it: Scientific knowledge, technological capability and external linkages in biotechnology," *Journal of Economic Behavior and Organization*, 24(1): 361–379.

Arora, Ashish and Gambardella, Alfonso. (2010). "Ideas for rent: an overview of markets for technology," *Industrial and Corporate Change*, 19(3): 775–803.

Arora, Ashish and Nandkumar, Anand. (2007). "Securing their future? Entry and survival in the information security industry," NBER Working Paper 13634, Cambridge, Mass.: National Bureau of Economic Research.

Arrow, Kenneth J. (1962). "Economic welfare and the allocation of resources for invention." In *National Bureau of Economic Research (Ed.), The Rate and Direction of Inventive Activity: Economic and Social Factors*: 609–625. Princeton, NJ: Princeton University Press.

Arthur, W. Brian. (1989). Competing technologies, increasing returns, and lock-in by historical events. *The Economic Journal*, 99 (394): 116–131.

Arundel, Anthony (2001). "The relative effectiveness of patents and secrecy for appropriation," *Research Policy*, 30(4): 611–624.

Asakawa, Kazuhiro. (2001). "Organizational tension in international R&D management: The case of Japanese firms," *Research Policy*, 30(5): 735–757.

Asakawa, Kazuhiro. (2004). "The coevolution of national innovation system and business strategy: The case of Japanese biotechnology industry," *Journal of Asian Business*, 20(4): 9–40.

Asakawa, Kazuhiro and Lehrer, Mark. (2003). "Managing local knowledge assets globally: The role of regional innovation relays," *Journal of World Business*, 38(1): 31–42.

Asakawa, Kazuhiro and Som, Ashok. (2008). "Internationalization of R&D in China and India: Conventional wisdom versus reality," *Asia Pacific Journal of Management*, 25(3): 375–394.

Asakawa, Kazuhiro and Un, Annique. (2012). *Types of R&D collaborations and process innovation*. Paper presented at the National University of Singapore, September.

Asakawa, Kazuhiro, Nakamura, Hiroshi and Sawada, Naohiro. (2010). "Firms' open innovation policies, laboratories' external collaborations, and laboratories' R&D performance," *R&D Management*, 40(2): 109–123.

Asheim, Bjørn T. and Gertler, Meric S. (2005). "The geography of innovation: regional innovation systems," in: Fagerberg, Jan, Mowery, David C., Nelson, Richard R. (Eds.), *The Oxford Handbook of Innovation*. Oxford: Oxford University Press.

Athreye, Suma and Cantwell, John. (2007). "Creating competition? Globalization and the emergence of new technology producers," *Research Policy*, 36(2): 209–226.

Azoulay, Pierre. (2004). "Capturing knowledge within and across firm boundaries: Evidence from clinical development," *American Economic Review*, 94(5): 1591–1612.

Azoulay, Pierre, Repenning, Nelson P. and Zuckerman, Ezra W. (2010). "Nasty, brutish, and short: Embeddedness failure in the pharmaceutical industry," *Administrative Science Quarterly*, 55(3): 472–507.

Bahemia, Hanna and Squire, Brian. (2010). "A contingent perspective of open innovation in new product development projects," *International Journal of Innovation Management*, 14(4): 603–627.

Bahlmann, Marc D. and Huysman, Marleen H. (2008). The emergence of a knowledge-based view of clusters and its implications for cluster governance. *The Information Society*, 24(5), 304–318.

Baker, Ted and Nelson, Reed E. (2005). "Creating something from nothing: resource construction through entrepreneurial bricolage," *Administrative Science Quarterly*, 50(3): 329–366.

Baldwin, Carliss Y. (2012). "Organization design for business ecosystems," *Journal of Organizational Design* 1(1): 20–23.

Baldwin, Carliss Y. and Henkel, Joachim. (2011). *The impact of modularity on intellectual property and value appropriation*, HBS Working Paper 12-040, Harvard Business School.

Baldwin, Carliss Y., Hienerth, Christoph, and von Hippel, Eric. (2006). "How user innovations become commercial products: A theoretical investigation and case study," *Research Policy*, 35(9): 1291–1313.

Baldwin, Carliss Y. and von Hippel, Eric. (2011). "Modeling a paradigm shift: from producer innovation to user and open collaborative innovation," *Organization Science*, 22(6): 1399–1417.

Barge-Gil, Andrés. (2010). "Open, semi-open and closed innovators: Towards an explanation of the degree of openness," *Industry & Innovation* 17(6): 577–607.

Barney, Jay B. (1986). "Strategic factor markets: expectations, luck, and business strategy," *Management Science*, 32(10): 1231–1241.

Barney, Jay B. (1991). "Firm resources and sustained competitive advantage," *Journal of Management*, 17(1): 99–120.

Bartlett, Chrisopher A. and Ghoshal, Sumantra. (1990). "Managing innovation in the transnational corporation," in Chrisopher A. Bartlett, Yves Doz, and Gunnar Hedlund, eds., *Managing the Global Firm*, London: Routledge, 215–255.

Basberg, Bjørn L. (1987). "Patents and the measurement of technological change: a survey of the literature," *Research Policy*, 16(2–4): 131–141.

Bathelt, Harald. (2008). "Knowledge-based clusters: Regional multiplier models and the role of 'buzz' and 'pipelines,'" in: Karlsson, Charlie (Ed.), *Handbook of Research on Cluster Theory*. Cheltenham, UK: Edward Elgar.

Baum, Joel A. C., Calabrese, Tony and Silverman, Brian S. (2000). "Don't go it alone: Alliance network composition and startups performance in Canadian biotechnology," *Strategic Management Journal*, 21(3): 267–294.

Becker, Wolfgang and Dietz, Jürgen. (2004). "R&D cooperation and innovation activities of firms—evidence for the German manufacturing industry," *Research Policy*, 33(2): 209–223.

Behrman, Jack N. and Fischer, William A. (1980). *Overseas R&D Activities Of Transnational Companies*, Cambridge, Mass.: Oelgeschlager, Gunn and Hain.

Bekkers, Rudi. (2001). *Mobile Telecommunications Standards: GSM, UMTS, TETRA, and ERMES*. Boston: Artech House.

Bekkers, Rudi and West, Joel. (2009) "The limits to IPR standardization policies as evidenced by strategic patenting in UMTS," *Telecommunications Policy*, 33 (1–2): 80–97.

Belderbos, René and Carree, Martin. (2002). "The location of Japanese investments in China: Agglomeration effects, keiretsu, and firm heterogeneity," *Journal of the Japanese and International Economies*, 16(2): 194–211.

Belderbos, René, Carree, Martin and Lokshin, Boris. (2004). "Cooperative R&D and firm performance," *Research Policy*, 33(10): 1477–1492.

Belderbos, René, Carree, Martin and Lokshin, Boris. (2006). "Complementarity in R&D cooperation strategies," *Review of Industrial Organization*, 28(4): 401–426.

Benkler, Yochai. (2006). *The Wealth of Networks: How Social Production Transforms Markets and Freedom*. New Haven, Conn.: Yale University Press.

Berchicci, Luca. (2013). "Towards an open R&D system: Internal R&D investment, external knowledge acquisition and innovative performance," *Research Policy*, 42(1): 117–127.

Bergek, Anna, Tell, Fredrik, Berggren, Christian and Watson, Jim (2008). "Technological capabilities and late shakeouts: Industrial dynamics in the advanced gas turbine industry, 1987–2002," *Industrial and Corporate Change*, 17(2): 335–392.

Bessant, John. (1999). "The rise and fall of "Supernet": A case study of technology transfer policy for smaller firms," *Research Policy* 28(6): 601–614.

Bianchi, Mattia, Campo dall'Orto, Sergio, Frattini, Federico and Vercesi, Paolo. (2010). "Enabling open innovation in small- and medium-sized enterprises: How to find alternative applications for your technologies," *R&D Management*, 40(4): 414–431.

Bianchi, Mattia, Cavaliere, Alberto, Chiaroni, Davide, Frattini, Federico and Chiesa, Vittorio. (2011). "Organisational modes for Open Innovation in the bio-pharmaceutical industry: An exploratory analysis," *Technovation*, 31(1): 22–33.

Bierly, Paul and Chakrabarti, Alok. (1996). "Generic knowledge strategies in the U.S. pharmaceutical industry," *Strategic Management Journal*, 17(winter): 123–135.

Bikhchandani, Sushil, Hirshleifer, David and Welch, Ivo. (1992). "A theory of fads, fashion, custom, and cultural change as informational cascades," *Journal of Political Economy*, 100(5): 992–1026.

Bikhchandani, Sushil, Hirshleifer, David and Welch, Ivo. (1998). "Learning from the Behavior of Others: Conformity, Fads, and Informational Cascades," *Journal of Economic Perspectives*, 12(3):151–170.

Bingham, Alpheus and Spradlin, Dwayne. (2011). *The Open Innovation Marketplace: Creating Value in the Challenge Driven Enterprise*. Upper Saddle River, NJ: FT Press.

Bird, Barbara. (1988). "Implementing entrepreneurial ideas: the case for intention," *Academy of Management Review*, 13(3): 442–453.

Birkinshaw, Julian and Hood, Neil. (1998). "Multinational subsidiary evolution: Capability and charter change in foreign owned subsidiary companies," *Academy of Management Review*, 23(4): 773–795.

Birley, Sue. (1985). "The role of networks in the entrepreneurial process," *Journal of Business Venturing* 1(1): 107–117.

Block, Zenas and Ornati, Oscar A. (1987). "Compensating corporate venture managers," *Journal of Business Venturing*, 2(1): 41–51.

Blomqvist, Kirsimarja, Hurmelinna, Pia and Seppänen, Risto. (2005). "Playing the collaboration game right—balancing trust and contracting," *Technovation*, 25(5): 497–504.

Bogers, Marcel. (2011). "The open innovation paradox: Knowledge sharing and protection in R&D collaborations," *European Journal of Innovation Management*, 14(1): 93–117.

Bogers, Marcel. (2012). "Knowledge sharing in open innovation: An overview of theoretical perspectives on collaborative innovation," in C. de Pablos Heredero and D. López (Eds.), *Open Innovation at Firms and Public Administrations: Technologies for Value Creation*: 1–14. Hershey, PA: IGI Global.

Bogers, Marcel and Lhuillery, Stephane. (2011). "A functional perspective on learning and innovation: Investigating the organization of absorptive capacity," *Industry and Innovation*, 18(6): 581–610.

Bogers, Marcel and West, Joel. (2012). "Managing distributed innovation: Strategic utilization of open and user innovation," *Creativity and Innovation Management*, 21(1): 61–75.

Bogers, Marcel, Afuah, Allan and Bastian, Bettina. (2010). "Users as innovators: A review, critique, and future research directions," *Journal of Management*, 36(4): 857–875.

Bogers, Marcel, Bekkers, Rudi and Granstrand, Ove. (2012). "Intellectual property and licensing strategies in open collaborative innovation," in Carmen de Pablos Heredero and David López Berzosa (Eds.), *Open Innovation at Firms and Public Administrations: Technologies for Value Creation*, Hershey, Penn.: IGI Global, 37–58.

Borod, Ronald S. (2005). "An update on intellectual property securitization," *Journal of Structured Finance*, 10 (4): 65–72.

Boscherini, Lorenzo, Chiaroni, Davide, Chiesa, Vittorio and Frattini, Federico. (2010). "How to use pilot projects to implement open innovation," *International Journal of Innovation Management*, 14(6): 1065–1097.

Boudreau, Kevin (2010). "Open platform strategies and innovation: granting access vs. devolving control," *Management Science*, 56 (10): 1849–1872.

Boudreau, Kevin J. and Lakhani, Karim. (2009). "How to manage outside innovation: competitive markets or collaborative communities?," *MIT Sloan Management Review*, 50(4): 69–75.

Boudreau, Kevin J., Lacetera, Nicola and Lakhani, Karim. (2011). "Incentives and problem uncertainty in innovation contests: An empirical analysis," *Management Science*, 57(5): 843–863.

Bougrain, Frédéric and Haudeville, Bernard. (2002). "Innovation, collaboration and SMEs internal research capacities," *Research Policy*, 31(5): 735–747.

Brandenburger Adam and Nalebuff, Barry J. (1996). *Co-opetition*, New York: Doubleday.

Brem, Alexander and Tidd, Joseph. (Eds.). (2012). *Perspectives on Supplier Innovation: Theories, Concepts and Empirical Insights on Open Innovation and the Integration of Suppliers*. London: Imperial College Press.

Breschi, Stefano, Lissoni, Francesco and Malerba, Franco. (2003) "Knowledge-relatedness in firm technological diversification." *Research Policy*, 32 (1): 69–87.

Bresnahan, Timothy F. and Trajtenberg, M. (1995). "General purpose technologies: Engines of growth?," *Journal of Econometrics*, 65(1): 83–108.

Bröring, Stefanie and Herzog, Philip, (2008). "Organising new business development: open innovation at Degussa," *European Journal of Innovation Management*, 11(3): 330–348.

Bross, Matt. (2009). "Innovation at the speed of life," *BT Innovation Journal*, 1: 3–11.

Brown, Shona L. and Eisenhardt, Kathleen M. (1995). "Product development: past research, present findings, and future directions," *Academy of Management Review*, 20(2): 343–378.

Brüderl, Joseph and Preisendörfer, Peter. (1998). "Network support and the success of newly founded business," *Small Business Economics*, 10(3), 213–225.

Brunswicker, Sabine. (2011). *An empirical multivariate examination of the performance impact of open and collaborative innovation strategies*, unpublished dissertation, Universität Stuttgart.

Brunswicker, Sabine. (2013). "Crowdsourcing in SMEs: the Ocean Optics case," working paper, Universität Stuttgart.

Brunswicker, Sabine and Vanhaverbeke, Wim. (2010). "Beyond open innovation in large enterprises: How do small and medium-sized enterprises (SMEs) open up to external innovation sources?," working paper, Universität Stuttgart.

Brunswicker, Sabine and Ehrenmann, Frank. (2013). Managing open innovation in SMEs: A good practice example of German software firm, *International Journal of Industrial Engineering and Management*, 4(1): 33–41.

Brusoni, Stefano, Prencipe, Andrea and Pavitt, Keith (2001). "Knowledge specialization, organizational coupling and the boundaries of the firm: why firms know more than they make?," *Administrative Science Quarterly*, 46(4): 597–621.

Büchel, Bettina and Raub, Steffen. (2002). "Building knowledge-creating value networks," *European Management Journal*, 20(6): 587–596.

Buganza, Tommaso, Chiaroni, Davide, Colombo, Gabriele and Frattini, Federico, (2011). "Organisational implications of open innovation: An analysis of inter-industry patterns," *International Journal of Innovation Management*, 15(2): 423–455.

Bughin, Jacques, Chui, Michael and Johnson, Brad. (2008). "The next step in open innovation" *McKinsey Quarterly*, (4): 112–122.

Campbell-Kelly, Martin. (2003). *From Airline Reservations to Sonic the Hedgehog: A History of the Software Industry*. Cambridge, Mass.: MIT Press.

Campbell, Alexandra J. and Cooper, Robert G. (1999). "Do customer partnerships improve new product success rates?," *Industrial Marketing Management*, 28(5): 507–519.

Cantwell, John and Mudambi, Ram. (2005). "MNE competence-creating subsidiary mandates," *Strategic Management Journal*, 26(12): 1109–1128.

Cantwell, John and Santangelo, Grazia D. (2006). "Evolution of markets, technology and M&A," in Bruno Cassiman and Massimo G. Colombo (eds.) *Mergers & Acquisitions: The Innovation Impact*, Cheltenham, UK: Edward Elgar, pp. 28–36.

Capron, Laurence and Mitchell, Will. (2000). "Internal versus external knowledge sourcing: evidence from telecom operators in Europe," working paper, INSEAD.

Carlile, Paul R. (2002). "A pragmatic view of knowledge and boundaries: boundary objects in new product development," *Organization Science*, 13(4): 442–455.

Casper, Steven. (2007). "How do technology clusters emerge and become sustainable? Social network formation and inter-firm mobility within the San Diego biotechnology cluster," *Research Policy*, 36(4): 438–455.

Cassiman, Bruno and Veugelers, Reinhilde. (2002). "Spillovers and R&D cooperation: Some empirical evidence for Belgium," *American Economic Review*, 92(4): 1169–1184.

Cassiman, Bruno and Veugelers, Reinhilde. (2006). "In search of complementarity in innovation strategy: Internal R&D and external knowledge acquisition," *Management Science*, 52(1): 68–82.

Cassiman, Bruno, Veugelers, Reinhilde and Zuniga, Pluvia. (2008). "In search of performance effects of (in)direct industry-science links," *Industrial and Corporate Change*, 17(4): 611–646.

Ceccagnoli, Marco, Graham, Stuart J. H., Higgins, Matthew J. and Lee, Jeongsik. (2010). "Productivity and the role of complementary assets in firms' demand for technology innovations," *Industrial and Corporate Change*, 19 (3): 839–869.

Ceci, Federica and Iubatti, Daniela. (2012). "Personal relationships and innovation diffusion in SME networks: A content analysis approach," *Research Policy*, 41(3): 565–579.

Chandler, Alfred D., Jr. (1962). *Strategy and Structure*, Cambridge, Mass.: MIT Press.

Chandler, Alfred D., Jr. (1990). *Scale and Scope: The Dynamics of Industrial Capitalism*, Cambridge, Mass.: Harvard University Press.

Che, Yeon-Koo, and Gale, Ian. (2003). "Optimal design of research contests," *American Economic Review*, 93(1): 646–671.

Chen, J., Chen, Y. and Vanhaverbeke, Wim (2011). "The influence of scope, depth, and orientation of external technology sources on the innovative performance of Chinese firms," *Technovation*, 31(8): 362–373.

Cheng, Joseph L.C. and Bolon, Douglas S. (1993). "The management of multinational R&D: A neglected topic in international business research," *Journal of International Business Studies*, 24(1): 1–18.

Chesbrough, Henry W. (1999). "The organizational impact of technological change: a comparative theory of national institutional factors," *Industrial and Corporate Change*, 8(3): 447–485.

Chesbrough, Henry W. (2000). "Designing corporate ventures in the shadow of private venture capital," *California Management Review*, 42 (3): 31–49.

Chesbrough, Henry W. (2002). "Graceful exits and missed opportunities: Xerox's management of its technology spin-off organizations," *Business History Review*, 76 (4): 803–837.

Chesbrough, Henry W. (2003a). *Open Innovation: The New Imperative for Creating and Profiting from Technology*. Boston, MA: Harvard Business School Press.

Chesbrough, Henry W. (2003b). "The era of open innovation," *Sloan Management Review*, 44(3): 35–41.

Chesbrough, Henry W. (2003c). "The logic of open innovation: managing intellectual property," *California Management Review*, 45(3): 33–58.

Chesbrough, Henry W. (2006a). *Open Business Models: How to Thrive in the New Innovation Landscape*. Boston: Harvard Business School Press.

Chesbrough, Henry W. (2006b). "Open innovation: A new paradigm for understanding industrial innovation," in H. Chesbrough, W. Vanhaverbeke, and J. West (eds.), *Open Innovation: Researching a New Paradigm*. Oxford: Oxford University Press, 1–12.

Chesbrough, Henry W. (2006c). "New puzzles and new findings," in H. Chesbrough, W. Vanhaverbeke, and J. West (eds.), *Open Innovation: Researching a New Paradigm*. Oxford: Oxford University Press, 15–34.

Chesbrough, Henry W. (2006d). "Emerging secondary markets for Intellectual Property: US and Japan Comparisons," research report to National Center for Industrial Property Information and Training, Lafayette, CA: Open Innovation Corporation.

Chesbrough, Henry W. (2007a). "Business model innovation: It's not just about technology anymore," *Strategy & Leadership*, 35(6): 12–17.

Chesbrough, Henry W. (2007b). "Why companies should have open business models," *Sloan Management Review*, 48(2): 22–28.

Chesbrough, Henry W. (2010). "Business model innovation: Opportunities and barriers," *Long Range Planning*, 43 (2–3): 354–363.

Chesbrough, Henry W. (2011). *Open Services Innovation: Rethinking Your Business to Grow and Compete in a New Era*. San Francisco: Jossey-Bass.

Chesbrough, Henry W. (2012a). "GE's ecomagination challenge: an experiment in open innovation," *California Management Review* 54 (3): 140–154.

Chesbrough, Henry W. (2012b). "Open innovation: Where we've been and where we're going," *Research Technology Management*, 55 (4): 20–27.

Chesbrough, Henry W. and Appleyard, Melissa M. (2007). "Open innovation and strategy," *California Management Review*, 50(1): 57–76.

Chesbrough, Henry W. and Brunswicker, Sabine. (2013). *Managing Open Innovation In Large Firms*, Stuttgart: Fraunhofer Verlag.

Chesbrough, Henry W. and Chen, Eric. (2013). "Recovering abandoned compounds through expanded external IP licensing," *California Management Review*, 55(4): 83–101.

Chesbrough, Henry W. and Crowther, Adrienne Kardon (2006). "Beyond high tech: Early adopters of open innovation in other industries," *R&D Management*, 36(3): 229–236.

Chesbrough, Henry W. and Garman, Andrew R. (2009). "How open innovation can help you cope in lean times," *Harvard Business Review*, 87(12): 68–76.

Chesbrough, Henry W. and Kusonoki, Ken. (2001), "The modularity trap: innovation, technology phases shifts and the resulting limits of virtual organizations," in Ikujiro Nonaka and David J. Teece (eds.), *Managing Industrial Knowledge: creation, transfer and utilization*, London: Sage, pp. 202–231.

Chesbrough, Henry W. and Rosenbloom, Richard S. (1999). "The dual-edged role of the business model in leveraging corporate technology investments," paper delivered at the *NIST conference on Managing Technical Risk*, the John F. Kennedy School of Government, Harvard University, September.

Chesbrough, Henry W. and Rosenbloom, Richard S. (2002). "The role of the business model in capturing value from innovation: Evidence from Xerox Corporation's technology spin-off companies," *Industrial and Corporate Change*, 11(3): 529–555.

Chesbrough, Henry W. and Schwartz, Ken. (2007). "Innovating business models with co-development partnerships," *Research—Technology Management*, 50(1): 55–59.

Chesbrough, Henry W., and Socolof, Stephen J. (2000). "Creating new ventures from Bell Labs Technologies," *Research-Technology Management*, 43 (2): 13–17.

Chesbrough, Henry W. and Teece, David J. (1996). "When is virtual virtuous: organizing for innovation," *Harvard Business Review*, January–February, 65–74.

Chesbrough, Henry W. and Vanhaverbeke, Wim. (2012). *Open Innovation and Public Policy in Europe*. Commissioned by ESADE and the Science Business Innovation Board, Brussels: Science Business Publishing Ltd.

Chesbrough, Henry W., Vanhaverbeke, Wim, and West, Joel. (eds.). (2006). *Open Innovation: Researching a New Paradigm*. Oxford: Oxford University Press.

Chiaroni, Davide, Chiesa, Vittorio and Frattini, Federico. (2010). "Unraveling the process from Closed to Open Innovation: evidence from mature, asset-intensive industries," *R&D Management* 40 (3): 222–245.

Chiaroni, Davide, Chiesa, Vittorio and Frattini, Federico. (2011). "The open innovation journey: How firms dynamically implement the emerging innovation management paradigm," *Technovation*, 31(1): 34–43.

Christensen, Clayton. (1997). *The Innovator's Dilemma: When New Technologies Cause Great Firms To Fail*. Boston: Harvard Business School Press.

Christensen, Clayton. (2012). *Open Innovation and Getting Things Right*, weblog article, 19 Sept., URL: http://www.claytonchristensen.com/open-innovation, accessed 26 Sept 2013.

Christensen, Jens Frøslev (1995). "Asset profiles for technological innovation," *Research Policy*, 24 (5): 727–745.

Christensen, Jens Frøslev, (2006). "Wither core competency for large corporation in an open innovation world?," in: Chesbrough, H., Vanhaverbeke, Wim, West, Joel (eds.),

Open Innovation: Researching a New Paradigm. Oxford University Press: Oxford, 35–61.

Christensen, Jens Frøslev (2011), "Industrial evolution through complementary convergence: the case of IT security," *Industrial and Corporate Change*, 20 (1), 57–89.

Christensen, Jens Frøslev, Olesen, Michael Holm, & Kjaer, Jonas Sorth (2005). "The industrial dynamics of Open Innovation: Evidence from the transformation of consumer electronics," *Research Policy*, 34(10): 1533–1549.

Christiansen, Christian A., Burke, Brian E., Kolodgy, Charles J., and Hudson, Sally (2003). "The Big Picture: IT Security Products and Services Forecast and Analysis, 2002–2006," International Data Corporation, Report #29137.

Churchill, Joan, von Hippel, Eric and Sonnack, Mary. (2009). *Lead User Project Handbook. A practical guide for lead user project teams*, Cambridge, Mass.: MIT.

Clark Kim B. and Wheelwright, Steven C. (1990). *Managing New Product and Process Development: Text and Cases*, Boston: Harvard Business School.

Classen, Nicolas, van Gils, Anita, Bammens, Yannick and Carree, Martin. (2012). "Accessing resources from innovation partners: the search breadth of family SMEs," *Journal of Small Business Management*, 50(2): 191–215.

Clay, Alexa, and Paul, Roshan. (2012). "Open innovation: a muse for scaling," *Stanford Social Innovation Review*, Fall, 17–18.

Cloyd, Gil, Euchner, James. (2012). "Building open innovation at P&G: An interview with Gil Cloyd," *Research Technology Management*, 55 (4), 14–19.

Cohen, Wesley M. and Levinthal, David A. (1989). "Innovation and learning: the two faces of R&D," *Economic Journal*, 99 (397): 569–596.

Cohen, Wesley M. and Levinthal, David A. (1990). "Absorptive capacity: A new perspective on learning and innovation," *Administrative Science Quarterly*, 35(1): 128–152.

Collinson, Simon, and Gregson, Geoff. (2003). "Knowledge networks for new technology-based firms: an international comparison of local entrepreneurship promotion." *R&D Management*, 33(2), 189–208.

Colombo, Massimo G. and Paola Garrone (2006). "The impact of M&A on innovation: empirical results," in Bruno Cassiman and Massimo G. Colombo (eds.) *Mergers & Acquisitions: The Innovation Impact*, Cheltenham, UK: Edward Elgar, pp. 104–133.

Cooke, Philip. (2005). "Regionally asymmetric knowledge capabilities and open innovation exploring 'Globalisation 2': A new model of industry organisation," *Research Policy*, 34(8): 1128–1149.

Cooke, Philip. (2006). "Regional knowledge capabilities and open innovation: regional innovation systems and clusters in the asymmetric knowledge economy," in Stefano Breschi & Franco Malerba (eds.), *Clusters, Networks & Innovation*, Oxford: Oxford University Press: 80–112.

Cooke, Philip. (2007). "Social capital, embeddedness, and market interactions: An analysis of firm performance in UK regions." *Review of Social Economy*, 65 (1), 79–106.

Cooper, Robert G. (1990). "Stage-gate systems: A new tool for managing new products," *Business Horizons*, 33(3): 44–54.

Cooper, Robert G. (1999). "The invisible success factors in product innovation," *Journal of Product Innovation Management*, 16(2): 115–133.

Cooper, Robert G. (2008). "Perspective: the stage-gate idea-to-launch process-update, what's new, and nexgen systems," *Journal of Product Innovation Management*, 25(3): 213–232.

Cooper, Robert G., Edgett, Scott J. and Kleinschmidt, Elko J. (2004). "Benchmarking best NPD practices—I," *Research-Technology Management*, 47 (1): 31–43.

Cooper, Robert G., Edgett, Scott J. (2012). "Best practices in the idea-to-launch process and its governance," *Research-Technology Management,* 55(2): 43–53.

Cordón-Pozo, Euloio, Garcia-Morales, Victor J., and Aragón-Correa, J. Alberto. (2006), "Inter-departmental collaboration and new product development success: a study on the collaboration between marketing and R&D in Spanish high-technology firms," *International Journal of Technology Management*, 35(1): 52–79.

Cornelissen, Joep, Christensen, Lars Thøger and Kinuthia, Kendi. (2012). "Corporate brands and identity: Developing stronger theory and a call for shifting the debate," *European Journal of Marketing*, 46 (7–8): 1093–1102.

Cosh, Andy and Zhang, Joanne Jin. (2011). "Open innovation choices—What is British enterprise doing?," London: UK-Innovation Research Centre.

Cottrell, Tom and Nault, Barrie R. (2004). "Product variety and firm survival in the microcomputer software industry," *Strategic Management Journal*, 25 (10): 1005–1025.

Dahlander, Linus, Frederiksen, Linus and Rullani, Francesco. (eds.). (2011). *Online Communities and Open Innovation: Governance and Symbolic Value Creation*. London: Routledge.

Dahlander, Linus, and Gann, David M. (2010). "How open is innovation?," *Research Policy*, 39(6): 699–709.

Dahlander, Linus and Magnusson, Mats. (2008). "How do firms make use of open source communities?," *Long Range Planning*, 41(6): 629–649.

Dahlander, Linus, and Wallin, Martin W. (2006). "A man on the inside: Unlocking communities as complementary assets," *Research Policy*, 35(8): 1243–1259.

Dang, Rani J., Mortara, Letizia, Thomson, Ruth and Minshall, Tim. (2011). "Developing a technology intelligence strategy to access knowledge of innovation clusters," in Hülsmann, M., Pfeffermann, N. (eds.), *Strategies and Communications for Innovations. An Integrative Management View for Companies and Networks*. Berlin: Springer-Verlag, 51–68.

Daniel, Caroline. (2000). "Psion plans Symbian flotation," *Financial Times*, August 9.

Davis, Lee. (2008), "Licensing strategies of the new 'intellectual property vendors,'" *California Management Review*, 50(2): 6–30.

de Jong, Jereon P. J. and Marsili, Orietta. (2006). "The fruit flies of innovations: A taxonomy of innovative small firms," *Research Policy*, 35 (2): 213–229.

de Jong, Jereon P. J., Kalvet, Tarmo, and Vanhaverbeke, Wim. (2010). "Exploring a theoretical framework to structure the public policy implications of open innovation," *Technology Analysis & Strategic Management*, 22(8): 877–896.

De Meyer, Arnoud. (1991). "Tech talk: How managers are stimulating global R&D communication," *Sloan Management Review*, 32 (3): 49–58.

De Meyer, Arnoud and Mizushima, Atsuo. (1989). "Global R&D management," *R&D Management*, 19 (2): 135–146.

de Pablos Heredero, Carmen and López, David. (eds.). (2012). *Open Innovation at Firms and Public Administrations: Technologies for Value Creation*. Hershey, PA: IGI Global.

Diener, Kathleen, and Piller, Frank. (2008), "Facets of open innovation: development of a conceptual framework," paper presented at the Open User Innovation Conference, Boston.

Diener, Kathleen, and Piller, Frank. (2010). *The Market for Open Innovation: Increasing the Efficiency and Effectiveness of the Innovation Process,* Raleigh, NC: Lulu.

Diener, Kathleen, and Piller, Frank. (2013). *The Market for Open Innovation: A Survey of Open Innovation Accelerators*, 2nd edition, Raleigh, NC: Lulu.

Dimancescu, Dan and James Botkin. (1986). *The New Alliance: America's R&D Consortia*, Cambridge, Mass.: Ballinger.

Di Minin, Alberto, Frattini, Federico, Piccaluga, Andrea. (2010). "Fiat: open innovation in a downturn (1993–2003)," *California Management Review* 52(3): 132–159.

Di Minin, Alberto and Bianchi, Mattia. (2011). "Safe nests in global nets: Internationalization and appropriability of R&D in wireless telecom," *Journal of International Business Studies*, 42 (7): 910–934.

Dittrich, Koen and Duysters, Geert. (2007). "Networking as a means to strategy change: The case of open innovation in mobile telephony," *Journal of Product Innovation Management*, 24 (6): 510–521.

Doan, Anhai, Ramakrishnan, Raghu, and Halevy, Alon Y. (2011). "Crowdsourcing systems on the world-wide web," *Communications of the ACM*, 54(4): 86–96.

Dodgson, Mark, Gann, David, and Salter, Ammon. (2006). "The role of technology in the shift towards open innovation: The case of Procter & Gamble," *R&D Management*, 36(3): 333–346.

Doppen, Casper S., (2008). "Research on the role of MiPlaza within the Open Innovation R&D community," unpublished thesis, Twente University, School of Management and Governance.

dos Santos, Ricardo and Spann, Martin. (2011). "Collective entrepreneurship at Qualcomm: combining collective and entrepreneurial practices to turn employee ideas into action," *R&D Management*, 41 (5): 443–456.

Dosi, Giovanni. (1988) "Sources, procedures, and microeconomic effects of innovation." *Journal of Economic Literature*, 26 (3): 1120–1171.

Doz, Yves L., Santos, Jose and Williamson Peter. (2001). *From Global to Metanational: How Companies Win in the Knowledge Economy*, Boston: Harvard Business School Press.

Doz, Yves L., Olk, Paul M. and Smith Ring, Peter. (2000). "Formation processes of R&D consortia: Which path to take where does it lead?," *Strategic Management Journal*, 21(3): 239–266.

Drayton, Bill. (2011). "Collaborative Entrepreneurship: How social entrepreneurs can tip the world by working in global teams," *Innovations*, 6(2): 35–38.

Drayton, Bill, and Budinich, Valeria. (2010). "A new alliance for global change," *Harvard Business Review*, 88(9), 56–64.

Drechsler, Wenzel and Natter, Martin. (2012). "Understanding a firm's openness decisions in innovation," *Journal of Business Research*, 65(3): 438–445.

Drees, Johannes M. and Heugens, Pursey. (2013). "Synthesizing and extending resource dependence theory: A meta-analysis," *Journal of Management*, 39(6): 1666–1698.

Dreyfuss, Rochelle C. (2011). "Evaluating the public impact of open innovation," *The Australian Economic Review*, 44(1): 66–72.

Droge, Cornelia, Stanko, Michael A., and Pollitte, Wesley A. (2010), "Lead users and early adopters on the Web: The role of new technology product blogs," *Journal of Product Innovation Management*, 27(1): 66–82.

Drucker, Peter. (1974). *Management: Tasks, Responsibilities, Practices.* New York: Harper & Row.

Drucker, Peter. (1993). *Post-Capitalist Society.* New York: Harper Business.

Drucker, Peter. (2001). "The Next Society—Special Report," *The Economist*, Nov 3.

Du, Jingshu, Leten, Bart, and Vanhaverbeke, Wim. (2014). "How to make open innovation more relevant for multinational enterprises?," working paper, Hasselt University.

Du Chatenier, Elise, Biesmans, Harm J. A., Verstegen, Jos A.A.M and Mulder, Martin. (2007). Collaborative knowledge creation in open innovation teams. Paper presented at the Eighth International Conference on HRD Research and Practice across Europe.

Du Chatenier, Elise, Verstegen Jos, Biemans Harm, Mulder Martin and Omta Onno (2010). "Identification of competencies for professionals in open innovation teams." *R & D Management*, 40 (3), 271–280.

Duff, Karen, (2011). "Unilever's open innovation learning programmes," *Innovation UK,* London: Maritime Media.

Dushnitsky, Gary, and Klueter, Thomas. (2011). "Is there an eBay for ideas? Insight from online knowledge marketplaces," *European Management Review*, 8(1): 17–32.

Dyer, Jeffrey H. (1997). "Effective interfirm collaboration: How firms minimize transaction cost and maximize transaction value," *Strategic Management Journal*, 18(7): 535–556.

Dyer, Jeffrey H. and Singh Harbir. (1998). "The relational view: Cooperative strategy and sources of interorganizational competitive advantage," *Academy of Management Review*, 23(4), 660–679.

Dyer, Jeffrey H. and Nobeoka, Kentaro. (2000). "Creating and managing a high-performance knowledge sharing network: The Toyota case," *Strategic Management Journal*, 21(3): 345–367.

Dyer, Davis, Dalzell, Frederick and Olegario, Rowena. (2004). *Rising Tide: Lessons from 165 Years of Brand Building at Procter & Gamble*, Boston: Harvard Business School Press.

Edwards, Tim, Delbridge, Rick and Munday, Max. (2005). "Understanding innovation in small and medium-sized enterprises: A process manifest," *Technovation*, 25(10): 1119–1127.

Eisenhardt, Kathleen M. and Tabrizi, Behnam N. (1995). "Accelerating adaptive processes: Product innovation in the global computer industry," *Administrative Science Quarterly*, 40(1): 84–110.

Eisenman, Thomas R. (2008). "Managing proprietary and shared platforms," *California Management Review*, 50(4): 31–53.

Eisenmann, Thomas R., Parker, Geoffrey and Van Alstyne, Marshall. (2011). "Platform Envelopment," *Strategic Management Journal*, 32(12): 1270–1285.

Elmquist, Maria, Fredberg, Tobias and Ollila, Susanne. (2009). "Exploring the field of open innovation," *European Journal of Innovation Management*, 12(3): 326–345.

Enkel, Ellen and Gassmann, Oliver. (2007) "Driving open innovation in the front end. The IBM case," paper presented at the EURAM conference, Paris.

Enkel, Ellen and Gassmann, Oliver, and Chesbrough, Henry. (2009) Open R&D and open innovation: exploring the phenomenon, *R&D Management*, 39(4): 311–316.

Enkel, Ellen, Bell, John and Hogenkamp, Hannah. (2011). "Open innovation maturity framework," *International Journal of Innovation Management*, 15(6): 1161–1189.

Etzkowitz, Harry. (2003). "Innovation in innovation: the triple helix of university-industry-government relations," *Social Science Information* 42(3): 293–337.

European Commission (2003). "Commission Recommendation of 6 May 2003 concerning the definition of micro, small and medium-sized enterprises" *Official Journal of the European Union*, L124 (46): 36.

European Commission (2005). *The New SME Definition: User Guide and Model Declaration*. Brussels: European Commission.

Fabrizio, Kira. (2006). "The use of university research in firm innovation," in: Chesbrough, H., Vanhaverbeke, Wim and West, Joel (eds.), *Open Innovation: Researching a New Paradigm*. Oxford: Oxford University Press, 134–160.

Faems, Dries, De Visser, Matthias, Andries, Petra, and Van Looy, Bart. (2010). "Technology alliance portfolios and financial performance: Value-enhancing and cost-increasing effects of open innovation," *Journal of Product Innovation Management*, 27(6): 785–796.

Fasnacht, Daniel. (2009). *Open Innovation in the Financial Services: Growing Through Openness, Flexibility and Customer Integration*. Berlin: Springer.

Fast, Norman D. (1978). *The Rise and Fall of Corporate New Venture Divisions*, Ann Arbor, MI: UMI Research Press.

Feldman, Maryann P. and Florida, Richard. (1994). "The geographic sources of innovation: technological infrastructure and product innovation in the United States," *Annals of the Association of American Geographers*, 84(2): 210–229.

Felin, Teppo and Foss, Nicolai J. (2009). "Organizational routines and capabilities: Historical drift and a course-correction toward microfoundations," *Scandinavian Journal of Management*, 25(2): 157–167.

Ferrary, Michel. (2011). "Specialized organizations and ambidextrous clusters in the open innovation paradigm," *European Management Journal*, 29(3): 181–192.

Fitzgerald, Brian. (2006). "The transformation of open source software," *MIS Quarterly*, 30 (3): 587–598.

Fleming, Lee and Sorenson, Olav. (2001). "Technology as a complex system. evidence from patent data," *Research Policy*, 30 (7): 1019–1039.

Fleming, Lee and Sorenson, Olav. (2004). "Science as a map in technological search," *Strategic Management Journal*, 25(8–9): 909–928.

Fleming, Lee, and Waguespack, David M. (2007). "Brokerage, boundary spanning, and leadership in open innovation communities," *Organization Science*, 18(2): 165–180.

Florida, Richard and Kenney, Martin. (1990). *The Breakthrough Illusion*. New York: Basic Books.

Flowers, Stephen. (2008), "Harnessing the hackers: The emergence and exploitation of Outlaw Innovation," *Research Policy*, 37(2): 177–193.

Fontana, Roberto, Geuna, Aldo and Matt, Mireille. (2006). "Factors affecting university–industry R&D projects: The importance of searching, screening and signaling," *Research Policy*, 35(2): 309–323.

Ford, Simon, Garnsey, Elizabeth and Probert, David. (2010). "Evolving corporate entrepreneurship strategy: Technology incubation at Philips," *R&D Management*, 40(1): 81–90.

Ford, Simon and Probert, David. (2010). "Trial by market: the Brightstar incubation experiment," *International Journal of Entrepreneurial Venturing*, 2 (2): 185–200.

Foss, Nicolai J., Laursen, Keld, and Pedersen, Torben (2011). "Linking customer interaction and innovation: the mediating role of new organizational practices," *Organization Science*, 22(4): 980–999.

Franke, Nikolaus and Shah, Sonali. (2003), "How communities support innovative activities: An exploration of assistance and sharing among end-users," *Research Policy*, 32(1): 157–178.

Franke, Nikolaus, and Piller, Frank. (2004), "Toolkits for user innovation and design: an exploration of user interaction and value creation," *Journal of Product Innovation Management*, 21(6): 401–415.

Freel, Mark S. (2000). "Barriers to product innovation in small manufacturing firms," *International Small Business Journal*, 18(2): 60–80.

Frenz, Marion, and Ietto-Gillies, Grazia. (2009). "The impact on innovation performance of different sources of knowledge: Evidence from the UK Community Innovation Survey," *Research Policy*, 38(7): 1125–1135.

Frey, Karsten, Lüthje, Christian and Haag, Simon. (2011). "Whom should firms attract to open innovation platforms? The role of knowledge diversity and motivation," *Long Range Planning*, 44(5): 397–420.

Frost, Tony S. (2001) "The geographic sources of foreign subsidiaries' innovations," *Strategic Management Journal*, 22(2): 101–123.

Frost, Tony S. and Zhou, Changhui. (2005). "R&D co-practice and 'reverse' knowledge integration in multinational firms," *Journal of International Business Studies*, 36(6): 676–687.

Füller, Johann, Bartl, Michael, Ernst, Holger, and Mühlbacher, Hans. (2006). "Community based innovation: How to integrate members of virtual communities into new product development," *Electronic Commerce Research*, 6(1): 57–73.

Füller, Johann, and Matzler, Kurt. (2007). "Virtual product experience and customer participation—A chance for customer-centred, really new products," *Technovation*, 27(6): 378–387.

Füller, Johann, Matzler, Kurt, and Hoppe, Melanie. (2008). "Brand community members as a source of innovation," *Journal of Product Innovation Management*, 25(6): 608–619.

Füller, Johann, Mühlbacher, Hans, Matzler, Kurt, and Jawecki, Gregor. (2009). "Consumer empowerment through internet-based co-creation," *Journal of Management Information Systems*, 26(3): 71–102.

Füller, Johann, Schroll, Roland, and von Hippel, Eric. (2013). "User generated brands and their contribution to the diffusion of user innovations," *Research Policy*, 42(6–7): 1197–1209.

Gallagher, Scott, and West, Joel. (2009). "Reconceptualizing and expanding the positive feedback network effects model: A case study," *Journal of Engineering and Technology Management*, 26(3): 131–147.

Gambardella, Alfonso and Giarratana, Marco S. (2004). "Fingerprints of the visible hand. Chandlerian organizations and their inward looking malaise," LEM Working paper, 2004/16, Available at SSRN: http://ssrn.com/abstract=578302.

Gambardella, Alfonso and Giarratana, Marco S. (2013). "General technological capabilities, product market fragmentation, and markets for technology," *Research Policy*, 42(2): 315–325.

Gambardella, Alfonso and Giarratana, Marco S. (2007). "Technological Breadth, Product Market Fragmentation and the Market for Technology: Evidence from the Software Security Industry," Working Paper, Bocconi, Milan.

Gans, Joshua S., and Stern, Scott. (2003) "The product market and the market for "ideas": commercialization strategies for technology entrepreneurs," *Research Policy*, 32(2), 333–350.

Gans, Joshua S., and Stern, Scott. (2010). "Is there a market for ideas?," *Industrial and Corporate Change*, 19(3): 805–837.

Gardet, Elodie and Fraiha, Shady. (2012). "Coordination modes established by the hub firm of an innovation network: The Case of an SME bearer," *Journal of Small Business Management*, 50(2): 216–238.

Garnsey, Elizabeth. (1998). "A Theory of the Early Growth of the Firm," *Industrial & Corporate Change*, 7(3): 523–556.

Gassmann, Oliver. (2006). "Opening up the innovation process: Towards an agenda," *R&D Management*, 36(3): 223–228.

Gassmann, Oliver and Enkel, Ellen. (2004). "Towards a theory of open innovation: Three core process archetypes," *Proceedings of the R&D Management Conference (RADMA)*, Lisbon, Portugal, July 6–9.

Gassmann, Oliver, Enkel, Ellen. and Chesbrough, Henry. (2010). "The future of open innovation," *R&D Management*, 40(3): 213–221.

Gassman, Oliver, Kausch, Christoph and Enkel, Ellen. (2010). "Negative side effects of customer integration," *International Journal of Technology Management*, 50(1): 43–63.

Gatzweiler, Alexandra, Blazevic, Vera, and Piller, Frank. (2013). "Deviant participant behavior in ideation contests," *Proceedings of the 2013 PDMA Research Forum*.

Gaule, Andrew. (2006). *Open Innovation in Action: How to Be Strategic in the Search for New Sources of Value*. London: H-I Network.

Gawer, Annabelle. (2009). "Platform dynamics and strategies: from products to services," in Annabelle Gawer (ed.), *Platforms, Markets and Innovation*, Cheltenham, UK: Edward Elgar, 45–76.

Gawer, Annabelle. (2010). "The organization of technological platforms," in Nelson Phillips, Graham Sewell and Dorothy Griffiths (eds.), *Research in the Sociology of Organizations*, 29: 287–296.

Gawer, Annabelle. and Cusumano, Michael A. (2002). *Platform Leadership: How Intel, Microsoft, and Cisco Drive Industry Innovation*. Boston: Harvard Business School Press.

Gawer, Annabelle. and Cusumano, Michael A. (2008). "How Companies Become Platform Leaders," *Sloan Management Review*, 49(2): 28–35.

Germeraad, Paul. (2010). "Integration of intellectual property strategy with innovation strategy," *Research-Technology Management*, 53(3): 10–18.

Gerwin, Donald. (2004). "Coordinating new product development in strategic alliances," *Academy of Management Review*, 29(2): 241–257.
Geykens, Inge, Steemkamp, Jan-Benedict E.M. and Kumar, Nirmalya. (2006). "Make, buy, or ally: A transaction cost theory meta-analysis," *Academy of Management Journal*, 49(3): 519–543.
Ghafele Roya and Gibert, Benjamin. (2011a). "The transaction cost benefits of electronic patent licensing platforms: a discussion at the example of the patentbooks model," MPRA Working Paper 36010.
Ghafele Roya and Gibert, Benjamin. (2011b). "The transformative impact of business models." MPRA Working Paper 38346.
Ghafele, Roya and Gibert, Benjamin. (2012a). "promoting intellectual property monetization in developing countries: a review of issues and strategies to support knowledge-driven growth." Policy Research Working Paper WPS 6143, World Bank.
Ghafele Roya and Gibert, Benjamin. (2012b). "Efficiency through openness: the economic value proposition of open source software." MPRA Working Paper 38088.
Ghafele, Roya, and O'Brian, Robert D. (2012). "Open innovation for sustainability: lessons from the GreenXchange experience." *Policy Brief No. 13. ICTSD*: 1–10.
Ghafele Roya, Gibert Benjamin and Malackowski, James. (2012). "Emerging IP monetization techniques. The institutionalization of an intellectual property exchange," *International Journal of Intellectual Property Management*, 5(2): 115–133.
Ghemawat, Pankaj. (2001). "Distance still matters: The hard reality of global expansion," *Harvard Business Review*, 79(8): 137–147.
Giannopoulou, Eleni, Yström, Anna, Ollila, Susanna, Fredberg, Tobias and Elmquist, Maria, (2010). "Implications of openness: A study into (all) the growing literature on open innovation," *Journal of Technology Management and Innovation*, 5(3), 162–180.
Giarratana, Marco S. (2004). "The birth of a new industry: entry by start-ups and the drivers of firm growth: The case of encryption software," *Research Policy*, 33(5), 787–806.
Giarratana, Marco S. and Fosfuri, Andrea (2007). "Product strategies and survival in schumpeterian environments: evidence from the security software," *Organization Studies*, 28 (6): 909–929.
Giuri, Paola, Hagedoorn, John and Mariani, Myriam. (2004). "Technological diversification and strategic alliances," in Cantwell, John, Gambardella, Alfonso and Ove Granstrand (eds.), *The Economics and Management of Technological Diversification*, London: Routledge, 116–152.
Gollin, Michael A. (2008). *Driving Innovation: Intellectual Property Strategies for a Dynamic World*. Cambridge, UK: Cambridge University Press.
Gomes-Casseres, Benjamin. (1996). *The Alliance Revolution: The New Shape of Business Rivalry*, Boston: Harvard Business School Publishing.
Gort, Michael and Klepper, Steven. (1982). "Time paths in the diffusion of product innovations," *Economic Journal*, 92(367): 630–653.
Gould, David M. and Gruben William C. (1996). "The role of intellectual property rights in economic growth," *Journal of Development Economics*, 48 (2): 323–350.
Graham, Bradley. (1978). "World of venture capitalists becomes more complicated," *Washington Post*, October 1, M1.
Granovetter, Mark. (1985). "Economic action and social structure: The problem of embeddedness," *American Journal of Sociology*, 91(3): 481–510.

Granstrand, Ove, Patel, Pari and Pavitt, Keith. (1997). "Multi-technology corporations: why they have "distributed" rather than "distinctive" core competences," *California Management Review*, 39(4): 8–25.

Grant, Robert M. (1996). "Toward a knowledge-based theory of the firm," *Strategic Management Journal*, 17(Winter): 109–122.

Grant, Robert M., (2012). "Eastman Kodak: meeting the digital challenge," *Contemporary Strategy Analysis: Text and Cases*. Hoboken, NJ: Wiley, pp. 591–612.

Grant, Robert M. and Baden-Fuller, Charles. (2004). "A knowledge accessing theory of strategic alliances," *Journal of Management Studies*, 41(1): 61–84.

Griffin, Abbie. (1997). "PDMA research on new product development practices: updating trends and benchmarking best practices," *Journal of Product Innovation Management*, 14(7): 429–458.

Griffiths, Dorothy, Boisot, Max, and Mole, Veronica. (1998). "Strategies for managing knowledge assets: A tale of two companies," *Technovation* 18(8–9): 529–539.

Griliches, Zvi. (1990). "Patent statistics as economic indicators: a survey," *Journal of Economic Literature*, 28(4): 1661–1707.

Grimpe, Christoph, and Sofka, Wolfgang. (2009). "Search patterns and absorptive capacity: Low- and high-technology sectors in European countries," *Research Policy*, 38(3): 495–506.

Grindley, Peter C. and Teece, D. (1997). "Managing intellectual capital: licensing and cross-licensing in semiconductors and electronics," *California Management Review*, 39(2): 8–41.

Groen, Aard J., and Linton, Jonathan D. (2010). "Is open innovation a field of study or a communication barrier to theory development?," *Technovation,* 30(11–12): 554.

Grönlund, Johann, Rönnberg Sjödin, David and Frishammar, Johann. (2010). "Open innovation and the Stage-Gate process: A revised model for new product development," *California Management Review*, 52(3): 106–131.

Gronum, Sarel, Verreynne, Martie-Louise, and Kastelle, Tim. (2012). "The role of networks in small and medium-sized enterprise innovation and firm performance," *Journal of Small Business Management*, 50(2): 257–282.

Grove, Andrew S. (1996). *Only the Paranoid Survive: How to Exploit the Crisis Points That Challenge Every Company and Career*. New York: Doubleday.

Gruber, Mark, and Henkel, Joachim. (2006). "New ventures based on open innovation—an empirical analysis of start-up firms in embedded Linux," *International Journal of Technology Management*, 33(4): 356–372.

Gruner, Kjell E. and Homburg, Christian. (2000). "Does customer interaction enhance new product success?," *Journal of Business Research*, 49(1): 1–14.

Grunwald, Roman and Kieser, Alfred. (2007). "Learning to reduce interorganizational learning: an analysis of architectural product innovation in strategic alliances," *Journal of Product Innovation Management*, 24(4): 369–391.

Gulati, Ranjay and Nickerson, Jack A. (2008). "Interorganizational trust, governance choice, and exchange performance," *Organization Science*, 19(5): 688–708.

Gulati, Ranjay (1995). "Does familiarity breed trust? The implications of repeated ties for contractual choice in alliances," *Academy of Management Journal*, 38(1): 85–112.

Gupta, Anil K., Tesluk, Paul E., and Taylor, M. Susan. (2007). "Innovation at and across multiple levels of analysis," *Organization Science*, 18(6): 885–897.

Gutierrez, Horacio. (2008). "Microsoft's Collaboration Imperative," *Intellectual Asset Management*, 29, 9–15.

Hackett, Sean M. and Dilts, David M., (2004). "A systematic review of business incubation research," *Journal of Technology Transfer*, 29(1): 55–82.

Hafkesbrink, Joachim, Hoppe, H. Ulrich, and Schlichter, Johann. (eds.) (2010). *Competence Management for Open Innovation.* Lohmar, Germany: Eul Verlag.

Hagedoorn, John. (1993). "Understanding the rationale of strategic technology partnering: interorganizational modes of cooperation and sectoral differences," *Strategic Management Journal*, 14 (5): 371–385.

Hagelin, Ted. (2002). "New method to value intellectual property," *AIPLA Quarterly Journal*, 30(3): 353–403.

Håkanson, Lars, and Nobel, Robet. (2001). "Organizational characteristics and reverse technology transfer," *Management International Review*, 41(4): 395–420.

Hallawell, Arabella, MacDonald, Neil, and Firstbrook, Peter. (2009). "Microsoft frees up its consumer security software product," Gartner, 26 June, http://www.gartner.com/id=1045412.

Hambrick, Donald C. and Mason, Phyllis A. (1984). "Upper echelons: the organization as a reflection of its top managers," *Academy of Management Review*, 9(2): 193–206.

Hamel, Gary, Doz, Yves L., and Prahalad, C. K. (1989). "Collaborate with your competitors-and win," *Harvard Business Review*, 67(1): 133–139.

Hamel, Gary, and Välikangas, Lisa. (2003). "The Quest for Resilience," *Harvard Business Review*, 81(9): 1–13.

Hamm, Steve. (2009). "The future of tech—Big Blue's Global Lab," *Business Week*, 27 August, http://www.businessweek.com/magazine/content/09_36/b4145040683083.htm.

Harhoff, Dietmar, Henkel, Joachim, and von Hippel, Eric. (2003). "Profiting from voluntary information spillovers: How users benefit by freely revealing their innovations," *Research Policy*, 32(10): 1753–1769.

Harrison, Debbie and Waluszewski, Alexandra. (2008). "The development of a user network as a way to re-launch an unwanted product," *Research Policy*, 37(1): 115–130.

Harryson, Sigvald J. (2008). "Entrepreneurship through relationships—navigating from creativity to commercialisation," *R&D Management*, 38(1): 290–310.

Hars, Alexander, and Ou, Shaosong. (2002). "Working for free? Motivations for participating in open-source projects," *International Journal of Electronic Commerce*, 6(3): 25–39.

Hasan, Iftekhar, and Tucci, Christopher L. (2010). "The innovation–economic growth nexus: Global evidence," *Research Policy*, 39(10): 1264–1276.

Hayek, Friedrich A. (1945). "The use of knowledge in society," *American Economic Review*, 35(4): 519–530.

Helfat, Constance, Finkelstein, Sydney, Mitchell, Will, Peteraf, Margaret A., Singh, Harbir, Teece, David J., and Winter, Sydney G. (2007). *Dynamic Capabilities: Understanding Strategic Change in Organizations.* Oxford: Blackwell.

Henkel, Joachim. (2006). "Selective revealing in open innovation processes: The case of embedded Linux." *Research Policy*, 35(7): 953–969.

Henkel, Joachim, Schöberl, Simone, and Alexy, Oliver. (2014). "The Emergence of Openness: How firms learn selective revealing in open innovation," *Research Policy*, 43(5): 879–890.

Herzog, Phillip, and Leker, Jens. (2010). "Open and closed innovation—Different innovation cultures for different strategies," *International Journal of Technology Management*, 52 (3–4): 322–343.

Heyman, James, and Ariely, Dan. (2004), "Effort for payment: a tale of two markets," *Psychological Science*, 15(11): 787–793.

Hillman, Amy J., Withers, Michael C. and Collins, Brian J. (2009). "Resource dependence theory: A review," *Journal of Management*, 35 (6): 1404–1427.

Hiltzik, Michael. (1999). *Dealers of Lightning: Xerox PARC and the Dawn of the Computer Age*, New York: HarperCollins.

Hoang, Ha, and Rothaermel, Frank T. (2005) "The effect of general and partner-specific alliance experience on joint R&D project performance," *Academy of Management Journal*, 48(2): 332–345.

Hobday, Michael. (2000). "The project-based organization: an ideal form for managing complex products and systems?," *Research Policy*, 29(7–8): 871–893.

Hollenbeck, John R., Ellis, Alexsander P.J., Humphrey, Stephen E., Garza, Adela S. and Ilgen, Daniel R. (2011). "Asymmetry in structural adaptation: The differential impact of centralizing versus decentralizing team decision-making structures," *Organizational Behavior and Human Decision Processes*, 114(1): 64–74.

Holmes, Joseph S., Jr. (2009). "Societal and economic valuation of technology-transfer deals," *Acta Astronautica*, 65(5–6): 834–840.

Hopkins, Michael M., Tidd, Joe, Nightingale, Paul, and Miller, Roger. (2011). "Generative and degenerative interactions: positive and negative dynamics of open, user-centric innovation in technology and engineering consultancies," *R&D Management*, 41(1): 44–60.

Hossain, Mokter. (2012). "Performance and potential of open innovation intermediaries," *Procedia—Social and Behavioral Sciences*, 58: 754–764.

Howe, Jeff. (2006), "The Rise of Crowdsourcing," *Wired*, 14(6): 176–183.

Howe, Jeff. (2008). *Crowdsourcing: How The Power of the Crowd is Driving the Future of Business*. London: Random House Business.

Howells, Jeremy. (2006). "Intermediation and the role of intermediaries in innovation," *Research Policy*, 35(5): 715–728.

Huang, Fang. and Rice, John. (2009). "The role of absorptive capacity in facilitating "open innovation" outcomes: A study of Australian SMEs in the manufacturing sector," *International Journal of Innovation Management*, 13(2): 201–220.

Huizingh, Eelko K., (2011). "Open innovation: State of the art and future perspectives," *Technovation*, 31(1): 2–9.

Hurmelinna, Pia, Kyläheiko, Kalevi and Jauhiainen, Tiina. (2007). "The Janus face of the appropriability regime in the protection of innovations: Theoretical re-appraisal and empirical analysis," *Technovation*, 27 (3): 133–144.

Huston, Larry, and Sakkab, Nabil. (2006). "Connect and develop: inside Procter and Gamble's new model for innovation," *Harvard Business Review*, 84(3): 58–66.

Huston, Larry, and Sakkab, Nabil. (2007). "Implementing open innovation," *Research-Technology Management*, 50(2): 21–25.

Iansiti, Marco, and Levien Roy. (2004a). *The Keystone Advantage: What The New Dynamics of Business Ecosystems Mean for Strategy, Innovation, and Sustainability*. Boston: Harvard Business School Press.

Iansiti, Marco, and Levien Roy. (2004b). "Strategy as ecology," *Harvard Business Review*, 82 (3): 68–78.

Ihl, Christoph, Piller, Frank T., & Wagner, Phillip. (2012). "Organizing for open innovation: Aligning internal structure and external knowledge sourcing," Open

Innovation: New Insights and Evidence conference, Imperial College London, June 26. Available at SSRN: http://ssrn.com/abstract=2164766.

Ihl, Christoph, Vossen, Alexander, and Piller, Frank T. (2012). "All for the money? the ambiguity of monetary rewards in firm-initiated ideation with users," SSRN Working Paper, http://ssrn.com/abstract=2164763.

Immelt, Jeffrey R., Govindarajan, Vijay, and Trimble, Chris. (2009). "How GE is disrupting itself," *Harvard Business Review*, 87(10): 56–65.

Inauen, Matthias and Schenker-Wicki, Andrea. (2011). "The impact of outside-in open innovation on innovation performance," *European Journal of Innovation Management*, 14(4): 496–520.

Inkpen, Andrew W. and Tsang, Eric W.K. (2005). "Social capital, networks, and knowledge transfer," *Academy of Management Review*, 30(1): 146–165.

Ireland, R. Duane, Hitt, Michael A., and Vaidyanath, Deepa. (2002). "Alliance management as a source of competitive advantage," *Journal of Management*, 28(3): 413–441.

Iwasa, Tomoko and Odagiri, Hiroyuki. (2004). "Overseas R&D, knowledge sourcing, and patenting: an empirical study of Japanese R&D investment in the US," *Research Policy*, 33(5): 807–828.

Jacobides, Michael. (2003). "How do markets emerge? Organizational unbundling and vertical dis-integration in mortgage banking," working paper, Centre for the Network Economy. London Business School. http://www.london.edu/facultyandresearch/research/docs/sim18.pdf.

Jacobides, Michael G., and Billinger, Stephan. (2006). "Designing the boundaries of the firm: From "Make, buy, or ally" to the dynamic benefits of vertical architecture," *Organization Science*, 17(2): 249–261.

Jacobides, Michael G., Knudsen, Thorbjørn, and Augier, Mie. (2006). "Benefiting from innovation: Value creation, value appropriation and the role of industry architectures," *Research Policy*, 35(8): 1200–1221.

Jaffe, Adam B. (1986). "Technological opportunity and spillovers of R&D: evidence from firms' patents, profits, and market value," *American Economic Review*, 76(5): 984–1001.

Jaffe, Adam B., Trajtenberg, Manuel, and Henderson. Rebecca. (1993) "Geographic localization of knowledge spillovers as evidenced by patent citations," *Quarterly Journal of Economics*, 108(3): 577–598.

Jansen, Justin J.P., van den Bosch, Frans A. J., and Volberda, Henk W. (2005). "Managing potential and realized absorptive capacity: how do organizational antecedents matter?," *Academy of Management Journal*, 48(6): 999–1015.

Jansen, Karen J., (2000). "The emerging dynamics of change: resistance, readiness, and momentum," *Human Resource Planning*, 23(2): 53–55.

Jansen, Karen J., (2004). "From persistence to pursuit: a longitudinal examination of momentum during the early stages of strategic change," *Organization Science*, 15(3): 276–294.

Jensen, Michael C. (1986). "Agency costs of free cash flow, corporate finance, and takeovers," *American Economic Review*, 76(2): 323–329.

Jensen, Michael C. (1989). "Eclipse of the public corporation," *Harvard Business Review*, 67(5): 61–75.

Jensen, Richard, and Thursby, Marie. (2001). "Proofs and prototypes for sale: The licensing of university inventions," *American Economic Review*, 91(1): 240–259.

Jeppesen, Lars Bo, and Frederiksen, Lars. (2006). "Why do users contribute to firm-hosted user communities? the case of computer-controlled music instruments," *Organization Science*, 17(1): 45–63.

Jeppesen, Lars Bo, and Lakhani, Karim R. (2010). "Marginality and problem solving effectiveness in broadcast search," *Organization Science*, 21(4): 1016–1033.

Jones, Candace, Hesterly, William S. and Borgatti Stephen P. (1997). "A general theory of network governance: exchange conditions and social mechanisms," *Academy of Management Review*, 22(4): 911–945.

Kahn, Kenneth B., Barzak, Gloria, and Moss, Roberta. (2006). "Perspective: establishing an NPD best practices framework," *Journal of Product Innovation Management*, 23(2): 106–116.

Karim, Samina, and Mitchell, Will. (2000). "Path-dependent and path-breaking change: reconfiguring business resources following acquisitions in the US medical sector 1978–1995," *Strategic Management Journal*, 21(10–11): 1061–1081.

Katila, Rita, Rosenberger, Jeff D. and Eisenhardt, Katkhleen M. (2008). "Swimming with sharks: technology ventures, defense mechanisms and corporate relationships," *Administrative Science Quarterly*, 53(2): 295–332.

Katz, Ralph, and Allen, Thomas J. (1982). "Investigating the Not Invented Here (NIH) syndrome: A look at the performance, tenure, and communication patterns of 50 R&D Project Groups," *R&D Management*, 12(1): 7–20.

Kawasaki, Guy. (1990). *The Macintosh Way*, Glenview, Ill.: Scott, Foresman.

Keil, Thomas. (2004). "Building external corporate venturing capability," *Journal of Management Studies*, 41(5), 799–825.

Kennedy, Jason. (2012). "Linux unites with android, adds business-friendly features," *PCWorld*, March 19, URL: http://www.pcworld.com/article/252137/linux_unites_with_android_adds_business_friendly_features.html

Kenney, Martin and Patton, Donald. (2009). "Reconsidering the Bayh-Dole act and the current university invention ownership model, *Research Policy*, 38(9): 1407–1422.

Kenney, Martin and Pon, Bryan. (2011). "Structuring the smartphone industry: is the mobile internet os platform the key?," *Journal of Industry, Competition and Trade*, 11 (3): 239–261.

Kerr, Clive IV., Mortara, Letizia, Phaal, R. and Probert, D.R., (2006). "A conceptual model for technology intelligence," *International Journal of Technology Intelligence and Planning*, 2(1): 73–93.

Kessler, Eric H., Bierly, Paul E., and Gopalakrishnan, Shanthi. (2000). "Internal vs. external learning in new product development: effects on speed, costs and competitive advantage," *R&D Management*, 30(3): 213–224.

King, William R. and Cleland, David I. (1983). "Life cycle management," in David I. Cleland and William R. King (eds.), *Project Management Handbook*, New York: Van Nostrand Reinhold Co.

Kinoshita, Yumiko. (2011). *Service Entities in Open-Closed Innovation*. New York: Nova Science Publishers Inc.

Kirschbaum, Robert. (2005). "Open innovation in practice," *Research-Technology Management*, 48(4): 24–28.

Kitching, John and Blackburn, Robert. (1998). "Intellectual property management in the small and medium enterprise (SME)," *Journal of Small Business and Enterprise Development*, 5(4): 327–335.

Klein, Benjamin, Crawford, Robert G. and Alchian, Armen A. (1978). "Vertical integration, appropriable rents, and the competitive contracting process," *Journal of Law and Economics*, 21(2): 297–326.

Klepper, Steven (1996), "Entry, exit, growth, and innovation over the product life cycle," *American Economic Review*, 86(3): 562–583.

Klepper, Steven (1997), "Industry Life Cycles," *Industrial and Corporate Change* 6(1): 145–182.

Knudsen, Mette Praest (2007). "The relative importance of interfirm relationships and knowledge transfer for new product development success," *Journal of Product Innovation Management*, 24(2): 117–138.

Knudsen, Mette Praest and Mortensen, Thomas Bøtker (2011). "Some immediate—but negative—effects of openness on product development performance," *Technovation*, 31(1): 54–64.

Koberg, Christine S., Uhlenbruck, Nikolaus and Sarason, Yolanda. (1996). "Facilitators of organizational innovation: The role of life-cycle stage," *Journal of Business Venturing*, 11(2): 133–149.

Koch, Felix and Coates, Nick. (2010). "Rulemaking or playmaking? Implications of the emerging co-creation landscape," *ESOMAR 2010 Collection on Online Research*, Part 3: Online Cosmos: Panels, Communities and Social Networks.

Kogut, Bruce. (1988). Joint ventures: theoretical and empirical perspectives, *Strategic Management Journal*, 9(4), 319–332.

Krugman, Paul. (1991). *Geography and Trade*, Cambridge, MA: MIT Press.

Kuemmerle Walter. (1997) "Building effective R&D capabilities abroad," *Harvard Business Review*, 75(2): 61–70.

Kuhn, Thomas S. (1962). *The Structure of Scientific Revolutions*. Chicago: University of Chicago Press.

Kuschel, Jonas, Remneland, Bjorn and Kuschel, Magnus Holmqvist. (2011). "Open innovation and control: A case from Volvo," *International Journal of Networking and Virtual Organisations*, 9(2): 123–139.

Laffan, Liz. (2011). "Open Governance Index: Measuring the true open source projects form Android to WebKit," Vision Mobile, URL: http://www.visionmobile.com/rsc/researchreports/Open%20Governance%20Index%20(VisionMobile).pdf

Lakhani Karim R. and von Hippel, Eric. (2003). "How open source software works: "Free" user-to-user assistance," *Research Policy*, 32(6): 923–943.

Lakhani Karim R., and Jeppesen, Lars Bo. (2007). "Getting unusual suspects to solve R&D puzzles," *Harvard Business Review*, 85(5): 30–32.

Lakhani Karim R., Jeppesen, Lars Bo, Lohse, Peter A., and Panetta, Jill A. (2006). "The value of openness in scientific problem solving," HBS Working Paper Number 07-050, Harvard University.

LaMonica, Martin. (2010) "Fast EV charging stations plug in," CNET Oct 13. http://news.cnet.com/8301-11128_3-20019438-54.html.

Lampel, Joseph, Jha, Pushkar P., and Bhalla, Ajay. (2012). "Test-driving the future: How design competitions are changing innovation," *Academy of Management Perspectives*, 26(2): 71–85.

Lane, Peter J., Koka, Balaji R., and Pathak, Seemantini. (2006). "The reification of absorptive capacity: a critical review and rejuvenation of the construct," *Academy of Management Review*, 31(4): 833–863.

Langlois, Richard N. (2003). "The vanishing hand: the changing dynamics of industrial capitalism," *Industrial and Corporate Change*, 12(2): 351–385.

Langlois, Richard N. and Garzarelli, G. (2008). "Of hackers and hairdressers: modularity and the organizational economics of open—source collaboration," *Industry & Innovation*, 15(2): 125–143.

Lanjouw, Jean O. and Schankerman, Mark. (2004). "Patent quality and research productivity: Measuring innovation with multiple indicators." *Economic Journal*, 114(495): 441–465.

Lanjouw, Jean O., Pakes, Ariel, and Putnam Jonathan. (1998). "How to count patents and value intellectual property: the uses of patent renewal and application data," *Journal of Industrial Economics*, 46(4): 405–432.

Laursen, Keld and Salter, Ammon. (2004). "Searching high and low: What types of firms use universities as a source of innovation?," *Research Policy*, 33: 1201–1215.

Laursen, Keld and Salter, Ammon. (2006). "Open for innovation: The role of openness in explaining innovation performance among UK manufacturing firms," *Strategic Management Journal*, 27(2): 131–150.

Laursen, Keld and Salter, Ammon. (2012). "The paradox of openness: Appropriability and the use of external sources of knowledge for innovation," Open Innovation: New Insights and Evidence conference, Imperial College London, June 26.

Lazzarotti, Valentina and Manzini, Raffaella. (2009). "Different modes of open innovation: a theoretical framework and an empirical study," *International Journal of Innovation Management*, 13(4): 615–636.

Lee, Keun, Park, Kyooho, Oh, Jun-Byoung, and Kim, Jinyoung. (2009). "Economics of IP in the context of shifting innovation paradigm." *WIPO Annual Report*.

Lee, Sungjoo, Park, Gwangman, Yoon, Byungun, and Park, Jinwoo. (2010). "Open innovation in SMEs—An intermediated network model," *Research Policy*, 39(2): 290–300.

Lehrer, Mark, and Asakawa, Kazuhiro. (2002). "Offshore knowledge incubation: The third path for embedding R&D labs in foreign systems of innovation," *Journal of World Business*, 37(4): 297–306.

Lehrer, Mark, and Asakawa, Kazuhiro. (2003). "Managing intersecting R&D social communities: A comparative study of European "knowledge incubators" in Japanese and American firms," *Organization Studies*, 24(5): 771–792.

Lehrer, Mark, Asakawa, Kazuhiro, and Behnam, Michael. (2011). "Home base-compensating R&D: Indicators, public policy, and ramifications for multinational firms," *Journal of International Management*, 17(1): 42–53.

Leiponen, Aija, and Byma, Justin. (2009). "If you cannot block, you better run: Small firms, cooperative innovation, and appropriation strategies," *Research Policy*, 38(9): 1478–1488.

Lemley, Mark A. (2002). "Intellectual property rights and standard-setting organizations," *California Law Review*, 90(6): 1889–1980.

Lenox, Michael and King, Andrew. (2004). "Prospects for developing absorptive capacity through internal information provision," *Strategic Management Journal*, 25(4): 331–345.

Leonard-Barton, Dorothy. (1992). "Core capabilities and core rigidities: a paradox in managing new product development," *Strategic Management Journal*, 13: (S1), 111–125.

Leonard-Barton, Dorothy. (1995). *Wellsprings of knowledge: Building and sustaining the source of innovation*. Boston: Harvard Business School Press.

Lerner, Josh. (1995). "Venture capitalists and the oversight of private firms," *Journal of Finance*, 50(1): 301–318.

Leten, Bart, Vanhaverbeke Wim, Roijakkers Nadine, Clerix, Andre, and Van Helleputte, Johan. (2012). "IP models to orchestrate innovation ecosystems: imec, a public research institute in nano-electronics," *California Management Review*, 55(4): 51–64.

Lettl, Christopher, Hienerth, Christoph, and Gemünden, Hans Georg. (2008). "Exploring how lead users develop radical innovation: opportunity recognition and exploitation in the field of medical equipment technology," *IEEE Transactions on Engineering Management*, 55(2): 219–233.

Levin, Richard C., Cohen, Wesley M. and Mowery, David C. (1985). "R&D appropriability, opportunity and market structure: new evidence on some Schumpeterian hypotheses," *American Economic Review*, 75(2): 20–24.

Levinthal, Daniel A. and March, James G. (1993). "The myopia of learning," *Strategic Management Journal*, 14(S2): 95–112.

Lewin, Kurt. (1947). "Frontiers in group dynamics." *Human Relations*, 1(1): 5–41.

Liao, Jianwen Jon, and Welsch, Harold. (2008). "Patterns of venture gestation process: Exploring the differences between tech and non-tech nascent entrepreneurs," *Journal of High Technology Management Research*, 19(2): 103–113.

Lichtenthaler, Ulrich. (2007). "The drivers of technology licensing: an industry comparison," *California Management Review*, 49(4): 67–89.

Lichtenthaler, Ulrich. (2011). "Open innovation: past research, current debates, and future directions," *The Academy of Management Perspectives*, 25(1): 75–93.

Lichtenthaler, Ulrich, and Ernst, Holger. (2006). "Attitudes to externally organising knowledge management tasks: a review, reconsideration and extension of the NIH syndrome," *R&D Management*, 36(4): 367–386.

Lichtenthaler, Ulrich, and Ernst, Holger. (2008). "Innovation intermediaries: Why internet marketplaces for technology have not yet met the expectations," *Creativity and Innovation Management*, 17(1), 14–25.

Lilien, Gary L., Morrison, Pamlea D., Searls, Kathleen, Sonnack, Mary, and Von Hippel, Eric. (2002). "Performance assessment of the lead user idea-generation process for new product development," *Management Science*, 48(8): 1042–1059.

Lindegaard, Stefan. (2010). *The Open Innovation Revolution: Essentials, Roadblocks, and Leadership Skills*. Hoboken, NJ: Wiley.

Linton, Jonathan D., (2002). "Implementation research: state of the art and future directions," *Technovation*, 22(2): 65–79.

Livne-Tarandach, Reut, and Bartunek, Jean M., (2009). "A new horizon for organizational change and development scholarship connecting planned and emergent change," *Research in Organizational Change and Development*, 17, 1–35.

Lopez-Vega, Henry. (2009). "How demand-driven technological systems of innovation work? The role of intermediary organizations," *DRUID-DIME Academy PhD Conference*, Economics and Management of Innovation, Technology and Organizational Change, 1–35.

Lopez, Henry, and Vanhaverbeke, Wim. (2009). "How innovation intermediaries are shaping the technology market? An analysis of their business model," working paper, available at http://mpra.ub.uni-muenchen.de/27016.

Lundvall, Bengt-Åke, ed. (1992). *National Systems of Innovation: Towards a Theory of Innovation and Interactive Learning.* London: Pinter.

Lunn, John and Martin, Stephen. (1986). "Market structure, firm structure, and research and development," *Quarterly Review of Economic and Business*, 26(1): 31–44.

Lüttgens, Dirk, Pollok, Patrick, Antons, David, and Piller, Frank. (2014). "Wisdom of the crowd and capabilities of the few: internal success factors of crowdsourcing for innovation," *Journal of Business Economics*, 84(3): 339–374.

Lyons, Andrew C., Coronado Mondragon, Adrian E., Piller, Frank, and Poler, Raúl. (2012). *Customer-Driven Supply Chains: From Glass Pipelines to Open Innovation Networks.* London: Springer.

Ma, Chaoqun, Yang, Zhi, Yao, Zheng, Fisher, Greg, Fang, Eric. (2012). "The effect of strategic alliance resource accumulation and process characteristics on new product success: Exploration of international high-tech strategic alliances in China," *Industrial Marketing Management*, 41(3): 469–480.

Maarse, Johann Henk, and Bogers, Marcel. (2012). "An integrative model for technology-driven innovation and external technology commercialization," in Carmen de Pablos Heredero, & David López (Eds.), *Open Innovation at Firms and Public Administrations: Technologies for Value Creation*: Hershey, PA: IGI Global, 59–78.

Macpherson, Allan, and Holt, Robin. (2007). "Knowledge, learning and small firm growth: A systematic review of the evidence," *Research Policy*, 36(2): 172–192.

Madhok, Anoop. (1997). "Cost, value and foreign market entry mode: the transaction and the firm," *Strategic Management Journal*, 18(1): 39–61.

Madhok, Anoop and Tallman, Stephen B. (1998). "Resources, transactions and rents: Managing value through interfirm collaborative relationships," *Organization Science*, 9(3): 326–339.

Malerba, Franco, and Orsenigo, Luigi (1997). "Technological regimes and sectoral patterns of innovative activities," *Industrial and Corporate Change*, 6(1): 83–118.

Malone, Michael S. (1999). *Infinite Loop: How the World's Most Insanely Great Computer Company Went Insane.* New York: Currency/Doubleday.

Manceau, Delphine, Moatti, Valérie, Fabbri, Julie, Kaltenbach, Pierre-François and Bagger-Hansen, Line. (2011). "Open Innovation. What's behind the buzzword?" working paper, i7 Institute for Innovation and Competitiveness, ESCP Europe.

Markham, Stephen K., Gentry, Stuart T., Hume, David, Ramachandran, Ram, and Kingon, Angus I., (2005). "Strategies and Tactics for External Corporate Venturing," *Research Technology Management,* 48(2), 49–59.

Markus, M. Lynne. (2007). "The governance of free/open source software projects: monolithic, multidimensional, or configurational?," *Journal of Management & Governance*, 11 (2): 151–163.

Maskus, Keith E. and Reichman Jerome H. (2004). "The Globalization of private knowledge goods and the privatization of global public goods," *Journal of International Economic Law*, 7(2): 279–320.

Mathews, John A. (2002). *Dragon Multinational: A New Model of Global Growth.* New York: Oxford University Press.

Maula, Markku V.J., Autio, Erkko, and Murray, Gordon C. (2009). "Corporate venture capital and the balance of risks and rewards for portfolio companies," *Journal of Business Venturing*, 24(3): 274–286.

Maula, Markku, Keil, Thomas, and Salmenkaita, Jukka-Pekka. (2006). "Open innovation in systemic innovation contexts," in Henry Chesbrough, Wim Vanhaverbeke, and Joel West (eds.), *Open Innovation: Researching a New Paradigm*. Oxford: Oxford University Press, 241–257.

Mazzoleni, Roberto & Nelson, Richard R. (1998). "The benefits and costs of strong patent protection: a contribution to the current debate," *Research Policy*, 27(3): 273–284.

McGrath, Rita and Nerkar, Atul. (2004). "real options reasoning and a new look at the r&d investment strategies of pharmaceutical firms," *Strategic Management Journal*, 25(1): 1–21.

McGrath, Rita and MacMillan, Ian C. (2009), *Discovery Driven Growth: A Breakthrough Process to Reduce Risk and Seize Opportunity*, Harvard Business Press, Boston: MA.

Meehan, William F., III, Koehane, Georgia Levenson. (2012). "Ashoka: Innovators for the Public," Stanford Graduate School of Business, case number SM203.

Meijer, Elise. (2012). *Team performance in R&D alliances: A micro-level perspective*. Unpublished doctoral dissertation, Eindhoven University of Technology.

Melese, Teri, Lin, Salima M., Chang, Julia L., and Cohen, Neal H. (2009). "Open innovation networks between academia and industry: an imperative for breakthrough therapies," *Nature Medicine,* 15(5): 502–507.

Ménière, Yann. (2012). "The market for patents: a quantitative analysis for patent ownership transfers based on the European and French Registers." Presentation given at the 7th Annual EPIP Conference. IP in Motion. Leuven, September 28.

Merges, Robert P. (1999). "Institutions for intellectual property transactions." Working paper, Bolt School of Law, U.C. Berkeley, https://2048.berkeley.edu/files/pools.pdf

Mesquita, Luiz F. (2007). "Starting over when the bickering never ends: Rebuilding aggregated trust among clustered firms through trust facilitators." *Academy of Management Review*, 32(1), 72–91.

Millien, Raymond, and Laurie, Ron. (2007). "A summary of established and emerging IP business models," URL: https://vcexperts.com/buzz_articles/592.

Minbaeva, Dana B., (2005). "HRM practices and MNC knowledge transfer," *Personnel Review*, 34(1): 125–144.

Minshall, Tim, Kouris, Stefan, Mortara, Letizia, and Weiss, David (2014). "Developing infrastructure to support open innovation: Analysis of case studies in the East of England," *International Journal of Innovation & Technology Management*, 11(1).

Möller, Krisitan, Rajala, Risto, and Westerlund, Mika (2008). "Service innovation myopia? a new recipe for client-provider value creation," *California Management Review*, 50(3): 31–48.

Moore, James F. (1993). "Predators and prey: a new ecology of competition," *Harvard Business Review*, 71(3): 75–86.

Moore, James F. (1996). *The Death of Competition: Leadership and Strategy in the Age of Business Ecosystems*, New York: HarperBusiness.

Morgan Keegan (2006). *Navigating the 7 C's of Security As Industry Shifts from Infrastructure to Information Security*, Morgan Keegan & Company, July 21.

Morgan Stanley (2005), "Security Software Data—The Next Perimeter of Defence," Morgan Stanley, January 5.

Mortara, Letizia. (2010a). *Getting Help With Open Innovation*. Cambridge: University of Cambridge Press.

Mortara, Letizia (2010b), "The role of intermediaries," white paper. Center for Technology Management. Cambridge University.

Mortara, Letizia, Slacik, Imke, Napp, Johann J., and Minshall, Tim. (2010). "Implementing Open Innovation: cultural issues," *International Journal of Entrepreneurship and Innovation Management*, 11(4): 369–397.

Mortara, Letizia, Thomson, Ruth, Moore, Chris, Armara, Kalliopi, Kerr, Clive, Phaal, Robert, and Probert, David. (2010). "Developing a technology intelligence strategy at Kodak european research: scan and target," *Research—Technology Management*, 53: 27–38.

Mortara, Letizia, and Minshall, Tim. (2011). "How do large multinational corporations implement Open Innovation?" *Technovation*, 31(10–11): 586–597.

Mortara, Letizia, Ford, Simon J., Jaeger, Manuel, (2013). "Idea competitions under scrutiny: acquisition, intelligence or public relation mechanism?," *Technological Forecasting & Social Change*, 80(8): 1563–1578.

Moschella, David C. (1997). *Waves of Power: Dynamics of Global Technology Leadership, 1964–2010*. New York: AMACOM.

Mowery, David C. (2009). "Plus ca change: Industrial R&D in the 'third industrial revolution," *Industrial and Corporate Change*, 18(1): 1–50.

Mowery, David C., Oxley, Joanne E., and Silverman, Brian S. (1996). "Strategic alliances and interfirm knowledge transfer," *Strategic Management Journal*, 17(Winter): 77–91.

Mudambi, Susan M. and Tallman, Stephen. (2010). "Make, buy or ally? Theoretical perspectives on knowledge process outsourcing through alliances," *Journal of Management Studies*, 47(8): 1434–1456.

Müller-Seitz, Gordon, Sydow, Jorg (2012). "Open innovation at the interorganizational network level—Collaborative Practices in a Semiconductor Industry Consortium," Open Innovation: New Insights and Evidence conference, Imperial College London, June 25.

Müller-Stewens, Günther. and Lechner, Christoph. (2005) *Strategisches Management*, 3rd ed. Stuttgart: Schäffer-Poeschel Verlag.

Muller, Amy, and Hutchins, Nate. (2012). "Open innovation helps Whirlpool Corporation discover new market opportunities," *Strategy and Leadership*, 40(4): 36–42.

Murray, Fiona, & O'Mahony, Siobhán. (2007). "Exploring the foundations of cumulative innovation: Implications for organization science," *Organization Science*, 18(6): 1006–1021.

Murray, Robin, Caulier-Grice, Julie, and Mulgan, Geoff. (2010). *The Open Book of Social Innovation*. NESTA, The Young Foundation.

Nakagaki, Paul, Aber, Josh, and Fetterhoff, Terry. (2012). "The challenges in implementing open innovation in a global innovation-driven corporation," *Research Technology Management*, 55(4): 32–38.

Nambisan, Satish, and Sawhney, Mohanbir. (2007). *The Global Brain: Your Roadmap for Innovating Faster and Smarter in a Networked World*. Philadelphia: Wharton School Publishing.

Nambisan, Satish, and Sawhney, Mohanbir. (2011). "Orchestration process in network-centric innovation: Evidence from the field," *Academy of Management Perspectives*, 25(3): 40–57.

Nambisan, Satish, Bacon, John, and Throckmorton, James. (2012). "The role of the innovation capitalist in open innovation a case study and key lessons learned," *Research-Technology Management*, 55(3): 49–57.

Narula, Rajneesh. (2004). "R&D collaboration by SMEs: New opportunities and limitations in the face of globalization," *Technovation*, 24(2): 153–161.

NSF (2006). *Survey of Industrial Research and Development*, Division of Science Resource Studies, National Science Foundation.

Nelson, Richard R. (1959). "The simple economics of basic scientific research," *Journal of Political Economy*, 67(3): 297–306.

Nelson, Richard R., (1993). *National Innovation Systems: A Comparative Analysis*. New York: Oxford University Press.

Nelson, Richard R. and Phelps, Edmund S. (1966). "Investment in humans, technological diffusion, and economic growth," *American Economic Review*. 56(1–2): 69–75.

Ng, Irene C.L, and Yip, Nick. (2010). "Theoretical foundations in the pricing of intermediating services: the case of mobile phone payments," *Journal of Revenue and Pricing Management*, 9(3): 1–16.

Nieto, Maria Jesús and Santamaría, Lluis. (2007). "The importance of diverse collaborative networks for the novelty of product innovation," *Technovation*, 27(6–7): 367–377.

Nobel, Robert, and Birkinshaw, Julian. (1998). "Innovation in multinational corporations: Control and communication patterns in international R&D operations," *Strategic Management Journal*, 19(5): 479–496.

Nonaka, Ikujiro and Takeuchi, Hirotaka. (1995). *The Knowledge Creating Company: How Japanese Companies Create the Dynamics of Innovation*. New York: Oxford University Press.

Nooteboom, Bart, Vanhaverbeke, Wim, Duysters, Geert, Gilsing, Victor and van den Oord, Ad. (2007). "Optimal cognitive distance and absorptive capacity," *Research Policy*, 36(7): 1016–1034

Normann, Richard, and Ramirez, Rafael (1993), "From value chain to value constellation: designing interactive strategy," *Harvard Business Review*, 71: 65–65.

Northam, Phil. (ed.) (2006). *How Smartphones Work: Symbian and the Mobile Phone Industry*, Chichester, UK: Wiley.

O'Hern, Matthew S. and Rindfleisch, Aric. (2009) "Customer co-creation: a typology and research agenda," in: Naresh K. Malhotra (ed.): *Review of Marketing Research*, Vol. 6., Armonk, NY: M.E. Sharpe, 84–106.

OECD (2008). *Open Innovation in Global Networks*. Paris: Organisation for Economic Co-operation and Development.

OECD (2009). *The Impact of the Global Crisis on SME and Entrepreneurship Financing and Policy Response*. Paris: Organisation for Economic Co-operation and Development.

Orlowski, Andrew. (2011). "Symbian's secret history: Davies on what went right (and wrong)," *The Register* Jan. 12, URL: http://www.theregister.co.uk/2011/01/12/symbian_history_part_three_charles_davies_interview/p

Osterwalder, Alexander, and Pigneur, Yves. (2009). *Business Model Generation. A Handbook for Visionaries, Game Changers, and Challengers*. Hoboken, NJ: Wiley.

Oxley, Joanne E. (1997). "Appropriability hazards and governance in strategic alliances: A transaction cost approach," *Journal of Law, Economics, and Organization*, 13(2): 387–409.

Paasi, Jaako, Luoma, Tuja, Valkorari, Katri and Lee, Nari. (2010). "Knowledge and intellectual property management in customer-supplier relationship," *International Journal of Innovation Management*, 14 (4): 629–654.

Page, Albert L. (1993). "Assessing new product development practices and performance: establishing crucial norms," *Journal of Product Innovation Management*, 10(4): 273–290.

Parida, Vinit, Westerberg, Mats, and Frishammar, Johan. (2012). "Inbound open innovation activities in high-tech SMEs: The impact on innovation performance," *Journal of Small Business Management*, 50(2): 283–309.

Park, Choelsoon (2002). "The effects of prior performance on the choice between related and unrelated acquisitions: implications for the performance consequences of diversification strategy," *Journal of Management Studies*, 39(7): 1003–1019.

Parker, Geoffrey G. and van Alstyne, Marshall. (2005). "Two-sided network effects: A theory of information product design," *Management Science*, 51(10): 1494–1504.

Patel, Pari, and Pavitt, Keith. (1991). "Large firms in the production of the world's technology: An important case of 'Non-Globalisation,'" *Journal of International Business Studies*, 22(1): 1–21.

Pavitt, K. (2002). "Innovating routines in the business firm: What corporate tasks should they be accomplishing?," *Industrial and Corporate Change*, 11(1): 117–133.

Pellegrini, Luisa, Lazzarotti, Valentina, and Pizzurno, Emanuele, (2012). "From outsourcing to Open Innovation: A case study in the oil industry," *International Journal of Technology Intelligence and Planning*, 8(2): 182–196.

Penin, Julien. (2005). "Patents versus ex post rewards: A new look," *Research Policy*, 34(5), 641–656.

Penner-Hahn, Joan, and Shaver, J. Myles. (2005). "Does international research and development increase patent output? An analysis of Japanese pharmaceutical firms," *Strategic Management Journal*, 26(2): 121–140.

Perkmann, Marcus, and Walsh, Kathryn. (2007). "University-industry relationships and open innovation: Towards a research agenda," *International Journal of Management Reviews*, 9(4): 259–280.

Persson, Magnus (2006). "The impact of operational structure, lateral integrative mechanisms and control mechanisms on intra-MNE knowledge transfer," *International Business Review*, 15(5): 547–569.

Petrash, Gordon. (1997) "Intellectual asset management at Dow Chemical," in Patrick H. Sullivan, ed. *Profiting from Intellectual Capital: Extracting Value from Innovation*. New York: Wiley.

Petroni, Giorgio, Venturini, Karen, and Verbano, Chiara, (2012). "Open innovation and new issues in R&D organization and personnel management." *International Journal of Human Resource Management*, 23(1): 147–173.

Pettigrew, Andrew M., (1990). "Longitudinal field research on change: theory and practice," *Organization Science*, 1(3): 267–292.

Pfeffer, Jeffrey. (1987). "A resource dependence perspective on intercorporate relations," in Mizruchi, Mark S. and Schwartz, Michael, *Intercorporate*

Relations: The Structural Analysis of Business, Cambridge: Cambridge University Press, 25–55.

Pfeffer, Jeffrey and Salancik, Gerald R. (1978). *The External Control of Organizations*. New York: Harper and Row.

Pfeffermann, Nicole. (2011a). "The scent of innovation: towards an integrated management concept for visual and scent communication of innovation," in: Michael Hülsmann and Nicole Pfeffermann (eds.), *Strategies and Communications for Innovations: An Integrative Management View for Companies and Networks*. Berlin: Springer, 163–181.

Pfeffermann, Nicole. (2011b). "Innovation communication as a cross-functional dynamic capability: strategies for organisations and networks," in: Michael Hülsmann and Nicole Pfeffermann (eds.), *Strategies and Communications for Innovations: An Integrative Management View for Companies and Networks* Berlin: Springer, 257–290.

Philpott, Kevin, Dooley, Lawrence, O'Reilly, Caroline, and Lupton, G., (2011). "The entrepreneurial university: Examining the underlying academic tensions," *Technovation*, 31(4): 161–170.

Piller, Frank T. and Walcher, Dominik. (2006). "Toolkits for idea competitions: A novel method to integrate users in new product development," *R&D Management*, 36(3): 307–318.

Piller, Frank, Ihl, Christoph, and Vossen, Alexander. (2011), "customer co-creation: open innovation with customers," in Volker Wittke and Heidemarie Hanekop (eds.) *New Forms of Collaborative Innovation and Production on the Internet: An Interdisciplinary Perspective*, Göttingen: Universitätsverlag Göttingen, 31–61.

Piller, Frank, Vossen, Alexander, and Ihl, Christoph. (2012). "From social media to social product development: the impact of social media on co-creation of innovation," *Die Unternehmung*, 65(1): 7–27.

Pinto, Jeffrey K. and Prescott, John E. (1988). "Variations in critical success factors over the stages in the project life cycle," *Journal of Management*, 14(1): 5–18.

Pisano, Gary P. (1990). "The R&D boundaries of the firm: An empirical analysis," *Administrative Science Quarterly*, 35(1): 153–176.

Pisano, Gary P. and Teece, David J. (2007). "How to capture value from innovation: shaping intellectual property and industry architecture," *California Management Review*, 50 (1): 278–296.

Pisano, Gary P. and Verganti, Roberto. (2008). "Which kind of collaboration is right for you?," *Harvard Business Review*, 86 (12): 78–86.

Podmetina Daria, Volchek Daria, Dabrowska Justina and Fiegenbaum Irina (2013). "Human resource practices and open innovation," *International Journal of Innovation Management*, 17(5).

Poetz, Marion K. and Prügl, Reinhard. (2010). "Crossing domain-specific boundaries in search of innovation: exploring the potential of pyramiding," *Journal of Product Innovation Management*, 27(6): 897–914.

Poetz, Marion K. and Schreier, Martin. (2012). "The value of crowdsourcing: can users really compete with professionals in generating new product ideas?," *Journal of Product Innovation Management*, 29(2): 245–256.

Poot, Tom, Faems, Dries, and Vanhaverbeke, Wim. (2009). "Toward a dynamic perspective on open innovation: a longitudinal assessment of the adoption of internal and external innovation strategies in the Netherlands," *International Journal of Innovation Management*, 13, 177–200.

Porter, Michael E. (1980), *Competitive Strategy*, New York: Free Press.

Porter, Michael E. (1983), *Cases in Competitive Strategy*, New York: Free Press.

Porter, Michael E. (1985). *Competitive Advantage: Creating and Sustaining Superior Performance*, New York: Free Press.

Porter, Michael E (1998). "Clusters and the new economics of competition," *Harvard Business Review*, 76(6): 77–90.

Porter, Michael E. and Kramer, Mark R. (2011). "Creating shared value," *Harvard Business Review*, 89(1–2): 62–77.

Powell, Walter W. (1990). "Neither market nor hierarchy: Network forms of organization," *Research in Organizational Behavior*, 12: 295–336.

Prahalad, C. K. and Ramaswamy, Venkat. (2003). *The Future of Competition: Co-Creating Unique Value with Customers*. Boston, MA: Harvard Business School Press.

Prahalad, C. K. and Ramaswamy, Venkat. (2004a). "Co-creation experiences: The next practice in value creation," *Journal of Interactive Marketing*, 18(3): 5–14.

Prahalad, C. K. and Ramaswamy, Venkat. (2004b). "Co-creating unique value with customers," *Strategy & Leadership*, 32(3): 4–9.

Prud'homme van Reine, Peter (2010). Open innovation and regional growth, in: Philip Cooke, Bjørn T. Asheim, Ron Boshma, Ron Martin, Dafna Schwartz and Franz Tödtling (eds.), *Handbook of Regional Innovation and Growth*, Cheltenham: UK, Edward Elgar, 391–405.

Pye, Annie, and Pettigrew, Andrew. (2006). "Strategizing and organizing: change as a political learning process, enabled by leadership," *Long Range Planning*, 39(6): 583–590.

Quigley, Narda R., Tesluk, Paul E., Locke, Edwin A., & Bartol, Kathryn M. (2007). "A multilevel investigation of the motivational mechanisms underlying knowledge sharing and performance," *Organization Science*, 18(1): 71–88.

Raasch, Christina, Herstatt, Cornelius, and Balka, Kerstin. (2009). "On the open design of tangible goods," *R&D Management*, 39(4): 382–393.

Rahman, Hakikur and Ramos, Isabel. (eds.). (2011). *SMEs and Open Innovation: Global Cases and Initiatives*. Hershey, Penn.: IGI Global.

Ragatz, Gary L., Handfield, Robert B., and Scannell, Thomas V. (1997). "Success factors for integrating suppliers into new product development," *Journal of Product Innovation Management*, 14(3): 190–202.

Ragatz, Gary L., Handfield, Robert B., and Petersen, Kenneth J., (2002). "Benefits associated with supplier integration into new product development under conditions of technology uncertainty," *Journal of Business Research*, 55(5): 389–400.

Ramaswamy, Venkat and Gouillart, Francis. (2010), *The Power of Co-Creation*. New York: Free Press.

Reilly, Robert F., and Robert P Schweihs. (2004). *The Handbook of Business Valuation and Intellectual Property Analysis*. New York: McGraw-Hill.

Rhoten, Diana and Powell, Walter W. (2007). "The Frontiers of Intellectual Property: Expanded Protection versus New Models of Open Science," *Annual Review of Law and Social Science*, 3: 345–373.

Rigby, Darrell and Zook, Chris. (2002) "Open-market innovation," *Harvard Business Review*, 80(10): 80–89.

Rind, Kenneth W. (1981). "The Role of Venture Capital in Corporate Development," *Strategic Management Journal*, 2(2): 169–180.

Ring, Peter Smith and van de Ven, Andrew H. (1992). "Structuring cooperative relationships between organizations," *Strategic Management Journal*, 13(7), 483–498.

Ringberg, Torsten and Reihlen, Markus. (2008), "Communication assumptions in consumer research: An alternative socio-cognitive approach," *Consumption, Markets and Culture*, 11(3): 173–189.

Rivette, Kevin G. and David Kline. (2000). *Rembrandts in the Attic. Unlocking the Hidden Value of Patents*. Boston: Harvard Business Press.

Robertson, Paul L., Casali, Gian Luca. and Jacobson, David. (2012). "Managing open incremental process innovation: Absorptive capacity and distributed learning," *Research Policy*, 41(5): 822–832.

Rochet, Jean-Charles and Tirole, Jean. (2003). "Platform competition in two-sided markets," *Journal of the European Economic Association*, 1 (4): 990–1029.

Rochet, Jean-Charles and Tirole, Jean. (2006). "Two-sided markets: A progress report," *The RAND Journal of Economics*, 37(3): 645–667.

Rogers, Everett M. (1995). *Diffusion of Innovations*, 4th ed. Free Press, New York, NY.

Rogers, Mark. (2004). "Networks, firm size and innovation," *Small Business Economics*, 22(2): 141–153.

Rohrbeck, René, (2010). "Harnessing a network of experts for competitive advantage: technology scouting in the ICT industry," *R&D Management*, 40(2): 169–180.

Rohrbeck, René, Hölzle, Katharina, and Gemünden, Hans Georg. (2009). "Opening up for competitive advantage—How Deutsche Telekom creates an open innovation ecosystem," *R&D Management*, 39 (4): 420–430.

Roper, Stephen. (1999). "Modeling small business growth and profitability," *Small Business Economics*, 13(3): 235–252.

Rosenberg, Nathan. (1990). "Why do firms do basic research (with their own money)?," *Research Policy*, 19(2): 165–174.

Rosenkopf, Lori, Metiu, Anca, and George, Varghese, P. (2001). "From the bottom up? Technical committee activity and alliance formation," *Administrative Science Quarterly*, 46(4): 748–772.

Roser, Thorsten, Samson, Alain, Humphreys, Patrick, and Cruz-Valdivieso, Eidi. (2009). "New pathways to value: Co-creating products by collaborating with customers," working paper, London School of Economics.

Rothaermel, Frank T., and Deeds, David L. (2004). "Exploration and exploitation alliances in biotechnology: A system of new product development," *Strategic Management Journal*, 25(3): 201–222.

Rothaermel, Frank T. and Hess, Andrew M. (2007). "Building dynamic capabilities: Innovation driven by individual-, firm-, and network-level effects," *Organization Science*, 18(6): 898–921.

Rothaermel, Frank T., Agung, Shanti D., and Jiang, Lin. (2007). "University entrepreneurship: A taxonomy of the literature," *Industrial and Corporate Change*, 16(4): 691–791.

Rothwell, R., & Dodgson, M. (1991). "External linkages and innovation in small and medium-sized enterprises," *R&D Management*, 21(2): 125–138.

Ryu, Christopher J., (2011). "How LG electronics is transforming itself into an innovation company," in: Paul Sloane (ed.), *A Guide to Open Innovation and Crowdsourcing*. London: Kogan Page, 85–90.

Salge, Torsten Oliver, Bohné, Thomas Marc, Farchi, Tomas, and Piening, Erk Peter, (2012). "Harnessing the value of open innovation: The moderating role of innovation management," *International Journal of Innovation Management*, 16(3): 1–26.

Sahlman, William A. (1992). "Insights from the American Venture Capital Organization," working paper, Harvard Business School, 92–047.

Sahlman, William A. (1990). "The structure and governance of venture capital organizations," *Journal of Financial Economics*, 27(2): 473–521.

Sakakibara, Mariko (1997). "Heterogeneity of firm capabilities and cooperative research and development: an empirical examination of motives." *Strategic Management Journal*, 18 (S1): 143–164.

Sanchez, Ron. (2008). "Modularity in the mediation of market and technology change," *International Journal of Technology Management*, 42(4): 331–364.

Santamaría, Lluís, Nieto, María Jesús, and Barge-Gil, Andrés. (2009). "Beyond formal R&D: Taking advantage of other sources of innovation in low- and medium-technology industries," *Research Policy*, 38(3): 507–517.

Savitskaya, Irina, Salmi, Pekka, Torkkeli, Marko, (2010). "Barriers to open innovation: Case China," *Journal of Technology Management and Innovation*, 5(4): 10–21.

Sawhney, Mohanbir, & Prandelli, Emanuela (2000), "Beyond customer knowledge management: customers as knowledge co-creators," in Yogesh Malhotra, ed, *Knowledge Management and Virtual Organizations*, Hershey, Penn.: Idea Group, 258–282.

Sawhney, Mohanbir, Verona, Gianmario, & Prandelli, Emanuela. (2005). "Collaborating to create: The Internet as a platform for customer engagement in product innovation," *Journal of Interactive Marketing*, 19(4): 4–17.

Schiele, Holger. (2010). "Early supplier integration: The dual role of purchasing in new product development," *R&D Management*, 40(2): 138–153.

Schildt, Henri, Maula, Markku, V. J., and Keil, Thomas. (2005). "Explorative and exploitative learning from external corporate ventures," *Entrepreneurship Theory and Practice*, 29(4): 493–515.

Schlegelmilch, Bodo B., Ambos, Björn, and Chini, Tina C. (2003). "Are you ready to learn from your offshore affiliates?," *European Business Forum*, 16: 50–54.

Schroll, Alexander, and Mild, Andreas. (2012). "A critical review of empirical research on open innovation adoption," *Journal für Betriebswirtschaft*, 62(2): 85–118.

Schulze, Anja, and Hoegl, Martin. (2008). "Organizational knowledge creation and the generation of new product ideas: A behavioral approach," *Research Policy*, 37(10): 1742–1750.

Schweitzer, Fiona M., Gassmann, Oliver, and Gaubinger, Kurt. (2011). "Open innovation and its effectiveness to embrace turbulent environments," *International Journal of Innovation Management*, 15(6): 1191–1207.

Selzer, Larry. (2006). "Vista, Security Check: This Time Microsoft Means Business," *eWeek*, May 29, (http://www.eweek.com/c/a/Security/Vista-Security-Check-This-Time-Microsoft-Means-Business/).

Senge, Peter M. (1990). *The Fifth Discipline: The Art and Practice of the Learning Organization*, New York: Doubleday.

Serrano, Carlos J. (2006). "The market for intellectual property: Evidence from the transfer of patents," working paper, University of Minnesota.

SG Cowen & Co. (2004). "McAfee—Initiating Coverage," SG Cowen Securities Corporation (November 18, 2004).

Shah, Sonali K. (2006), "Motivation, governance, and the viability of hybrid forms in open source software development," *Management Science*, 52(7): 1000–1014.

Shah, Sonali K., and Tripsas, Mary (2007), "The accidental entrepreneur: The emergent and collective process of user entrepreneurship," *Strategic Entrepreneurship Journal*, 1(1–2): 123–140.

Shan, Weijan, Walker, Gordon, and Kogut, Bruce. (1994). "Interfirm cooperation and startup innovation in the biotechnology industry," *Strategic Management Journal*, 15(5): 387–394.

Shane, Scott, Shane Venkataraman, and Ian MacMillan. (1995). Cultural differences in innovation championing strategies. *Journal of Management*, 21 (5): 931–952.

Shapiro, Carl, and Varian, Hal R. (1999). *Information Rules: A Strategic Guide to the Network Economy*. Boston: Harvard Business School Press.

Sieg, Jan Henrik, Martin W. Wallin, and Georg Von Krogh. (2010). "Managerial challenges in open innovation: a study of innovation intermediation in the chemical industry," *R&D Management*, 40(3): 281–291.

Siegel, Robin, Siegel, Eric, and MacMillan, Ian C. (1988). "Corporate venture capitalists: autonomy, obstacles and performance," *Journal of Business Venturing*, 3(3): 233–247.

Silveira, Rafael, and Wright, Randall. (2010). "Search and the market for ideas," *Journal of Economic Theory*, 145(4): 1550–1573.

Simcoe, Tim. (2006) "Open standards and intellectual property rights," in Henry Chesbrough, Wim Vanhaverbeke, and Joel West (eds.), *Open Innovation: Researching a New Paradigm*. Oxford: Oxford University Press, 161–183.

Simcoe, Tim. (2012). "Standard setting committees: Consensus governance for shared technology platforms," *American Economic Review*, 102(1): 305–336.

Simard, Caroline, and West, Joel. (2006). "Knowledge networks and the geographic locus of innovation," in: Henry Chesbrough, Wim Vanhaverbeke, and Joel West (eds.), *Open Innovation: Researching a New Paradigm*. Oxford: Oxford University Press, 220–240.

Simon, Herbert A. (1947). *Administrative Behavior*. New York: Macmillan.

Simon, Herbert A. (2002). "Near decomposability and the speed of evolution," *Industrial and Corporate Change*, 11(3): 587–599.

Sivadas, Eugene and Dwyer, F. Robert. (2000). "An examination of organizational factors influencing new product success in internal and alliance-based processes," *Journal of Marketing*, 64(1): 31–49.

Sloane, Paul. (ed.). (2011). *A Guide to Open Innovation and Crowdsourcing: Advice From Leading Experts*. London: Kogan Page.

Slowinski, Gene, Hummel, Edward, Gupta, Amitabh and Gilmont, Ernest R. (2009). "Effective practices for sourcing innovation," *Research Technology Management*, 52(1): 27–34.

Slowinsky, Gene, and Sahal, Matthew W. (2010). "Good practices in open innovation," *Research -Technology Management*, 53 (5): 38–45.

Sminia, Harry and de Rond, Mark, (2012). "Context and action in the transformation of strategy scholarship," *Journal of Management Studies*, 49(7): 1329–1349.

Sofka, Wolfgang and Grimpe, Christoph. (2010). "Specialized search and innovation performance—evidence across Europe," *R&D Management*, 40(3): 310–323.

Song, Jaeyong, Almeida, Paul, and Wu, Geraldine. (2003). "Learning-by-hiring: When is mobility more likely to facilitate inter-firm knowledge transfer?," *Management Science*, 49(4): 351–365.

Song, Jaeyong and Shin, Jongtae. (2008). "The paradox of technological capabilities: A study of knowledge sourcing from host countries of overseas R&D operations," *Journal of International Business Studies*, 39(2): 291–303.

Song, Jaeyong, Asakawa, Kazuhiro, and Chu, Youngeun. (2011). "What determines knowledge sourcing from host locations of overseas R&D operations? A study of global R&D activities of Japanese multinationals," *Research Policy*, 40(3): 380–390.

Southwick, Karen. (1999). *High Noon: The Inside Story of Scott McNealy and the Rise of Sun Microsystems*, New York: Wiley.

Spaeth, Sebastian, Stuermer, Matthias and von Krogh, Georg (2010), "Enabling knowledge creation through outsiders: towards a push model of open innovation," *International Journal of Technology Management*, 52(3–4): 411–431.

Spithoven, André, Clarysse, Bart, and Knockaert, Mirjam. (2010). "Building absorptive capacity to organise inbound open innovation in traditional industries," *Technovation*, 31(1): 130–141.

Spithoven, André, Teirlinck, Peter, and Frantzen, Dirk (2012). *Managing Open Innovation: Connecting the Firm to External Knowledge*. Cheltenham: Edward Elgar.

Spithoven, André, Vanhaverbeke, Wim and Rojiakkers, Nadine. (2013). "Open innovation practices in SMEs and large enterprises," *Small Business Economics*, 41(3): 537–552.

Spradlin, Dwayne. (2012) "Are you solving the right problem? Asking the right questions is crucial," *Harvard Business Review*, 90(9): 84–101.

Staber, Udo (2007): Contextualizing research on social capital in regional clusters. *International Journal of Urban and Regional Research*, 31(3), 505–521.

Staudenmayer, Nancy, Mary Tripsas and Chris L. Tucci. (2000). "Development Webs: A new paradigm for product development," in Rudi K. F. Bresser, Michael A. Hitt, Robert D. Nixon, Dieter Heuskel (eds.) *Winning Strategies in a Deconstructing World*, New York: Wiley, pp. 135–161.

Sullivan, Patrick H. (2000). *Value Driven Intellectual Capital: How to Convert Intangible Corporate Assets into Market Value*, New York: Wiley.

Swink, Morgan, Talluri, Srinivas, and Pandejpong, Temyos. (2006). "Faster, better, cheaper: A study of NPD project efficiency and performance tradeoffs," *Journal of Operations Management*, 24(5): 542–562.

Sydow, Jörg, Lindkist, Lars, and DeFillippi, Robert. (2004). "Project based organizations, embeddedness and repositories of knowledge: editorial," *Organization Studies*, 25(9): 1475–89.

Sykes, Hollister B. (1986). "The anatomy of a corporate venturing program: factors influencing success," *Journal of Business Venturing*, 1(3): 275–293.

Symbian. (2006). "Ownership," November 11, accessed via Archive.org, original URL: http://www.symbian.com/about/overview/ownership/ownership.html

Takahashi, Dean. (2002). *Opening the XBox: Inside Microsoft's Plan to Unleash an Entertainment Revolution*. Roseville, CA: Prima.

Takeishi, Akira. (2001). "Bridging inter- and intra-firm boundaries: Management of supplier involvement in automobile product development," *Strategic Management Journal*, 22(5): 403–433.

Tanriverdi, Hüseyin and Chi-Hyon Lee (2008). "Within-industry diversification and firm performance in the presence of network externalities: evidence from the software industry," *Academy of Management Journal*, 51(2): 381–397.

Tao, John and Magnotta, Vincent. (2006). "How air products and chemicals identifies and accelerates," *Research Technology Management*, 49(5): 12–18.

Tapscott, Don, and Williams, Anthony D. (2006). *Wikinomics: How Mass Collaboration Changes Everything*. London: Penguin.

Taptich, Brian E. (1998). "The New Startup," *Red Herring*, October, 52–56.

Taylor, Curtis R. (1995). "Digging for golden carrots: An analysis of research tournaments," *American Economic Review*, 85(4): 872–890.

Teece, David J. (1986). "Profiting from technological innovation: Implications for integration, collaboration, licensing and public policy," *Research Policy*, 15(6): 285–305.

Teece, David J. (1998). "Capturing value from knowledge assets: the new economy, markets for know-how, and intangible assets," *California Management Review*, 40(3): 55–79.

Teece, David J. (2006). "Reflections on 'profiting from innovation," *Research Policy*, 35 (8): 1131–1146.

Teece, David J. (2007). "Explicating dynamic capabilities: The nature and microfoundations of (sustainable) enterprise performance," *Strategic Management Journal*, 28(13): 1319–1350.

Teece, David J. (2008). "Dosi's technological paradigms and trajectories. Insights for economics and management," *Industrial and Corporate Change*, 17(3), 507–512.

Teece, David J., Pisano, Gary and Shuen, Amy. (1997). "Dynamic capabilities and strategic management," *Strategic Management Journal*, 18(7): 509–533.

Terwiesch, Christian, and Xu, Yi. (2008). "Innovation contests, open innovation, and multiagent problem solving," *Management Science*, 54(9): 1529–1543.

Tether, Bruce S. (2002). "Who co-operates for innovation, and why: An empirical analysis," *Research Policy*, 31(6): 947–967.

Tether, Bruce S. and Tajar, Abdelouahid (2008). "Beyond industry–university links: Sourcing knowledge for innovation from consultants, private research organisations and the public science-base," *Research Policy*, 37(6–7): 1079–1095.

Theyel, Nelli. (2012). "Open innovation—a gold mine or fool's gold for young firms?," *Academy of Management Proceedings 2012*.

Thomke, Stefan, Sinofsky, Steven Jay (1999). "Learning from projects: note on conducting a postmortem analysis," HBS Case #N9-600-021, September 3.

Thomke, Stefan, and von Hippel, Eric, (2002). "Customers as innovators: a new way to create value," *Harvard Business Review*, 80(4): 74–81.

Thrift, Nigel. (2006). "Re-inventing invention: New tendencies in capitalist commodification," *Economy and Society*, 35(2): 279–306.

Tirpak, Thomas M., Miller, Roger, Schwartz, Larry, Kashdan, David. (2006). "R&D structure in a changing world," *Research-Technology Management*, 49(5):19–26.

Tödtling, Franz, van Reine, Peter Prud'homme, and Dörhöfer, Steffen, (2011). "Open innovation and regional culture-findings from different industrial and regional settings," *European Planning Studies*, 19(11): 1885-1907.

Torkkeli, Marko T., Kock, Carl Joachim and Salm, Pekka A.S., (2009). 'The "Open Innovation" paradigm: A contingency perspective," *Journal of Industrial Engineering and Management*, 2(1): 176–207.

Tranekjer, Tina Lundø and Knudsen, Mette Præst (2012). "The (unknown) providers to other firms' new product development: What's in it for them?," *Journal of Product Innovation Management*, 29(6): 986–999.

Trott, Paul, and Hartmann, Dap. (2009). "Why 'open innovation' is old wine in new bottles," *International Journal of Innovation Management*, 13(4): 715–736.

Tsai, Kuen-Hung. (2009). "Collaborative networks and product innovation performance: Toward a contingency perspective," *Research Policy*, 38(5): 765–778.

Ulrich David, and Barney Jay B. (1984). "Perspectives in organizations: Resource dependence, efficiency, and population," *Academy of Management Review*, 9(3): 471–481.

Un, C. Annique, Cuervo-Cazurra, Alvaro, and Asakawa, Kazuhiro. (2010). "R&D collaborations and product innovation," *Journal of Product Innovation Management*, 27(5): 673–689.

Urban, Glen L., and Von Hippel, Eric, (1988). "Lead user analyses for the development of new industrial products," *Management Science*, 34(5): 569–582.

Utterback, James M. (1994), *Mastering the Dynamics of Innovation*, Boston: Harvard Business School Press.

Uzzi, Brian. (1996). "The sources and consequences of embeddedness for the economic performance of organizations: The network effect," *American Sociological Review*, 61(4): 674–698.

Van den Biesen, Jan (2008). "Open Innovation @ Philips Research, Business Symposium," Symposium on Open Innovation in Global Networks, Danish Enterprise and Construction Authority, Copenhagen, February 26, http://www.oecd.org/science/inno/40206366.pdf

van de Vrande, Vareska, Lemmens, Charmianne and Vanhaverbeke, Wim (2006). "Choosing governance modes for external technology sourcing," *R&D Management*, 36(3): 347–363.

van de Vrande, Vareska, Vanhaverbeke, Wim and Duysters, Geert. (2009a). "External technology sourcing: The effect of uncertainty on governance mode choice," *Journal of Business Venturing*, 24(1): 62–80.

van de Vrande, Vareska, de Jong, Jeroen P. J., Vanhaverbeke, Wim and de Rochemont, Maurice. (2009b). "Open innovation in SMEs: Trends, motives and management challenges," *Technovation*, 29(6–7): 423–437.

van de Vrande, Vareska, Vanhaverbeke, Wim, and Gassmann, Oliver. (2010). "Broadening the scope of open innovation: past research, current state and future directions," *International Journal of Technology Management*, 52(3): 221–235.

van der Borgh, Michel, Cloodt, Myriam and Romme, A. Georges L. (2012). "Value creation by knowledge-based ecosystems: evidence from a field study," *R&D Management*, 42(2): 150–169.

Van der Meer, Hans. (2007). "Open Innovation—The Dutch treat: Challenges in thinking in business models," *Creativity and Innovation Management*, 16(2): 192–202.

Van de Ven, Andrew H. (1989). "Nothing is quite so practical as a good theory," *Academy of Management Review*, 14(4): 486–489.

Van Steerthem, Angie, Delcour, Fauve and De Stobbeleir, Kathleen (2013). "De menselijke factor in open innovatie: hoe people management open innovatie kan stimuleren" (*"The human factor in open innovation: how people management can promote open innovation"*), white paper, Flanders DC Kennisstudie, Vlerick Business School.

Vanhaverbeke, Wim (2006). "The inter-organizational context of open innovation," in Henry Chesbrough, Wim Vanhaverbeke, and Joel West (eds.), *Open Innovation: Researching a New Paradigm*. Oxford: Oxford University Press, 205–219.

Vanhaverbeke, Wim and Cloodt, Myriam. (2006).

Vanhaverbeke, Wim (2012). *Open Innovation in SMEs: How can small companies and start-ups benefit from open innovation strategies?* Flanders District of Creativity: Leuven, Belgium.

Vanhaverbeke, Wim, and Bosch, Servaas. (2010). "Curana BVBA: Managing open innovation in SMEs," European Case Clearinghouse, case ECCH 810-062-1.

Vanhaverbeke, Wim and Cloodt, Myriam. (2006). "Open innovation in value networks," in Henry Chesbrough, Wim Vanhaverbeke, and Joel West (eds) *Open Innovation: Researching a new paradigm*, Oxford: Oxford University Press, 258–281.

Vanhaverbeke, Wim, Du, Jingshu, and von Zedtwitz, Maximilian. (2013). "Managing open innovation in multinational enterprises: combining open innovation and R&D globalization literature," in Joe Tidd, ed, *Open Innovation Research, Management and Practice*, Series on Technology Management: Volume 23, Imperial College Press, 213–233.

Vanhaverbeke, Wim, Gilsing, Victor, and Duysters, Geert. (2012). "Competence and governance in strategic collaboration: The differential effect of network structure on the creation of core and non-core technology," *Journal of Product Innovation Management*, 29(5): 784–802.

Vanhaverbeke, Wim, Gilsing, Victor and Duysters, Geert. (2012). "Competence and governance in strategic collaboration: The differential effect of network structure on the creation of core and non-core technology," *Journal of Product Innovation Management*, 29(5): 784–802.

Vanhaverbeke, Wim and Peeters, Nico. (2005). "Embracing innovation as strategy: corporate venturing, competence building and corporate strategy making," *Creativity and Innovation Management*, 14(3): 246–257.

Vanhaverbeke, Wim and Roijakkers, Nadine. (2013). "The role of strategy in open innovation," in: Nicole Pfeffermann, Tim Minshall, Letizia Mortara (eds.), *Strategy and Communication for Innovation: An Integrative Management View for Start-Ups, Companies and Collaborative Networks in the Open Innovation Economy*, Heidelberg: Springer Verlag, 15–25.

Vanhaverbeke, Wim, van de Vrande, Vareska, Chesbrough, Henry. (2008). "Understanding the advantages of open innovation practices in corporate venturing in terms of real options," *Creativity & Innovation Management*, 17(4): 251–258.

Van Oorschot, Kim, Sengupta, Kishore, Akkermans, Henk, and Van Wassenhove, Luk. (2010). "Get fat fast: Surviving Stage-Gate® in NPD," *Journal of Product Innovation Management*, 27(6): 828–839.

Van Wijk, Raymond, Jansen, Justin J.P. and Lyles, Marjorie A. (2008). "Inter-and intra-organizational knowledge transfer: a meta-analytic review and assessment of its antecedents and consequences," *Journal of Management Studies*, 45(4): 830–853.

Venture Economics (1998). "Venture capital funding statistics press release," October 1. http://www.secdata.com/vepressrlse/VEpressrlse.html accessed xxxxx

Venture One (1999). "3Q'99 *Venture Capital Financings Top 8 Billion Dollars*," November 2, http://www.v1.com/news/press/Q399PRFinancings.htm, accessed xxxxx

von Hippel, Eric, (1973). *An Exploratory Study of Corporate Venturing—A. New Product Innovation Strategy Used by Some Major Corporations*, unpublished dissertation, Carnegie Mellon University.
von Hippel, Eric, (1977). "Successful and failing internal corporate ventures: an empirical analysis," *Industrial Marketing Management*, 6(3): 163–174.
von Hippel, Eric, (1988). *The Sources of Innovation*. New York: Oxford University Press.
von Hippel, Eric (1994). "Sticky information and the locus of problem solving: implications for innovation," *Management Science*, 40(4): 429–439.
von Hippel, Eric. (2001). "User toolkits for innovation," *Journal of Product Innovation Management*, 18(4): 247–257.
von Hippel, Eric. (2005). *Democratizing Innovation*. Cambridge, Mass.: MIT Press.
von Hippel, Eric. (2007). "Horizontal innovation networks—by and for users," *Industrial & Corporate Change*, 16(2): 293–315.
von Hippel, Eric. (2010). "Open user innovation," *Handbook of the Economics of Innovation*, edited by Bronwyn H. Hall and Nathan Rosenberg, Volume 1: 411–427.
von Hippel, Eric, and Katz, Ralph. (2002). "Shifting innovation to users via toolkits," *Management Science*, 48(7): 821–834.
von Hippel, Eric, and de Jong, Jeroen P.J. (2010), *Open, Distributed and User-Centered: Towards a Paradigm Shift in Innovation Policy*, EIM Research Report, Number H201009, Zoetermeer, NL: EIM Business & Policy Research, URL: http://www.entrepreneurship-sme.eu/pdf-ez/H201009.pdf
von Hippel, Eric, Ogawa, Susumu, and de Jong, Jeroen P.J. (2012). "The age of consumer-innovator," *Sloan Management Review*, 53(1): 27–33.
von Hippel, Eric, and von Krogh, Georg, (2003). "Open source software and the "private-collective' innovation model: Issues for organization science," *Organization Science*, 14(2): 209–223.
von Hippel, Eric, and von Krogh, Georg, (2006). "Free revealing and the private-collective model for innovation incentives," *R&D Management*, 36(3): 295–306.
von Krogh, Georg, Spaeth, Sebasian and Lakhani, Karim R. (2003). "Community, joining, and specialization in open source software innovation: a case study," *Research Policy*, 32(7): 1217–1241.
von Krogh, Georg, Wallin, Martin, and Sieg, Jan Henrik (2012). "A problem in becoming: How firms formulate sharable problems for innovation contests," Working Paper. ETH Zürich.
Vossen, Robert W. (1988). "Relative strengths and weaknesses of small firms in innovation," *International Small Business Journal*, 16(3): 88–94.
Wagner, Marcus (2011). "Growth of university-based start-ups and acquisition as an exit strategy in academic entrepreneurship: evidence from software-based ventures," *International Journal of Entrepreneurship and Small Business*, 12(4): 395–412.
Wall Street Transcript (2001). "Internet security software," Report 327, Wall Street Transcript Corporation, April 23,
Wall Street Transcript (2004a). "Security/Internet security & Identity authentication," Report 904, Wall Street Transcript Corporation, April 26.
Wall Street Transcript (2004b). "Analyst interview: Gary Spivak, Network security," Report 18485, Wall Street Transcript Corporation, May 19.

Wall Street Transcript (2004c). "CEO interview: Peter Privateer, Internet Security Systems," Report 18652, Wall Street Transcript Corporation, June 2.

Wallin, Martin W. and von Krogh, Georg, (2010). "Organizing for open innovation: Focus on the integration of knowledge: Designing organizations for the 21st-century global economy," *Organizational Dynamics*, 39(2): 145-154.

Webb, David, and Pettigrew, Andrew. (1999). "The temporal development of strategy: patterns in the U.K. insurance industry," *Organization Science*, 10(5): 601-621.

Wernerfelt, Birger. (1984). "A resource based view of the firm," *Strategic Management Journal*, 5(2): 171-180.

West, Joel (2003). "How open is open enough? melding proprietary and open source platform strategies," *Research Policy*, 32 (7): 1259-1285.

West, Joel (2006). "Does appropriability enable or retard open innovation?," in Henry Chesbrough, Wim Vanhaverbeke, and Joel West (eds.), *Open Innovation: Researching a New Paradigm*. Oxford: Oxford University Press, 109-133.

West, Joel. (2007a). "The economic realities of open standards: black, white and many shades of gray," in Shane Greenstein and Victor Stango (eds.), *Standards and Public Policy*, Cambridge: Cambridge University Press, 87-122.

West, Joel. (2007b). "What is Open Innovation?," Open Innovation Blog, Aug. 28, URL: http://blog.openinnovation.net/2007/08/what-is-open-innovation.html.

West, Joel and Dedrick, J. (2001). "Open source standardization: the rise of Linux in the network era," *Knowledge, Technology & Policy*, 14(2): 88-112.

West, Joel and Gallagher, Scott (2006a). "Challenges of open innovation: The paradox of firm investment in open-source software," *R&D Management*, 36(3): 319-331.

West, Joel and Gallagher, Scott (2006b). "Patterns of open innovation in open source software," in Henry Chesbrough, Wim Vanhaverbeke and Joel West, eds., *Open Innovation: Researching a New Paradigm*, Oxford: Oxford University Press, pp. 82-106.

West, Joel, Vanhaverbeke, Wim, and Chesbrough, Henry. (2006). "Open innovation: a research agenda," in Henry Chesbrough, Wim Vanhaverbeke, and Joel West (eds.), *Open Innovation: Researching a New Paradigm*. Oxford: Oxford University Press, 285-307.

West, Joel, and Lakhani, Karim R. (2008). "Getting clear about communities in open innovation," *Industry and Innovation*, 15(2): 223-261.

West, Joel and O'Mahony, Siobhán. (2008). "The role of participation architecture in growing sponsored open source communities," *Industry & Innovation*, 15(2): 145-168.

West, Joel and Mace, Michael. (2010) "Browsing as the killer app: Explaining the rapid success of Apple's iPhone," *Telecommunications Policy*, 34(5-6): 270-286.

West, Joel and Sims, Jonathan. (2012). "Coupling firms and communities: how innovative are innovation communities, and when do firms care?," working paper, UT Austin, McCombs School of Business.

West, Joel and Bogers, Marcel. (2014). "Leveraging external sources of innovation: A review of research on open innovation," *Journal of Product Innovation Management*, 31(4): 814-831.

West, Joel and Wood, David. (2013). "Evolving an open ecosystem: the rise and fall of the Symbian platform," in Ron Adner, Joanne E. Oxley, Brian S. Silverman (ed.)

Advances in Strategic Management, Volume 30: Collaboration and Competition in Business Ecosystems, 27–67.

Westergren, Ulrika H. and Holmström, Jonny. (2012). "Exploring preconditions for open innovation: Value networks in industrial firms," *Information and Organization,* 22(4): 209–226.

Wijen, Frank, Noorderhaven, Niels and Vanhaverbeke, Wim. (2011). "Structural antecedents of corporate network evolution," *International Journal of Business Environment,* 4(3): 207–233.

Wikström, Solveig. (1996). "Value creation by company-consumer interaction," *Journal of Marketing Management,* 12(5): 359–374.

Williamson, Oliver E. (1975). *Markets and Hierarchies: Analysis and Antitrust Implications, Study in the Economics of Internal Organization,* New York: Free Press.

Williamson, Oliver E. (1985). *The Economic Institutions of Capitalism: Firms, Markets, Relational Contracting,* New York: Free Press.

Wynarczyk, Pooran, Piperopoulos, Panagiotis and McAdam, Maura. (2013). "Open innovation in small and medium-sized enterprises: An overview," *International Small Business Journal,* 31(2): 1–16.

Xu, Sean Xin, Zhu, Christina and Zhu, Kevin Xiaoguo (2012). "Why do firms adopt innovations in bandwagons? Evidence of herd behaviour in open standards adoption," *International Journal of Technology Management,* 59(1): 63–91.

Yin, Robert K. (2009). *Case Study Research: Design and Methods,* 4th ed., Thousand Oaks, CA: Sage.

Yoffie, David B. (1996). "Competing in the age of digital convergence," *California Management Review,* 38(4): 31–53.

Yost, Matthew. (1994). "The state of corporate venturing: the number of active programs levels off as corporations complete shifts back to core businesses," *Corporate Venturing,* June.

Zajac, Edward J. and Olsen, Cyrus P. (1993). "From transaction cost to transactional value analysis: implications for the study of interorganizational strategies," *Journal of Management Studies,* 30(1): 131–145.

Zenger, Todd R. (1994). "Explaining organizational diseconomies of scale in R&D: the allocation of engineering talent, ideas, and effort by firm size," *Management Science,* 40(6): 708–729.

Zott, Christoph and Amit, Rafael. (2007). "Business model design and the performance of entrepreneurial firms," *Organization Science,* 18(2): 181–199.

Zott, Christoph and Amit, Rafael. (2008). "The fit between product market strategy and business model: implications for firm performance," *Strategic Management Journal,* 29(1): 1–26.

Index

Page numbers in **bold** refer to figures and tables

Adams, Robert 212
Adner, Ron 61
Air Products and Chemicals (APD) 158
Amazon 54
Apple 74, 86, 164
   iPhone 54, 59–60, 82, 83–4, 87, **89**, 90, **92**, 92n4, 93n10
Ariely, Dan 36
Arora, Ashish 194–5, 246
Arrow, Kenneth J. 16, 198
Arts and Humanities Citation Index (A&HCI) 7
Asakawa, Kazuhiro 163
AT&T 195
   Unix 77, 88, **89**, 93
aviation industry 63–4, 68n17

Baldwin, Carliss Y. 35
BASF 157
Better Place **54**, 61–4, 67n14
Bianchi, Mattia 164
biotechnology 73, 106, 137, **140**, 148, 196
Bogers, Marcel 19, 25, 37, 75
Boudreau, Kevin J. 48, 49n4, 277n7
boundaries of the firm v, vi, 16, **258**
   permeable 17, 34, 35, 265
   see also business models; business theory; knowledge flows
Bowie, David 201
BP (née British Petroleum) 65, 231, 233–4
British Telecom 215, 218–20
   BrightStar 215–17, 228
broadcast search, *see* crowdsourcing
Brunswicker, Sabine 281
business ecosystems vi–vii, 20, 53, 63–4, 71–93, 102–12, **103**, 283–4
   see also networks
business models v, vi, 15, 19, 21, 170, 257–8, 260
   closed innovation 16, 22, 55–6
   closed/stand-alone 53, 54–5, 56–8, 64
   flexible 173
   hybrid 35, 44, 57
   inclusive vii
   linked/networked 53, 55, 61, 64–5, 68n19, 72–4
   static 51, 66, 224

   see also business theory; external innovation; open business models; open innovation; value creation and capture
management theory 9, 94–114, 256–78
   change management 225
   convergence/divergence frameworks 94–5, 102–3, 104–8
   convergence life cycle 95, 96, 98–102, 108–10, **109**
   ecosystem vs. industry frameworks 102–3
   firm–environment relationship 95, 102–4, **103**
   five forces framework 94–5, 103
   innovation life cycle 94–5, 101, 102, 106–7, 108–10, **109**, 112
   knowledge-based view of the firm v–vi, 259, 274
   make–buy–ally decisions 258
   open business dynamics 104
   product life cycle (PLC) model 94–5, 101, 106–7, 108–10, **109**, 112
   real options 259, 271–3
   relational view of the firm 265–6, 274
   resource-based view 256, 259, 264–6, 274, 277n8
   resource dependence theory 266, 274, 277n8
   upper echelon theory 225, 231
   see also product development; product innovation; value creation and capture

Cantwell, John 164
CAS Software AG 154
chemicals industry 56, 157–8, 195–7, 260
Chen, Eric 222
Chesbrough, Henry v, 4, 5, 14, 16, 19, 22–6, 29, 33, 50–3, 75, 117, 158, 167, 194, 222, 224, 281, 284, 287, 289
Chiaroni, Davide 154
China 85, 196
Christensen, Clayton M. 20
Chu, Youngeon 163
Cisco 100, 113n9
co-creation:
   external partners 44–8
   firm-initiated 39–41, **40**
   incremental vs. radical innovation 46, 94–5

co-creation: (*cont.*)
  interactive 30
  marketing concept 30, 39
  product/service development 39
  research 37, 44–8
  stakeholder collaboration 39
  see also collaboration; coupled open innovation; crowdsourcing; networks; user innovation
Cohen, Wesley M. 17, 152, 269
collaboration:
  barriers to 41
  bidirectional 38
  choosing/finding partners 40, 42–4, 48, 122
  community 37, **38**
  contractual 44
  dyadic 30, 37, **38**, 42, 72–4
  external actors/partners 37, **38**, **40**, 42–6
  firm–environment relationship 95, 102–3
  interactive 30, 36–7, 38–9
  leverage and commercialization 46–7
  motivating participants 42–4
  network 37, **38**
  process model 39–48, **40**
  resource allocation 42
  rules of cooperation 42
  strategic intent 38
  task formulation 41–2
  within firms 44
  see also communities; coupled open innovation; crowdsourcing; networks; strategic alliances; user innovation
Collabra Software 151
commercialization paths 16, 18, 37, 46–7, 143, 147–9, 268
communication and public relations (PR) 232, 233–5
communities:
  competitive vs. cooperative 49n4
  open source 35, 38, 42, 60, 73, 75, 84
  self-governing 73
  see also business ecosystems; networks; user innovation
competitive advantage 50, 52, 57, 91, 265–6
competitive bargaining:
  five forces framework 94–5, 103
ConAgra Foods 57
Cooper, Robert G. 9, 24
core competence v, vi, 111, 266–7
corporate ventures 208–22, 259
  complex spin-outs 217–22
  new venture divisions (NVDs) 210
  see also investment; R&D; venture capital
coupled open innovation 29–49
  contrast with user innovation 30–6, **31**, 34–6
  external actors/partners 37, **38**, **40**, 42–6
  governance of collaboration 45
  interactive 36–47
  process model of collaboration 39–48, **40**
  tools and infrastructures 45–6
crowdsourcing 19, **31**, 39, 248
  broadcast search 15, 36, 145
  IT-enabled 144–7, 285
  open call for participants 43, 145, 147
  tournaments and competitions 41–5, 49n6, 145, 234

Dahlander, Linus 19, 21, 34, 36
Davies, Charles 86
Dell 32, 264
Di Minin, Alberto 164
Dittrich, Koen 75
division of labor 20, 52–3, 97, 100–1, 124
Drayton, Bill 175, 176, 180, 188n5
Dreyfuss, Rochelle 194
Drucker, Peter 187n1, 208
DuPont 54, 56
Duysters, Geert 75
Dyer, Jeffrey H. 262

economic theory:
  bounded rationality 225
  evolutionary economics 224, 238
  industrial economics 94
  neo-classical economics vi
  Schumpeterian economics 94
  transaction cost economics 256, 261–4, 274, 277n5
ecosystems, *see* business ecosystems
emerging economies 165–6, 288–9
energy industry 50, 63–5, 211–12, 231, 248
Enkel, Ellen 36, 38, 47, 246
entrepreneurship 15, 32, 90
  collaborative 175–80
  see also open social innovation; social entrepreneurs
equity-based alliances 160
evolutionary theory 224, 238
external actors 37, **38**
  selection and recruitment of **40**, 42–4, 48
external innovation:
  effect on firm boundaries and activity 259
  inbound open innovation 19, 30, 33, 34, 140–2, **140**, 152, 291
  innovation research 72
  money markets 36
  sources 15, 75, 111, 152, 158, 159–60, 165–6
  transaction costs 261–3

Index

using intermediaries/"innomediaries" 46, 146, 242–56, **245**, 263–4, 276, 277nn3, 6
  see also co-creation; collaboration; coupled open innovation; crowdsourcing; external knowledge; inbound (outside-in) open innovation; open business models; open innovation; open social innovation; outbound (inside-out) open innovation; R&D; user innovation
external knowledge:
  absorptive capacity of the firm 14, 17, **31**, 46–7, 152–3, 160, 163, 236, 269–71
  balance with internal knowledge 267–71, 275
  business strategy 275–6
  coupling and commercialization 17, 27, 37
  leveraging 14–17, 25, 27, 29, **40**, 160
  optimal use of 9, 17
  role of intermediaries and brokers 242–56, 263–4, 277nn3, 6
  searching for 242–56, **245**
  sourcing 58, 65, 152, 157–8, 162
  sourcing overseas 163–4
  see also collaboration; communities; coupled open innovation; crowdsourcing; intellectual property; internal knowledge; networks; R&D
Exxon 209, 211–14, 217

Fast, Norman D. 210
finance, *see* investment; R&D
Fleming, Lee 14
food industry 57
free revealing, *see* outbound (inside-out) open innovation
Füller, Johann 43

Gallagher, Scott 75
Gann, David M. 19, 21, 36
Garman, Andrew R. 222
gas industry, *see* energy industry
Gassmann, Oliver 36, 38, 47
Gates, Bill 79
globalization vi, vii, 168, 264, 288
Google 3, 4, 5, 84, **92**
Groen, Aard J. 24
Gronum, Sarel 144
Grove, Andy 88

Hartmann, Dap 22–3
Hayek, F. A. 16
Heyman, James 36
high-tech vs. low-tech industries 286
Holmström, Jonny 235
horizontal integration, *see* value chains

human resource management (HRM) 232–3, 292–3

IBM 33, **54**, 60–1, 74, 88, **89**, **92**, 105, 158, 164, 193, 266, 284
ICT (information and communications technology) vi, 16, 75
  see also IT (information technology) security; social media
inbound (outside-in) innovation:
  absorptive capacity of the firm 14, 17, **31**, 46–7, 152–3, 160, 163, 236, 269–71
  benefits to firms 34
  business models **54**, 58, 61
  dyadic collaboration 42, 47, 75
  purposive knowledge flow 18, 33–34
  see also R&D; small and medium-sized enterprises
incentives and rewards **31**, 35–7, 43–4, 47–8, 123
"infomediation" 205
information technology, *see* IT (information technology) security
"innomediaries" 242–56, **245**, 277nn3, 6
innovation:
  context-embedding 99–101
  democratization of 34
  distributive model 29
  erosion factors 16, 22, 26, 158–9
  life cycle 94–5, 101
  "not invented here" (NIH) syndrome vi, 41, 46, 167
  private vs. collective models 35, 44
  public good nature of 20–1, 35
  shift from products to services vi
  systemic 98, 102, 106–7, 109
  theories of v–vi, 20–1
  "user-centric" 20–1
  see also open innovation
innovation merchants 33
Intel 33, 60, **78**, 88, **89**, **92**, 93n14,164, 195, 266, 284
intellectual property (IP) 15, 33, 34–6, 144–5, 158
  barriers to efficient IP markets 199–201
  business networks 198–9
  buy and sell motivators 194–9
  collaborative model 35
  creation paths **192**
  effect of open innovation 191–207
  in-licensing/out-licensing 19, **31**
  intermediate markets 194–201
  licensing 96–7, 99, 101, 107, 108, 191, 193–5, 200, 201, 206n3
  management strategy 21, 149–51, 192–4, 197–8, 204–6

intellectual property (IP) (*cont.*)
  monetarization mechanisms 201–5
  patents 120–1, 149–50, 192–3, **192**, 199, 200, 202–4
  TAEUS PatentBooks 203
  Unit License Right (ULR) contracts 202
  weak vs. strong regimes 191
Intellectual Property Exchange International (IPXI) 202–3
intermediaries and brokers 46, 146, 242–56, **245**, 263–4, 277nn3, 6
internal innovation:
  leveraging external knowledge 14, 29, 38–9
  purposive knowledge flow 4, 33–34, 52
  see also R&D; research laboratories
internal knowledge 7, 17, 27, 153, 247
  accessibility of **54**, 60–1
  balance with external knowledge 267–71, 275
  core rigidities 267
  licensing 115, 227, 258
  marketing of 65, 259, 260
  spin-off activities 53, 57, 115, 258, 270, 272
  unused 16, 17, 19, **54**, 56–7, 115, 186, 208, 243
  value creation 58
  see also intellectual property; outbound (inside-out) open innovation
Internet, *see* IT (information technology) security; social media
investment:
  capital vi
  platform development 79–83
  private vs. social return 16–17
  see also R&D; venture capital
Isobionics 148
IT (information technology) security 96–103, 105–6, 108–9, 111–13
ITRI (Industrial Technology Research Institute, Taiwan) 203–4
Iwasa, Tomoko 162, 163

Jacobides, Michael G. 196
Japan v, 56, **85**, 157
Jobs, Steve 59
joint ventures 19, 20, 57, 229
Joy, Bill 173, 175
Joy's Law ("most smart people work for someone else") 16, 173, 175

Kearns, David 212, 213
KLM 63–4
knowledge:
  accessive, adaptive, and integrative capacities 153

"black box" model vi, 116
complementary 107, 110
contribution to value chains vi
know-how vi, 60, 141, 171, 206n3, 269
local 160, 163–4
"not invented here" (NIH) syndrome vi, 41, 46, 167, 206, 235, 246–7
shared v–vii, 30, 32, 36, 38, 40, 198
useful 29, 38, 193–5
see also co-creation; collaboration; crowdsourcing; external knowledge; intellectual property; internal knowledge; knowledge flows; R&D
knowledge-based society v, 205
knowledge flows:
  external/internal 16, 38, 55–9
  knowledge transfer 31, 124–5, 159, 235, 270–1
  open innovation model **18**
  outside-in/inside-out/coupled 4, 14–15, **18**, 18–20, 33, **54**, 55–64, 75, 159, 171, 226
  purposive management 17, 19, 21, 33–4, 52
knowledge management vii, 153, 233, **240**, 291
Kodak 105, 193, 200, 226–7, 233
Korea 157
Kramer, Mark R. 187n1

Lafley, A. G. 231
Lakhani, Karim R. 25, 48, 49n4, 277n7
Laursen, Keld 9, 75
Lee, S. 38, 143, 289
Leibovitz, Annie 201
Lemley, Mark 198
Levinthal, David A. 17, 152, 269
Lhuillery, Stephane 25
Linton, Jonathan D. 24
Living Labs movement vii
Lucent 213–15

mergers and acquisitions 107–8, 110, 160
MacMillan, Ian C. 210, 213, 288
marketing:
  open innovation model **18**
  value creation 58
markets:
  intermediate 194–201
  money market vs. social market 31, 35–6, 44
  neo-classical view of vi
  open innovation model **18**
Mathews, John A. 165
Maula, Markku 75
Meijer, Elise 130

Mendel, Gregor 204
Microsoft 59, 79, 82, 100–1, 105, 113n11, 151, 164, 200, 221, 227
  Windows 60, 71, 77, **85**, 86, 88, **89**, **92**, 100, 113n10
Mild, Andreas 236
Millennium Pharmaceuticals 196
Mintzberg, Henry 38
mobile phone industry 59, 71–93, **78**, **80–1**, **83**, **85**, **87**, **89**, 109
Mortara, Letizia 235
Mowery, David C. 23–4
Mudambi, Ram 164
multinational corporations (MNCs) 157–68, 223–41

Nelson, Richard 16
NESTA 169, **170**
networks 38, 50, 72–4, 142–4, 147–9, 198–9, 229–30
  consortia 73, 92n1
  industry/technology clusters 158
  management 164–5
  market clusters 103
  platforms 20, 32, 60–1, 66, 68n19, 71–93, **78**, **80–1**, **83**, **85**, **87**, **89**, **92**, 105, 124
  R&D clusters 161–2, 165
  regional clusters 289
  see also business ecosystems; small and medium-sized enterprises; strategic alliances
Ninesigma 244, 245, 249–51, 254
Nokia 75, 78–87, **78**, **80–1**, **85**, **87**, 227, 229, 234
non-governmental organizations (NGOs) 286–7
  see also open social innovation
Nortel Networks 200

Ocean Optics 147
Odagiri, Hiroyuki 162, 163
oil industry, *see* energy industry
open business models 50–68
  combined with open innovation models 53–5, **54**
  definition 52–3
open call, *see* crowdsourcing
open collaborative innovation 4, 20, 21
open data 25
open distributed innovation 20, 35
open government 25, 288
open innovation (OI):
  contrast with open business models 50–68
  contrast with open collaborative innovation 4, 20, 21
  contrast with open distributed innovation 20, 35
  contrast with user innovation 30–6, **31**, 34–6
  critiques of OI 21–4
  definitions of OI v–vi, 4, 15–21, 24, 27, 51–2
  enablers and barriers 235–6
  firm-centric focus 29, 34, 289–90
  global research stream 157–68, **161**, **168**
  implementation framework 223–41, **237**
  implications for public policy 287–8
  IT-enabled OI 144, 146, 236, 285
  location decisions 161–2, 165, 228–9
  management theory 282–4, 291–2
  outside-in/inside-out/coupled 4, 14–15, **18**, 18–20, 33, **54**, 55–64, 75, 159, 171, 226
  research 3–28, **5–8**, **10–15**, 37, 49n2, 72–8, 115–17, **138–41**
  research agendas for the future 25–7, **26**, 144–56, **239–40**, 255, 275, 282–94
  role of leaders and managers 230–2, 292
  selective openness strategies 42
  types and mechanisms of OI **18**, 18–20
  see also business models; co-creation; collaboration; coupled open innovation; external innovation; inbound (outside-in) open innovation; knowledge flows; open social innovation; outbound (inside-out) open innovation; user innovation
open innovation communities 14, 35
open innovation funnel 52, 54, 257–60, **258**, 263, 270, 271, 273, 277n2
open innovation platforms 20, 32, 60–1, 66, 68n19, 71–93, **78**, **80–1**, **83**, **85**, **87**, **89**, **92**, 105, 124
Open Invention Network 204
open social innovation 169–88, **170**, 287
  Ashoka 171, 175–80, **177**, 182–6, **184**, 188n6
  Birmingham local government initiatives 180–3, **182**, **184**, 186, 188n7
  Emergency 171, 172–4, 182–5, **184**, 188nn1–2
  Hybrid Value Chain 177
open source software 35, 38, 42, 60, 73
  Linux 60–1, 88, 93n11, 204
outbound (inside-out) open innovation 4, 14–15, **54**, **184**, **239**
  business models 64–5
  commercialization paths 16, 18, 37, 46–7, 143, 147–9, 268, 269, 272
  complex ventures 208–22

outbound (inside-out) open innovation(*cont.*)
   free/selective revealing 11, 15, 19, 30, 35, 142, 150–1
   monetization of technology 166
   out-licensing 31, 115, 141, 227, 258, 270
   profit motive 33, 43
   spin-off activities 53, 57, 115, 258, 270, 272
outsourcing 15, 60, 103, 105, 112, 145, 258, 276

P&G (Procter & Gamble) 33, **54**, 57–9, 67nn4, 6, 8, 154, 157, 196, 227, 231
Pasteur, Louis 204
Paul, Roshan 179
Penner-Hahn, Joan 163
Petroni, Giorgio 232
pharmaceutical industry 150, 196, 222
Porter, Michael E. 94, 95, 187n1
process innovation 66, 95, 102, 106, 110
product development:
   four phases (initialization, planning, execution, termination) 126
   new product development (NPD) 116–17, 126, 259
   product life cycle (PLC) model 94–5, 101, 102, 106–7, 108–10, **109**, 112
   "scrum" approach v
   stage-gate model 9, 24, 131n1, 259
product innovation 50, 65
   bundling 96, 98, 101–2, 109, 111
   context-embedding 99–101, 111–12
   innovation life cycle (ILC) 94–5, 101, 102, 106–7, 108–10, **109**, 112
   market disintegration 104–5
   market formation 96–112, **109**
   market integration 96, 103–4
   product life cycle (PLC) model 94–5, 101, 102, 106–7, 108–10, **109**, 112
Psion 77–81, **78**, **80–1**, 87, 90–1, 105

R&D (research and development) 115–31
   absorptive capacity of the firm 14, 17, **31**, 46–7, 152–3, 160, 163, 236, 269–71
   alliances 44, 158
   centralization/decentralization cycles 228
   choice of organizational/governance mode 122, 127, **128**
   choice of partners 122
   closed innovation model 16, 22, 55–6, 257
   closed innovation projects 125
   first mover advantage 267
   global research stream 157–68, **161**, **168**, 264, 288

   individual and team composition 130
   internal 17, 123
   investment 16–17
   knowledge transfer 119–21, 124–5
   market-seeking vs. technology-seeking strategy 161
   measurement of innovation success 119–21, 123–9
   open innovation funnel 52, 54, 257–60, **258**, 263, 270, 271, 273, 277n2
   open innovation model 17
   open innovation projects 117–31, **121**, **128**, 259, 290
   performance effects of collaboration 115–18, 119–21
   phases of product development (initialization, planning, execution, termination) 126
   reverse innovation 166
   role of R&D in corporate strategy 129–30
   sourcing knowledge 58, 65, 126, 152, 157–8, 162–4
   timing of collaboration 125–7
   upstream and downstream activities 18, **121**, 167
   user innovation model **31**
Rambus 199
research laboratories 23, 33, **38**, 56, 75, 92n1, 127, 209, 228–9, 293
   "collaboratories" 158
   competence-leveraging vs. competence-creating 161
   home-base-exploiting vs. home-base-augmenting 161–2, 166
Rind, Kenneth W. 210
Rosenberg, Nathan 17
Royal Philips Electronics 119, 120, **121**, 130–1

Salomon Brothers 199, 200
Salter, Ammon 9, 75
Samsung 78–81, **78**, **80–1**, **85**, **87**, 157
Schroll, Alexander 236
Schwartz, Ken 167
Science Citation Index (SCI) 7
semiconductor industry 195–6, 199
Semmelweis, Ignaz 204
Shah, Sonali K. 32
Shane, Scott 288
Shaver, J. Myles 163
Shin, Jongtae 162, 163
Siegel, Eric 201, 213
Siegel, Robin 210, 213
Silveira, Rafael 247
Simard, Caroline 289

Index

SkyNRG-KLM **54**, 63–5, 68n18
small and medium-sized enterprises (SMEs) 135–56, 221, 273, 285–6
   crowdsourcing 145–7
   definition of 136, 156n1
   effect of ownership structure on open innovation 137, 142
   external focus of open innovation 135, 137
   importance of personal networks 137
   inbound and outbound innovation 141–2
   intellectual property (IP) management 149–51
   management of open innovation 152–5
   networks and firm performance 142–4, 147–9
   open innovation collaboration 38
   performance impact of open innovation 142
   R&D expenditure 136
   research on open innovation in SMEs **138–41**, 144–56
   see also start-up firms
social enterprise software 45
social entrepreneurs 175–80, 187, 188n5, 287
social media 16, 90–1, 145
   Facebook **54**, 90, 203
Social Science Citation Index (SSCI) 7
Song, Jaeyong 162, 163
spillovers:
   knowledge transfer 158–9
   purposive management of 4, 16–20, **18**
   R&D 4, 16–17, 273
spin-offs 24, 57, 212–13, 292
Spithoven, André 142
startup firms 16, 19, 22, 90–2, 136, 209, 217, 259
   see also venture capital
Strada, Cecilia 172, 173
Strada, Gino 172
strategic alliances/partnerships 20, 37, 65, 72–4, 137, 259
   see also coupled open innovation
strategic management 275–6
   collaborative relationships 95
   dynamic capabilities 9, **10–11**, 95, 111, 261, 266–9
   five forces framework 94–5, 103, 110
   industry-bounded paradigm 94, 110, 112
   see also business models; business theory
supply chains 24
   see also value chains
Symbian 71–93, **78**, **80–1**, **83**, **85**, **87**, **89**

Teece, David J. 9, 91, 219, 267, 268
technology 18, **31**, 52
   interface technologies 107, 108
   licensing 96–7, 99, 101, 107, 108, 191, 193–5, 200, 201, 206n3
   patents 120–1, 149–50, 192–3, **192**, 199, 200, 202–4
   platform technologies 124
   see also intellectual property; R&D
technology transfer organizations (TTOs) 146
Texas Instruments (TI) 193, 195
Thomson Reuters (formerly ISI) Web of Science **7–8**, 9, 28n7, 206n2
Tripsas, Mary 32
Trott, Paul 22–3

United States:
   Bayh–Dole Act (1980) 230
   declining hegemony 16, 22
   innovation system 23
university–industry relationships 9, 33
upstream and downstream activities 18, **121**
user innovation 29–36
   communities 32, 34, 44, 49n4
   contrast with open innovation 30–6, **31**, 34–6
   entrepreneurship 32
   freely revealed information **31**, 35, 142, 150–1, 205
   intellectual property (IP) 35
   knowledge sharing 32
   lead users 30–3, 42, 43, 45, 167
   open source/creative commons licensing **31**
   search fields 42
   social vs. monetary motivation 44
   "sticky information" 30, **31**, 44
   user–firm collaboration 45
   user firms 49n3

value chains:
   closed/linear vs. open/complex vi
   extraction vs. inclusion vi
   hybrid 177–9, **177**
   vertical vs. horizontal integration vi, 34, 142, 150, 196
value creation and capture 18, 28n8, 50, 91–2
   business models 52–5, 71, 76, 160
   division of labor 53
   event-based vi
   open approach 59–64
   role of "innomediaries" 242–56, 277nn3, 6
van de Vrande, Vareska 263, 273
Vanhaverbeke, Wim 25, 52, 117, 271, 287, 289

Venkataraman, S. [Sankaran] 288
venture capital 16, 19, 22, 90-2, 148, 158, 208-9, 211-13, 215, 219
vertical integration, *see* value chains
von Hippel, Eric 20-1, 28n5, 29, 30, 32, 35, 44, 210
von Krogh, Georg 44, 153

Waguespack, David M. 14
Wallin, Martin W. 34, 153

West, Joel 19, 25, 37, 52, 75, 117, 289
Westergren, Ulrika R. 235
worker mobility 16, 22, 158-9
World Economic Forum 169, 186
Wright, Randall 247

Xerox 66, 173, 209, 212-14, 233

Young Foundation 169, **170**

Printed and bound by CPI Group (UK) Ltd, Croydon, CR0 4YY